HUMAN LACTATION 3

The Effects of Human Milk on the Recipient Infant

HUMAN LACTATION 3

The Effects of Human Milk on the Recipient Infant

Edited by

Armond S. Goldman
University of Texas Medical Branch
Galveston, Texas

Stephanie A. Atkinson
McMaster University
Hamilton, Ontario, Canada

and

Lars Å. Hanson
University of Göteborg
Göteborg, Sweden

Springer Science+Business Media, LLC

Library of Congress Cataloging in Publication Data

International Conference on the Effects of Human Milk on the Recipient Infant (3rd: 1986: Konstanz, Germany)
Human lactation 3.

"Proceedings of the Third International Conference on the Effects of Human Milk on the Recipient Infant, held September 10-14, 1986, in Konstanz, West Germany"—T.p. verso.
Includes bibliographies and index.
1. Breast feeding—Physiological aspects—Congresses. 2. Infants—Growth—Congresses. 3. Milk, Human—Congresses. I. Goldman, Armond S. II. Atkinson, S. A. (Stephanie A.) III. Hanson, Lars Å. IV. Title. V. Title: Human lactation three. [DNLM: 1. Breast Feeding—congresses. 2. Infant, Newborn—physiology—congresses. 3. Lactation—metabolism—congresses. 4. Milk, Human—metabolism—congresses. W3 IN1948D 3rd 1986h/WS 125 I615 1986h]
RJ216.I555 1986 618.92'01 87-15359
ISBN 978-1-4899-0839-1

ISBN 978-1-4899-0839-1 ISBN 978-1-4899-0837-7 (eBook)
DOI 10.1007/978-1-4899-0837-7

Proceedings of the International Conference on the Effects of Human Milk on the Recipient Infant, held September 10-14, 1986, in Konstanz, West Germany

PREFACE

Four years ago the National Institutes of Child Health and Human Development (NICHD) brought together a group of scientists to Belmont, Maryland to examine the status of human milk banking. During those deliberations, the idea was generated to organize a series of research conferences concerning human lactation and the composition and biological effects of human milk. The first one, organized by Robert G. Jensen from the University of Connecticut and Margaret C. Neville from the University of Colorado, dealt with methodologic issues. An additional meeting to explore the effects of maternal and environmental factors upon human lactation and the composition of human milk was organized by Margit Hamosh from Georgetown University and me, and was held in January, 1986 in Oaxaca, Mexico.

Those meetings provided the foundation for the design of the present conference, 'The Effects of Human Milk Upon the Recipient Infant'. In addition to a grant from the NICHD, the conference was generously supported by Milupa AG from the Federal Republic of Germany; Wyeth Limited and Mead Johnson of Canada; and Ross Laboratories, Heinz USA, the Mead Johnson Nutritional Group, Wyeth International Limited, Gerber Products Company, the La Leche League International, Glaxo Incorporated and Sandoz Pharmaceutical Corporation from the United States. This allowed us to invite over sixty leading scientists from Australia, Belgium, Canada, the Federal Republic of Germany, France, Hong Kong, Italy, New Zealand, Norway, Pakistan, Sweden, Switzerland, the United Kingdom, and the United States to Konstanz, the Federal Republic of Germany to address the nutritional, epidemiologic, metabolic, hormonal, immunologic, and toxicologic aspects of this issue.

I am grateful to Stephanie A. Atkinson from Canada and Lars A. Hanson from Sweden for their important contributions to the design of the program and selection of participants. Atkinson organized the abstracts and prepared the discussants for their roles. Thorsten A. Fjellstedt from NICHD encouraged our efforts, reviewed the abstracts for the poster session, and arranged for the programs to be printed. Gerd Harzer from the Federal Republic of Germany deserves special recognition for selecting the meeting site, negotiating for the use of the University of Konstanz and our accomodations in the city, and aiding the participants during the conference. Robert Goldman designed a computer program that facilitated in the preparation of the index of this book. In addition, Susan Kovacevich, Stephanie Huery, and Elke Hutzenlaub should be acknowledged for their secretarial assistance.

June, 1987

Armond S. Goldman
Conference Chair

CONTENTS

SECTION II

GROWTH FACTORS, HORMONES AND INDUCERS

SECTION III

HOST RESISTANCE

SECTION IV

POTENTIALLY HARMFUL EFFECTS

SECTION V

EPIDEMIOLOGY

SECTION VI

POSTER SESSION

INTRODUCTION TO THE INTERNATIONAL CONFERENCE:

THE EFFECTS OF HUMAN MILK UPON THE RECIPIENT INFANT

Armond S. Goldman

Departments of Pediatrics and Human Biological
Chemistry and Genetics
University of Texas Medical Branch
Galveston, Texas, 77550

The background of the conference is a history of concepts concerning the role of human milk in infant nutrition. Until the past few decades, human milk was considered to be simply a source of calories, proteins, lactose, lipids, and minerals. It was conceded that human milk differed from cow's milk, but most clinicians believed that those differences were insignificant. Earlier reports of lower morbidity and mortality among breast-fed infants were attributed to infectious contaminants introduced into cow's milk formulations, rather than to protective effects of human milk. Human lactation was not emphasized and that was mirrored by a decline in breast-feeding in many countries during the mid-20th century. This vital area of human biology was also neglected because the technologies used to examine lactation performance, human milk composition, or the effects upon the infant were rudimentary. It was, therefore, understandable that most scientists shrank from the daunting task of conducting research on this subject. Finally, the paucity of studies may have been a reflection of the comparatively low priority regarding children in many countries.

In the early 1960's, the possibility that human milk had uniquely evolved for the needs of the infant began to be investigated. With the advent of new technologies, it was discovered that human milk was rich in not only an array of nutrient proteins, nonprotein nitrogen compounds, lipids, oligosaccharides, vitamins and certain minerals (1,2), but also in many types of defense agents (3,4), enzymes (5), hormones (6), inducers, growth factors (7-10), membranes and cells (3,4). Knowledge concerning the complexity of human milk is still increasing as new functions are ascribed to previously recognized factors and novel factors are uncovered. In addition, many factors in human milk are compartmentalized. Undoubtedly, these compartments are physiologically important. Perhaps they shield factors from the digestive, denaturing processes of the gastrointestinal tract or limit their actions to specific regions of the alimentary tract by selectively releasing them at those sites. There is little understanding, however, of the *in vivo* effects of many constituents of human milk or of the compartments in which they are housed.

In the main, clinical studies of human milk have been limited to examining the effects upon somatic growth or the frequency of intestinal or respiratory infections, and determining whether some constituents of human milk survive passage through the gastrointestinal tract or may be absorbed into the circulaton. The situation is, however, changing. New studies, some of which were presented at this meeting, are probing the differences between breast and artifically fed infants in greater depth. For example, Cutberto Garza and his co-workers (11) and Neville and Oliva-Rasbach (12) reported differences in energy intake and growth in breast-fed and cow milk formula-fed infants during the first year of life. It has also been found that the hormonal responses in the infant such as insulin secretion depend upon the type of feedings (13). As more sensitive, less invasive methods are adapted to investigations of the recipient infant, it will also become possible to detect changes in cellular, metabolic, and immunologic pathways in the infant.

The effects of human milk upon the premature infant are of special interest. Many low birth weight infants once considered to be unsalvagable are being saved, but the quality of their survival is in question (14,15). The issue has been raised repeately whether the type of feeding influences the ultimate development of those infants. Some questions that emanate from this central issue were discussed at this conference. 1) What should be the standard of growth in such infants? Rates of weight gain which approximate intrauterine patterns are obtainable in such infants fed mother's milk (16-20), but it remains unclear whether the intrauterine standard should pertain to extrauterine life. 2) What should be the optimal body composition of weight gain and what quantities of macronutrients and minerals are required to produce such an effect? Studies employing metabolic balances and indirect calorimetry demonstrate a partition of growth that is similar to the reference fetus when infants are fed their own mother's milk. The levels of proteins and minerals, especially sodium, also seem to be important determinants of weight gain and body composition (20). In infants fed cow's milk formulas containing a higher energy content, lipids seem to provide a greated proportion of weight gain (13). The participants at this meeting further addressed these matters. 3) Does banked human milk provide the same advantages as freshly secreted milk, and indeed does the use of banked milk introduce new hazards such as viruses (21), drugs or environmental pollutants (2)? This last point, which was addressed by Robert Pass (23) and John Wilson (24), is of particular concern since it is difficult for human milk banks to avoid or eliminate contaminants. 4) Are the concentrations of constituents of pooled human milk derived from mothers who deliver at term or before then appropriate for feeding premature infants? The use of human milk preparations in low birth weight infants has been reported (16-20), but it is unclear whether human milk used for that purpose should be obtained during the phase of lactation which corresponds to the postnatal age of the recipient (25-29). 5) Since the low birth weight infant would have been nourished via the placenta for a longer period if the pregnancy continued to term, will the alimentary tract of the premature infant be ready to handle the complex systems in human milk or foreign components from substitute feedings?

In regard to the last question, although much has been learned about the developing gastrointestinal tract (30), more information will be needed concerning changes in local defenses and the chemical structure of the surface of epithelial cells of the gastrointestinal tract during early life. Certain studies have, in fact, already shown that the organization of epithelial membranes and receptors for enterotoxic ligands change during the first several months of life (31,32). Undoubtedly, it will be important to determine whether human milk components influence the function of cells that first encounter the constituents of human milk or of substitute feedings.

There are other issues that were covered in this conference or that should be addressed in future studies. 1) Will detailed analyses of the physicochemical structure of the constituents of human milk lead to a discovery of new functions of those components? For instance, significant structural homologies between peptides found in human milk and other sites may provide new clues concerning the operation of certain proteins in human milk. (2) What are the interactive effects of the components of human milk? Heird (20) reported that the sodium balance in low birth weight infants fed cow's milk formula was dependent not only upon sodium intake but also upon nitrogen balance, but little is known whether similar interactive effects occur with human milk feedings. 3) Can suitable animal models or cultures of human epithelial cells be devised to test the effects of human milk? Even with new technologies, certain studies in human infants may be precluded because of their invasive nature. Therefore, animal or tissue culture models may be required to study certain questions. 4) Do intestinal bacteria in the breast-fed infant generate special nutrients or other important factors for the infant? In addition to the known contribution of vitamin K by common enteric bacteria of vitamin K and anti-bacterial organic acids by lactobacilli, are other important substances produced by the intestinal microflora of the breast-fed infant? 5) Are sites other than the respiratory and alimentary tracts protected by human milk feedings? The report by Randall Goldblum and his colleagues (33) suggests that may be the case. 6) Does part of the protection provided by human milk involve inhibition of inflammatory processes in the recipient infant? That hypothesis is supported by many pieces of evidence that were assembled in a recent publication (34). 7) What are the long term consequences of breast-feeding? The papers by Margit and Paul Hamosh (13) and by Ranjit Chandra (35) addressed the metabolic and atopic aspects of that problem. They pointed out that further studies will be needed to obtain definitive answers to these questions. 8) How can epidemiologic studies be designed to most effectively investigate these issues? The group lead by Jean-Pierre Habicht considered this point and prepared recommendations which should be useful in designing future epidemiologic investigations of the effects of human milk (36). Indeed a sharing of ideas between epidemiologists and basic and clinical scientists should advance the field more quickly.

Although this conference focused upon the recipient, it was evident that the controls over lactation, the complexity of human milk and maternal-infant interactions impact upon the questions that were posed. There is a dichotomy of paradigms regarding the syntheses and secretion of human milk. The first is that the initiation of lactation and the composition of human milk are governed by innate mechanisms, and the second is that the volume and constituents of human milk are modified by environmental factors such as neurogenic excitation of prolactin secretion, antigenic triggering of the entero-mammary gland pathway for the production of dimeric IgA antibodies (37) and the provision of certain components of human milk such as vitamin B-6 (38) or trans-fatty acids from the diet (39). A third mechanism should also be considered: feed-back systems may operate between the infant and the mother to control the synthesis and release of certain constituents in human milk by the mammary gland. Reciprocal relationship between the production of certain factors such as secretory IgA by the maternal mammary gland (40) and by mucosal tissues of the infant (4) suggest this possibility. Prime anatomic sites for the generation and receipt of possible feed-back signals are the mucosa of the mouth of the infant and the nipple and areola of the maternal breast. In that regard, it would behoove us to study these sites for receptors of principal components of the secretions to which they are exposed during the feeding process.

Each participant should be recognized for the light that their presentations shed upon this important field of human biology. Ultimately, this conference may point to a resolution of many complex problems concerning the effects of human milk upon the recipient infant. If that is the case, many children will benefit from our efforts.

REFERENCES

1. R. G. Jensen and C. Neville, "Human Lactation: Milk Components and Methodologies", Plenum Press, New York and London (1985).
2. M. Hamosh and A. S. Goldman, "Human Lactation 2: Maternal and Environmental Factors," Plenum Press, New York and London (1986).
3. A. S. Goldman and R. M. Goldblum, Protective properties of human milk, in : "Nutrition in Pediatrics-Basic Sciences and Clinical Application", W. A. Walker, J. B. Watkins, eds., Little, Brown, and Company, Boston, pg. 819 (1985).
4. A. S. Goldman, A. J. Ham Pong, and R. M. Goldblum, Host Defenses: Development and maternal contributions, in : "Advances in Pediatrics", L. A. Barness, ed., Yearbook Medical Publishers, Chicago, 32:71 (1985).
5. M. Hamosh, L. M. Freed, J. B. Jones, S. E. Berkow, J. Bitman, N. R. Mehta, B. Happ, and P. Hamosh, Enzymes in human milk, in : Human Lactation: Milk Components and Methodologies", R. G. Jensen and M. C. Neville, eds., Plenum Press, New York and London, pg. 251 (1985).
6. O. Koldovsky, A. Bedrick, P. Pollack, R. K. Rao, and W. Thornburg, Hormones in Milk: Their presence and possible physiological significance, in : "Human Lactation 3: Effects of Human Milk Upon the Recipient Infant," A. S. Goldman, S. A. Atkinson, and L. A. Hanson, eds., Plenum Press, New York and London, pg. 183 (1987).
7. F. H. Morris, Jr., Methods for investigating the presence and physiologic role of growth factors in milk, in : "Human Lacation: Milk Components and Methodologies", R. G. Jensen and M. C. Neville, eds., Plenum Press, New York and London, pg. 193 (1985).
8. Y. W. Shing, S. Dai, and M. Klagsbrun, Detection of three growth factors, HMGF I, HMGF II, and HMGF III in human milk, in : "Human Lactation: Milk Components and Methodologies", R. G. Jensen and M. C. Neville, eds., Plenum Press, New York and London, pg. 201 (1985).
9. W. R. Kidwell, M. Bano, K. Burdette, I. Lococzy, and D. Salomon, Human Lactation. Mammary derived growth factors in human milk, in : "Human Lactation: Milk Components and Methodologies", R. G. Jensen and M. C. Neville, eds., Plenum Press, New York and London, pg. 209 (1985).
10. W. R. Kidwell, D. S. Salomon, and S. Mohanam, Production of growth factors by normal human mammary cells in culture, in : "Human Lactation 3: Effects of Human Milk Upon the Recipient Infant", A. S. Goldman, S. A. Atkinson, and L. A. Hanson, eds., Plenum Press, New York and London, pg. 227 (1987).

11. C. Garza, J. Stuff, N. Butte, and K. Fraley, Growth patterns of the breast-fed infants, in : "Human Lactation 3: Effects of Human Milk Upon the Recipient Infant", A. S. Goldman, S. A. Atkinson, and L. A. Hanson, eds., Plenum Press, New York and London, pg. 109 (1987).
12. M. Neville and J. Oliva-Rasbach, Is maternal milk production limiting infant growth during first year life in breast-fed infants? in : "Human Lactation 3: Effects of Human Milk Upon the Recipient Infant", A. S. Goldman, S. A. Atkinson, and L. A. Hanson, eds., Plenum Press, New York and London, pg. 123 (1987).
13. M. Hamosh and P. Hamosh, Does nutrition in early life have long term metabolic effects? Can animal models be used to predict these effects in the human? in : "Human Lactation 3: Effects of Human Milk Upon the Recipient Infant", A. S. Goldman, S. A. Atkinson, and L. A. Hanson, eds., Plenum Press, New York and London, pg. 37 (1987).
14. P. Budetti, P. McManus, N. Barrand, and L. Heinen, The costs and effectiveness of neonatal intensive care. Washington, D. C., Congressional Office of Technology Assessment, U. S. Government Printing Office (1981).
15. P. Budetti and P. McManus, Assessing the effectiveness of neonatal intensive care, Medical Care 20:1027 (1982).
16. S. A. Atkinson, M. H. Bryan, and G. H. Anderson, Human milk feeding in premature infants: Protein and energy balance in the first two weeks of life, J. Pediatr. 99:61 (1981).
17. S. A. Atkinson, I. C. Radde, and G. H. Anderson, Macro-mineral balance in premature infants fed their own mother's milk or formula, J. Pediatr. 102:99 (1983).
18. R. K. Whyte, R. Haslam, H. S. Bayley, and J. C. Sinclair, Energy balance and nitrogen balance in growing low birthweight infants fed milk or formula, Pediatr. Res. 17:891 (1983).
19. R. J. Schanler, C. Garza, and E. O'Brian Smith, Fortified mother's milk for very low birth weight infants: Results of macromineral balance studies, J. Pediatr. 107:767 (1985).
20. W. C. Heird, S. Kashyap, K. F. Schulze, R. Ramakrishnan, C. L. Zucker, and R. B. Dell, Nutrient utilization and growth in LBW infants, in :"Human Lactation 3: Effects of Human Milk Upon the Recipient Infant", A. S. Goldman, S. A. Atkinson, and L. A. Hanson, eds., Plenum Press, New York and London, pg. 9 (1987).
21. S. Stagno, D. W. Reynolds, R. T. Pass, and C. A. Alford, Breast milk and the risk of cytomegalovirus infection, N. Engl. J. Med. 302:1073 (1980).
22. J. D. Wilson, J. L. Hinson, R. D. Brown, and I. J. Smith, Comprehensive assessment of drugs and chemical toxins excreted in breast milk, in "Human Lactation 2: Milk Components and Methodologies", M. Hamosh and A. S. Goldman, Plenum Press, New York and London, pg. 395 (1986).
23. R. F. Pass, Viral contamination of milk, in : "Human Lactation 3: Effects of Human Milk Upon the Recipient Infant", A. S. Goldman, S. A. Atkinson, and L. A. Hanson, eds., Plenum Press, New York and London, pg. 279 (1987).
24. J. T. Wilson, R. D. Brown, I. J. Smith, and J. L. Hinson, Potentially harmful effects of drugs and toxins in human milk, in : Human Lactation 3: Effects of Human Milk Upon the Recipient Infant", A. S. Goldman, S. A. Atkinson, and L. A. Hanson, eds., Plenum Press, New York and London, pg. 301 (1987).
25. S. A. Atkinson, G. H. Anderson, and M. H. Bryan, Human milk: Comparison of the nitrogen composition in milk from mothers of premature and full-term infants, Am. J. Clin. Nutr. 33:811 (1980).

26. A. S. Goldman, C. Garza, C. A. Johnson, B. L. Nichols, and R. M. Goldblum, Immunologic factors in human milk during the first year of lactation, J. Pediatr. 100:563 (1982).
27. A. S. Goldman, C. Garza, B. Nichols, C. A. Johnson, E. Smith, and R. M. Goldblum, The effects of prematurity upon the immunologic system in human milk, J. Pediatr. 101:901 (1982).
28. N. F. Butte, C. Garza, C. A. Johnson, E. O'Brian Smith, and B. L. Nichols, Longitudinal changes in milk composition of mothers delivering preterm and term infants, Early Hum. Develop. 9:153 (1984).
29. G. Harzer, M. Haug, I. Dieterich, and P. R. Gentner, Changing patterns of human milk lipids in the course of lactation and during the day, Am. J. Clinical Nutr. 37:612 (1983).
30. R. M. Klein, Models for the study of cell proliferation in the developing gastrointestinal tract, Pediatr. Gastroenterol. Nutr. 5:513 (1968).
31. J. L. Bresson, K. Y. Pang, and W. A. Walker, Microvillus membrane differentiation: Quantitative difference in cholera toxin binding to the intestinal surface newborn and adult rabbits, Pediatr. Res. 18:984 (1984).
32. K. Y. Pang, J. L. Bresson, and W. A. Walker, Development of the gastrointestinal mucosal barrier. Evidence for structural differences in microvillus membranes from newborn and adult rabbits, Biochemica. Biophysica. Acta. 727:204 (1983).
33. R. M. Goldblum, R. J. Schanler, C. Garza, and A. S. Goldman, Effect of fortified human milk feedings upon the development of immunity in low birthweight infants, in : "Human Lactation 3: Effects of Human Milk Upon the Recipient Infant", A. S. Goldman, S. A. Atkinson, and L. A. Hanson, eds., Plenum Press, New York and London, pg. 245 (1987)
34. A. S. Goldman, L. W. Thorpe, R. M. Goldblum, and L. A. Hanson, Anti-inflammatory properties of human milk, Acta Paediatr. Scand. 75:689 (1986).
35. R. Chandra, Prevention of atopic disease: Environmental engineering utilizing antenatal antigen avoidance and breast feeding, in : "Human Lactation. Effects of Human Milk Upon the Recipient Infant", A. S. Goldman, S. A. Atkinson, and L. A. Hanson, eds., Plenum Press, New York and London, pg. 269 (1987).
36. J-P. Habicht, et al., Report of epidemiology workshop: Recommendations regarding future research concerning the effects of human milk upon infant recipients in : "Human Lactation 3: Effects of Human Milk Upon the Recipient Infant", A. S. Goldman, S. A. Atkinson, and L. A. Hanson, eds., Plenum Press, New York and London, pg. 367 (1987).
37. A. S. Goldman and R. M. Goldblum, Immunoglobulins in human milk, in : "Natural Antimicrobial Systems", R. G. Board, ed., International Dairy Federation, Brussels, Belgium and Bath University Press, Bath, U. K., pg. 7 (1986).
38. A. Kirskey and S. A. Udipi, Analysis of water-soluble vitamins in human milk: Vitamin B-6 and Vitamin C, in : "Human Lactation: Milk Components and Methodologies", R. G. Jensen and M. C. Neville, eds., Plenum Press, New York and London, pg. 153 (1985).
39. B. Koletzko, M. Mrotzek, and H. J. Bremer, Trans-fatty acids in human milk and infant plasma and tissue, in : "Human Lactation 3: Effects of Human Milk Upon the Recipient Infant", A. S. Goldman, S. A. Atkinson, and L. A. Hanson, eds., Plenum Press, New York and London, pg. 325 (1987).

40. N. F. Butte, R. M. Goldblum, L. M. Fehl, K. Loftin, E. O. Smith, C. Garza, and A. S. Goldman, Daily ingestion of immunologic components in human milk during the first few months of life *Acta Paediatr. Scand.* 73:296 (1984).

NUTRIENT UTILIZATION AND GROWTH IN LBW INFANTS

William C. Heird, Sudha Kashyap,
Karl F. Schulze, Rajasekhar Ramakrishnan,
Christine L. Zucker, and Ralph B. Dell

Department of Pediatrics
College of Physicians and Surgeons
of Columbia University
630 West 168th Street
New York, NY 10032

Over the past decade, a number of studies concerning growth, nutrient retention and metabolic response of low birth weight (LBW) infants fed a variety of formulas as well as a variety of human milk preparations have been reported (1-21). These studies, in toto, provide important insights concerning both the nutrient requirements of LBW infants and the utilization of nutrients for growth. Some of the insights in both areas will be illustrated in this chapter, using primarily the authors' studies (21-23) as examples. These studies have the advantage of familiarity and, also, are among the few such studies in which protein and energy intakes were varied independently.

Summary of Studies All studies were conducted in LBW infants (birth weight, 900-1750g) who were able to tolerate the desired intake (180 ml/(kg.d)) by 28 days of age. These infants were assigned randomly to formulas providing different protein and energy intakes; intakes of all other nutrients were identical (Table I). Observations were made serially from the time desired intake was tolerated until body weight reached 2200g.

Table I. Range of Electrolyte and Mineral Intakes of Study Groups

Electrolyte/Mineral	Intake(mMoles/(kg.d))
Sodium	1.92 - 2.65
Potassium	3.94 - 4.24
Calcium	4.84 - 5.32
Magnesium	0.43 - 0.50
Chloride	2.50 - 3.57
Phosphorus	3.03 - 3.42

Initially, protein and energy intakes, respectively, of 2.24 g/(kg.d) and 115 kcal/(kg.d), 3.62 g/(kg.d) and 114 kcal/(kg.d) and 3.5 g/(kg.d) and 149 kcal/(kg.d) were studied (21). Some of the observed outcome variables related to growth, nutrient retention and metabolic response are summarized in Table II.

Table II. Growth, Nutrient Retention and Metabolic Response (Mean ± S.D.) of Infants Participating in Study I (21)

Variable	Group 1	Group 2	Group 3
Intake:			
Protein(g(kg.d))	2.24 ± 0.02	3.6 ± 0.03	3.5 ± 0.04
Energy(kcal/(kg.d))	115.0 ± 0.90	114.0 ± 1.0	149.0 ± 1.80
Growth:			
Δ Weight(g/(kg.d))	13.9 ± 2.8*	18.3 ± 2.8*	22.0 ± 3.1*
Δ Length(cm/wk)	0.94 ± 0.19	1.21 ± 3.32	1.24 ± 0.30
Δ Head circumference(cm/wk)	0.85 ± 0.15	1.22 ± 0.28	1.17 ± 2.24
Δ Triceps SF(mm/wk)	0.32 ± 0.14	0.38 ± 0.11	0.67 ± 0.34*
Δ Subscapular SF (mm/wk)	0.33 ± 0.20	0.36 ± 0.13	0.64 ± 0.26*
Nutrient Retention:			
Nitrogen(mg/(kg.d))	267.9 ± 12.0*	422.2 ± 21.7	425.3 ± 11.6
Energy(kcal/(kg.d))	47.0 ± 4.0	46.0 ± 6.0	69.0 ± 10.0*
Sodium(mMoles/(kg.d))	0.74 ± 0.27*	1.34 ± 0.35	1.40 ± 0.11
Potassium(mMoles/(kg.d))	1.10 ± 0.17	1.21 ± 0.38	1.54 ± 0.19+
Calcium(mMoles/(kg.d))	3.02 ± 0.60	2.99 ± 1.04	3.02 ± 0.56
Chloride(mMoles/(kg.d))	0.99 ± 0.31	1.33 ± 0.31	1.49 ± 0.21+
Phosphorus(mg/(kg.d))	2.17 ± 0.23	2.35 ± 0.16	2.39 ± 0.13
BUN(mg/dl)	1.2 ± 0.8*	2.9 ± 1.2	2.5 ± 0.5

*Difference from other groups statistically significant.
+Difference from Group 1 statistically significant.

Subsequently, protein and energy intakes, respectively, of 2.8 g/(kg.d) and 121 kcal/(kg.d), 3.8 g/(kg.d) and 119 kcal/(kg.d) and 3.9 g/(kg.d) and 144 kcal/(kg.d) were studied (22). Again, the intakes of all other nutrients were similar in the three groups (see Table I). Pertinent variables related to growth, nutrient retention and metabolic response observed in this study are summarized in Table III.

With respect to growth, it is obvious that the rate of weight gain is related to both protein and energy intakes. In both studies, for example, the rate of weight gain of Group 2, the intake of which differed from Group 1 only in protein intake, was greater than that of Group 1. Also in both studies, the rate of weight gain of Group 3, the intake of which differed from that of Group 2 only in energy intake, was greater than that of Group 2 (in Study II, this difference is not statistically significant). Moreover, in both studies, the greater rate of weight gain associated with the greater energy intake was accompanied by a greater rate of increase in skinfold thicknesses. In Study I, the higher protein intakes also were associated with a greater increase in both length and head circumference. In Study II, this effect was statistically significant only for the differences between Group 1 (protein intake, 2.8 g/(kg.d)) and Group 3 (protein intake, 3.9 g/(kg.d); energy intake, 144 kcal/(kg.d)).

In general, nitrogen retention was related directly to protein intake and energy retention was related directly to energy intake. In Study II, the greater energy intake also enhanced nitrogen retention, ie., nitrogen retention of Group 3 was greater than that of Group 2. A suggestive effect of energy intake on nitrogen retention also was apparent in Study I (eg., nitrogen retention of Group 3, the protein intake of which was slightly lower than that of Group 2 and the energy intake of which was greater, was almost exactly the same as that of Group 2); however, the difference in nitrogen retention between Groups 2 and 3 was not statistically significant.

Table III. Growth, Nutrient Retention and Metabolic Response (Mean ± S.D.) of Infants Participating in Study II (22)

Variable	Group 1	Group 2	Group 3
Intake:			
Protein(g/(kg.d))	2.8 ± 0.05	3.8 ± 0.04	3.9 ± 0.04
Energy(kcal/(kg.d))	120.9 ± 1.6	118.6 ± 1.2	143.7 ± 1.3
Growth:			
Δ Weight(g/(kg.d))	15.9 ± 1.8*	19.1 ± 3.2	21.5 ± 2.2
Δ Length(cm/wk)	1.04 ± 0.18	1.21 ± 0.34	1.28 ± 0.26
Δ Head circumference(cm/wk)	0.98 ± 0.11	1.21 ± 0.25	1.24 ± 0.47+
Δ Triceps SF(mm/wk)	0.30 ± 0.10	0.33 ± 0.19	0.52 ± 0.18*
Δ Subscapular SF(mm/wk)	0.39 ± 0.13	0.45 ± 0.14	0.70 ± 0.20*
Nutrient Retention:			
Nitrogen(g/(kg.d))	346.5 ± 12.5*	421.1 ± 27.6*	472.6 ± 20.4*
Sodium(mMoles/(kg.d))	0.94 ± 0.20	1.19 ± 0.27	1.26 ± 0.27+
Potassium(mMoles/(kg.d))	1.05 ± 0.20	1.39 ± 0.38	1.39 ± 0.31
Calcium(mMoles/(kg.d))	2.09 ± 1.02		2.99 ± 1.0
Chloride(mMoles/(kg.d))	0.82 ± 0.19	1.01 ± 0.11	1.10 ± 0.32+
Phosphorus(mMoles/(kg.d))	1.49 ± 0.34	1.72 ± 0.34	2.21 ± 0.32*
BUN(mg/dl)	2.48 ± 0.41	8.06 ± 2.30	5.54 ± 1.15

*Difference from other Groups statistically significant.
+Difference from Group 1 statistically significant.

In both studies, the retention of calcium reflected calcium intake; neither calcium intake nor calcium retention differed among the groups. Although the intakes of other electrolytes and minerals were similar in all groups, retention of these nutrients varied from group to group and was related to nitrogen retention rather than to the intakes of specific electrolytes and minerals (See below).

All intakes studied were well tolerated metabolically. For example, the highest mean blood urea nitrogen concentration observed was <10 mg/dl (Group 2 of Study II; Table III). Moreover, the mean plasma concentrations of most amino acids in all groups were within the 95% confidence limits (ie., ± 2 S.D.) of the plasma amino acid concentrations of LBW infants fed human milk (at least 40% of which was provided by the infant's mother (9)). Blood acid base status of all groups also was within normal limits throughout the study. Thus, protein intakes ranging from 2.24 g/(kg.d) to 3.9 g/(kg.d) appear to be well tolerated metabolically by LBW infants.

In toto, these results provide important insights concerning both the nutrient requirements of LBW infants and the relationship between nutrient intake, nutrient utilization and growth. These insights are discussed in more detail below.

Protein and Energy Requirements of LBW Infants As discussed above, a protein intake of 2.24 g/(kg.d), with an energy intake of 115 kcal/(kg.d), is associated with a slower rate of weight gain as well as a slower rate of increase in both head circumference and length than the higher protein intakes studied. More important, perhaps, the mean plasma albumin and transthyretin (prealbumin) concentrations of this group as well as the mean plasma transthyretin concentration of Group 1 of Study II were significantly lower than those of the groups that received higher protein intakes (Table IV). Moreover, in these groups, plasma albumin and transthyretin concentrations decreased from the mean values observed at birth (21), whereas, in all other groups, plasma albumin and transthyretin concentrations either increased from those observed at birth or did not change. This combination of results suggests that a protein intake of 2.24 g/(kg.d) is inadequate for the type of LBW infants studied (ie., a typical urban population). A protein intake of 2.8 g/(kg.d), on the other hand, may be marginally acceptable but it is clear that higher protein intakes result in greater rates of growth and that a protein intake as high as 3.9 g/(kg.d) is not associated with marked metabolic abnormalities. This protein intake, however, might require a concomitant energy intake greater than 115-120 kcal/(kg.d) for complete utilization.

Table IV. Mean Plasma Albumin (g/dl ± S.D) and Transthyretin (Prealbumin) Concentrations (mg/dl ± S.D) of All Study Groups

Group	Albumin	Transthyretin
Study I (21):		
Group 1	2.77 ± 0.28*	6.2 ± 2.0*
Group 2	3.36 ± 0.31	10.6 ± 2.4
Group 3	3.34 ± 0.54	10.0 ± 3.2
Study II (22):		
Group 1	3.23 ± 0.28	11.9 ± 1.96*
Group 2	3.47 ± 0.25	15.5 ± 2.40
Group 3	3.49 ± 0.19	14.8 ± 1.97

*Difference from other groups studied concurrently statistically significant.

Energy intakes of 140-150 kcal/(kg.d) result in slightly greater rates of weight gain than energy intakes in the range of 115-120 kcal/(kg.d). In Study II, the higher energy intake also was associated with better protein utilization. Other than this effect of energy intake on protein utilization, the major effect of energy intake on growth appears to be deposition of more fat. Such an effect is suggested both by the greater rate of increase in skinfold thicknesses of infants in Group 3 of both studies and by the estimated composition of weight gain derived from energy balance data of Study I (see Figures 1 and 2). The estimated daily fat deposition of Groups 1 and 2 (Study I) was quite similar whereas that of Group 3 (whose intake differed from that of Group 2 only in energy content) was considerably greater (Figure 1).

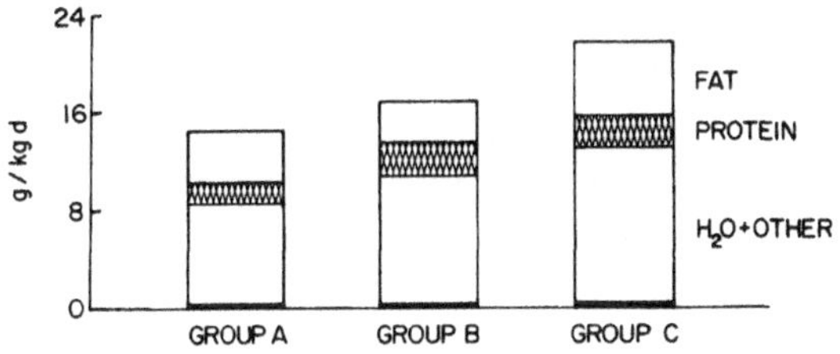

Figure 1. Absolute Composition of Weight Gained by Infants Enrolled in Study I as Estimated from Energy Balance Data (23).

However, the proportional intakes of protein and energy also appear to be important (see Figure 2). The relative composition of the weight gain of Groups 1 and 3, the intakes of which, although different in protein content, had a relatively high proportion of energy to protein, were quite similar;however,the relative composition of the weight gain of Group 2, the intake of which had a lower proportion of energy to protein, was proportionally higher in protein. In other words, the weight gained by this group was leaner than that gained by Groups 1 and 3.

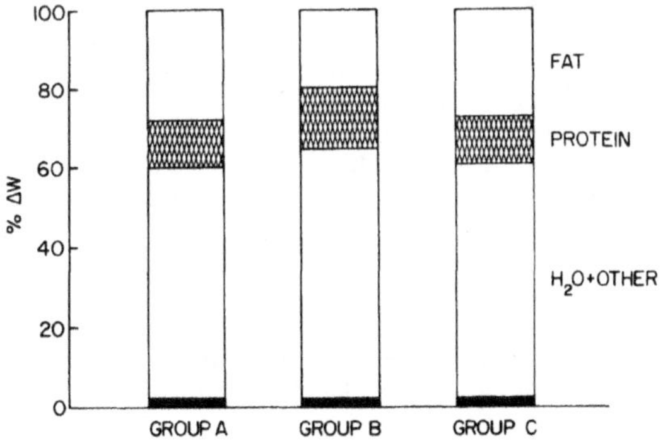

Figure 2. Relative Composition of Weight Gained by Infants Enrolled in Study I as Estimated from Energy Balance Data (23).

It is difficult to draw conclusions from these data concerning the optimal energy intake of LBW infants. On one hand, it is clear that the major effect of energy intakes above 115-120 kcal/(kg.d) is to increase fat deposition. In most settings, this effect would be considered undesirable (24). This is based, in part, on the concern that intakes during infancy may be associated with an "imprinting" effect on intake later in life (ie., that infants whose intake is greater during infancy continue to have greater intakes thoughout life) and, therefore, increase the likelihood of obesity later in life. Considering the many health problems associated with obesity, such an effect obviously may be undesirable. However, whether or not "imprinting" occurs remains to be proven. Also, it should be emphasized that the fat content of the weight gained by infants fed only 115-120 kcal/(kg.d) is roughly twice that retained by the fetus during the last trimester of gestation (25). Thus, should the goals of nutritional management of LBW infants include both an absolute rate of weight gain equal to or greater than the intrauterine rate as well as composition of weight gain similar to that in utero, even the lower energy intakes studied are "excessive".

On the other hand, the greater rate of fat deposition associated with higher energy intakes obviously enhance the rate of weight gain. Thus, since weight is the most frequently used criterion for discharge of LBW infants, it is obvious that those who gain weight more rapidly reach discharge weight sooner and, therefore, are hospitalized for a shorter period. This effect of energy intake on growth obviously has important implications with respect to decreasing the total cost of hospitalization of LBW infants.

Nutrient Utilization and Growth A number of studies in LBW infants suggest that both nitrogen retention and energy retention, respectively, are quite predictable from protein (nitrogen) and energy intakes. In fact, equations expressing the relationship between protein and energy intakes and nitrogen and energy retention have been reported. Those expressing the relationship between nitrogen(N) intake and retention of formula-fed infants include the following:

$$\text{N Balance} = 0.67\ (\text{N Intake}) - 0.03\ (2)$$
$$\text{N Balance} = 0.73\ (\text{Metabolizable N Intake}) - 0.07\ (14)$$

The relationship between protein intake and nitrogen retention derived from the studies discussed above (26) is:

$$\text{N Balance} = 0.7\ (\text{N Intake}) + 0.03$$

These relationships between nitrogen intake and retention of formula-fed infants are quite similar and also similar to the relationship reported by Reichman et al (14) for infants fed their own mother's milk (ie., Nitrogen Balance = 0.64 (Metabolizable N Intake)+0.28). Thus, it appears that the LBW infant fed protein intakes ranging from 2 to 4 g/(kg.d) can be expected to retain 65-75% of these intakes. Whether or not this relationship holds for higher protein intakes remains to be determined.

The equations that have been reported relating energy (E) balance of formula-fed infants to metabolizable energy intake include the following:

$$\text{E Balance} = 0.71\ (\text{E Intake}) - 28\ (2)$$
$$\text{E Balance} = 0.84\ (\text{E Intake}) - 41.8\ (14)$$

The relationship between energy balance and gross intake derived from the data reported above is:

$$\text{E Balance} = 0.65\ (\text{E Intake}) - 28$$

Thus, it appears that the amount of energy retained is a function, primarily, of energy intake.

Few investigators have attempted to derive multiple regression equations for weight gain. The equation which best describes the rate of weight gain of the 72 infants comprising the study populations of Study I and II plus 37 other infants studied by the authors is:

$$\Delta\ \text{Weight} = 3.6(\text{Protein}_{in}) + 0.095(\text{Energy}_{in}) - 0047(\text{Birthweight}) + 1.7 (r = 0.8).$$

This equation suggests that the rate of weight gain of LBW infants is highly predictable from protein and energy intakes.

To determine if this equation applies only to the data from which it was derived or if it is more universally applicable, weight gains of LBW infants reported in most recent studies (1-20) were compared with the weight gains predicted by this equation using mean protein and energy intakes as well as mean birthweights reported. As shown in Figure 3, the agreement between reported weight gains and weight gains predicted by the above equation is quite good.

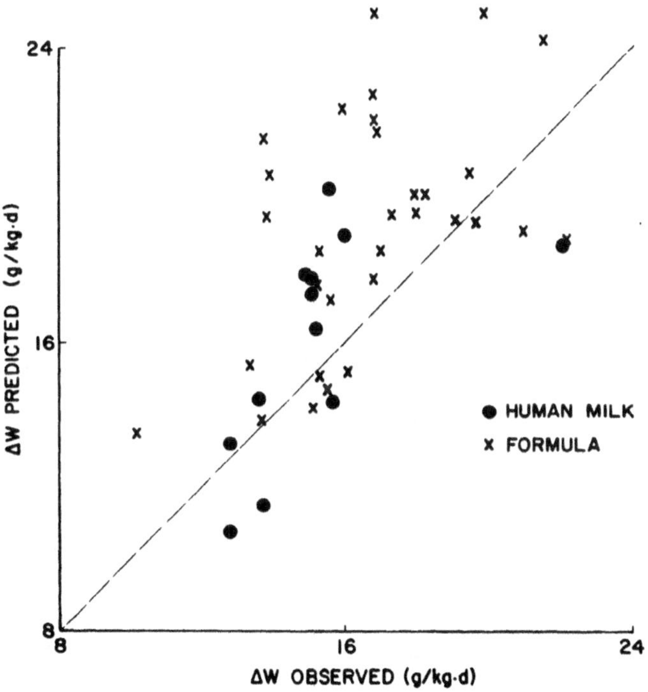

Figure 3. Comparison of Reported Weight Gain of LBW Infants (ΔW Observed) With Weight Gain Predicted (Δ W Predicted) Using Equations Summarizing Data of the Authors (See text).

Inconsistencies in expressing rate of weight gain, particularly when this rate is expressed as g/(kg.d) account for some of the deviation between observed and predicted rates of weight gain shown in Figure 3. In some reports, the mean daily rate of weight gain during the study period (g/d) was divided by birth weight. In other reports, the mean daily rate of weight gain during the study period (g/d) was divided by the mean weight over the period of study (eg, if an infant was studied from 1200 g to 2000 g, the mean daily rate of weight gain was divided by 1.6 kg). In the studies on which the equation is based, the rate of weight gain was expressed as the slope of the regression line relating daily weight gain (g/d) to daily weight (kg).

Another probable explanation for some of the discrepancies between predicted and observed rates of weight gain, particularly those associated with a predicted weight gain greater than observed weight gain, concerns the fact that retention of various electrolytes and minerals is related to nitrogen retention rather than to their respective intakes. This relationship, which is developed in more detail in the following section, suggests that inadequate intakes of one or more electrolytes and/or minerals may prevent utilization of nitrogen intake for growth; hence, predicted weight gain will be greater than observed weight gain.

Electrolyte/Mineral Retention and Nitrogen Retention In the studies reported above, the intakes of all electrolytes and minerals were similar. Yet, as noted, there is considerable variation among groups in retention of these nutrients. In fact, the retention of sodium, potassium, chloride and phosphorus corrected for bone deposition of phosphorus (ie., skeletal phosphorus, on a molar basis, is 60% of skeletal calcium (27)) appears to be related closely to nitrogen retention. In other words, the retention of each of these nutrients can be expressed in terms of nitrogen retention (26):

$$\text{Na Balance} = 2.87(\text{N Balance}) + 0.01 \quad (r=0.60)$$

$$\text{K Balance} = 2.22(\text{N Balance}) + 0.04 \quad (r=0.45)$$

$$\text{Cl Balance} = 1.72(\text{N Balance}) + 0.405 \quad (r=0.36)$$

$$\text{P Balance} = 21.9(\text{N Balance}) + 4.34 \quad (r=0.24)$$

These relationships, although having received little attention, are not unexpected. As shown in Figure 4, there is a quite predictable relationship, in utero, between depostion of each of these nutrients and deposition of protein. Moreover, the intrauterine relationships between retention of these nutrients and nitrogen retention are roughly the same as the relationships expressed by the above equations.

The predictable relationship between retention of various electrolytes and minerals and nitrogen retention suggests that an inadequate intake of one or more of the electrolytes and/or minerals may inhibit utilization of protein. In other words, if the intakes of these nutrients are not adequate, ingested protein may not be utilized for growth. Rudman, et al (28) demonstrated the feasability of this possibility in adults receiving parenteral nutrition regimens lacking one or more nutrients. Withdrawal of either sodium, potassium or phosphorus from an otherwise complete parenteral nutrition regimen resulted in a net loss of the withdrawn nutrient as well as lower retention of nitrogen and a lower rate of weight gain. In some cases, nitrogen retention became negative despite no change in nitrogen intake; weight gain, however, usually continued, albeit at a lower rate.

Some studies in LBW infants also suggest that inadequate electrolyte and/or mineral intakes may inhibit nitrogen retention and growth (See

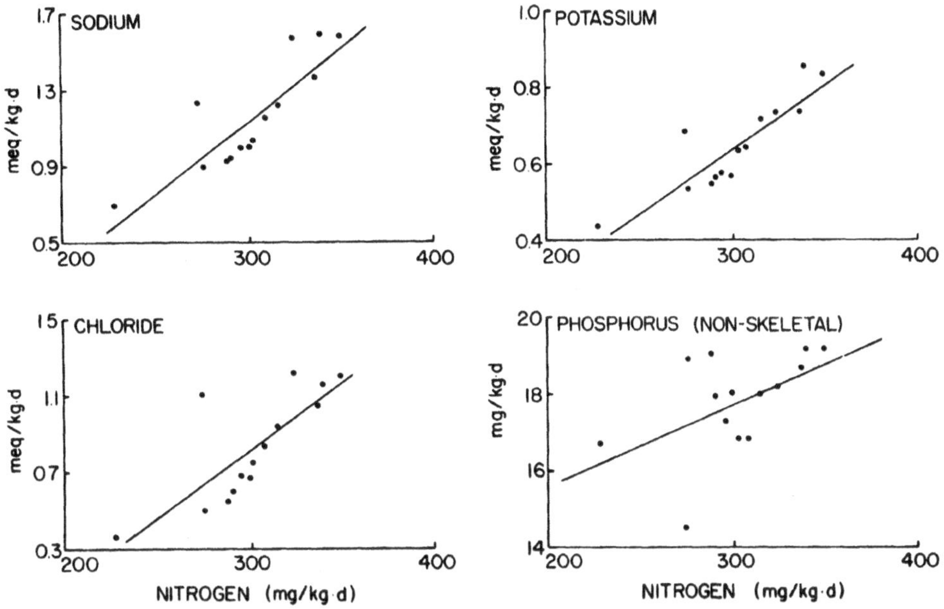

Figure 4. Intrauterine Relationship Between Nitrogen Accretion and Accretion of Selected Electrolytes and Minerals (Calculated from data of Ziegler, et al (25)).

Table V). For example, if the intake of sodium is limiting, as may have been the case for Group B (Table V (13)), the rate of weight gain is more closely predicted from sodium retention than from nitrogen retention whereas, if sodium intake is adequate, as was the case in the other groups (Table V), the rate of weight gain is equally predictable from either nitrogen or sodium retention.

Studies of LBW Infants Fed Human Milk Considering the intense interest in the the role of human milk in the nutritional management of the LBW infant, an obvious question arises: Is the predictable relationship between nutrient intake and growth observed in formula-fed infants true also for the infant fed human milk?

Studies by the authors over the past several years also provide insight into this question (30). Infants similar to those enrolled in the various formula studies were also enrolled into various human milk regimens. In one group, at least 80% of intake was human milk provided by the infant's mother. The predicted rate of weight gain as well as the predicted retentions of nitrogen, sodium, potassium and phosphorus of this group of infants, based on the equations derived from formula-fed infants, and the same variables observed in these infants are summarized in Table VI. From these data, it is apparent that the equations describing the predictable relationship between nutrient intake and growth of formula-fed infants are equally applicable to infants fed human milk. In other words, it appears that both growth and nutrient retention of infants fed human milk are entirely predictable from the nutrient content of that milk.

Whether or not the nutrient content of human milk is sufficient to permit growth at the intrauterine rate, it is clear that the rate of weight gain observed in infants fed human milk is that predicted from protein and energy intakes. More important, perhaps, the sodium retention observed

Table V. Example of Probable Effect of Inadequate Sodium Intake on Weight Gain.

	Group A*	Group B*	Group C+	Group D+
Protein Intake (g/(kg.d))	2.25	4.5	2.24	3.62
Energy Intake (kcal/(kg.d))	117	117	115	114
Sodium Intake (mEq/(kg.d))	1.2	1.2	2.7	2.6
Nitrogen Retention (mg/(kg.d))	272	550	270	422
Sodium Retention (mEq/(kg.d))	0.72	0.72	0.74	1.34
Δ Weight (g/(kg.d)), predicted from protein retention°	~13	~22	~14	~19
Δ Weight (g/(kg.d)), predicted from sodium retention°	~13	~13	~13	~18
Δ Weight (g/(kg.d)) observed	11.8	11.2	13.9	18.3

*Adapted from Raiha, et al (13). Nitrogen retention was estimated using equations expressing relationship between nitrogen intake and retention; (see text); sodium retention was estimated using measured urinary sodium concentration (29) and estimated urine volume.
+Data of Kashyap, et al (21).
°Estimated from intrauterine relationship between weight gain and accretion of protein and sodium (25).

also is quite similar to the retention predicted by the equation describing the relationship between sodium retention and nitrogen retention in formula-fed infants. In other words, although human milk may not contain sufficient sodium to permit deposition of this nutrient at the intrauterine rate, it contains sufficient sodium to permit deposition of both sodium and nitrogen at the rates permissible by the protein content of preterm human milk. The observed retention of phosphorus, on the other hand, is somewhat lower than predicted; however, the difference between

Table VI. Comparison of Growth Rates and Nutrient Retention observed in Infants Fed Human Milk With Those Predicted From Equations Derived From Data Obtained in Formula-Fed Infants.

Variable	Predicted	Observed
Δ Weight (g/(kg.d))	15.2	16.35 ± 2.20
N Balance (mg/(kg.d))	301	290 ± 55
Na Balance (mMoles/(kg.d))	0.8	0.99 ± 0.20
K Balance (mMoles/(kg.d))	1.0	1.02 ± 0.29
Cl Balance (mMoles/(kg.d))	0.9	1.02 ± 0.13
P Balance (mMoles/(kg.d))	10.3	7.80 ± 3.21

observed and predicted phosphorus retention is not statistically significant.

This predictable relationship between nutrient intake, nutrient utilization and growth in LBW infants fed human milk is also apparent from other studies of LBW infants fed human milk. From Figure 3, it is clear that the predicted and observed rates of weight gain of LBW infants fed human milk deviate no more from the line of identity than the predicted and observed rates of weight gain of formula-fed infants.

Conclusions From the preceeding discussion, it appears that the rate of weight gain of LBW infants can be predicted quite reasonably from their intakes of protein and energy, provided intakes of all other nutrients necessary for growth also are sufficient. Thus, despite a variety of physiologic and metabolic limitations, it appears that the LBW infant efficiently utilizes protein and energy intakes for growth and that this utilization is quite predictable. This implies that the LBW infant is capable of mounting the endocrine and metabolic responses necessary for utilization of the nutrient intakes provided. Moreover, the LBW infant appears to adjust its hormonal milieu in response to nutrient intake and this adjustment appears to favor utilization of the delivered nutrients for growth.

The predictable relationship between nutrient intakes and both nutrient retention and growth suggest that the protein and energy intakes required for both a specific rate and a specific composition of weight gain can be predicted. If true, it should eventually be possible to evaluate not only the currently recommended goal for growth of the LBW infant (ie., the intrauterine rate of weight gain) but also other goals concerning either the rate or quality of weight gain that may be proposed in the future.

References

1. O.G. Brooke, C. Wood and J. Barley, Energy Balance, Nitrogen Balance and Growth in Preterm Infants Fed Expressed Breast Milk, a Premature Infant Formula, and Two Low-Solute Adapted Formulae, Arch. Dis. Child. 57:898 (1982).
2. C. Catzeflis, Y. Schutz, J.L. Micheli, C. Welsch, M.J. Arnaud and E. Jequier, Whole Body Protein Synthesis and Energy Expenditure in Very Low Birth Weight Infant, Pediatr. Res 19:679 (1985).
3. P. Chessex, B.L. Reichman, G.J.E. Verellen, G. Putet, J.M. Smith, T. Heim and P.R. Swyer, Influence of Postnatal Age, Energy Intake, and Weight Gain on Energy Metabolism in the Very Low Birth Weight Infant, J. Pediatr. 99:761 (1981).
4. P. Darling, G. Lepage, P. Tremblay, S. Collet, L.C. Kien and C.C. Roy, Protein Quality and Quantity in Preterm Infants Receiving the Same Energy Intake, Am. J. Dis. Child. 139:186 (1985).
5. T. J. French, M. Colbeck, D. Burman, B.D. Speidel and R.A. Hendey, A Modified Cows' Milk Formula Suitable for Low Birth Weight Infants, Arch. Dis. Child. 57:507 (1982).
6. D. Freymond, Y. Schutz, J. Decombaz, J.L. Micheli and E Jequier, Energy Balance, Physical Activity, and Thermogenic Effect of Feeding in Premature Infants, Pediatr. Res. 20:638 (1986).
7. S.J. Gross, Growth and Biochemical Response of Preterm Infants Fed Human Milk or Modified Infant Formula, N. Eng. J. Med. 308:237 (1983)
8. R.K. Huston, J. W. Reynolds, C. Jensen and R.M. Buis, Nutrient and Mineral Retention and Vitamin D Absorption in Low Birth Weight Infants: Effect of Medium Chain Triglyceride, Pediatr. 72:44 (1983).

9. D.K. Rassin, G.E. Gaull, A.L. Jarvenpaa and N.C.R. Raiha, Feeding the Low Birth Weight Infant: II. Effects of Taurine and Cholesterol Supplementation on Amino Acids and Cholesterol, Pediatr. 71:179 (1983).
10. A. Lucas, S.M. Gore, T.J. Cole, M.F. Bamford, J.F.B. Dossetor, I Barr, L. Dicarlo, S. Cork and P.J. Lucas, Multicentre Trial on Feeding Low Birth weight Infants: Effects of Diet on Early Growth, Arch. Dis. Child. 59:702 (1984).
11. E. Okamoto, C.R. Muttart, C.L. Zucker and W.C. Heird, Use of Medium Chain Triglycerides in Feeding Low Birth Weight Infants, Am. J. Dis. Child. 136:428 (1982).
12. G. Putet, J. Senterre, J. Rigo and B. Salle, Nutrient Balance, Energy Utilization, and Composition of Weight Gain in Very Low Birth Weight Infants Fed Pooled Human Milk or a Preterm Formula, J. Pediatr. 105:79 (1984).
13. N.C.R. Raiha, K. Heinonen, D.K. Rassin and G.E. Gaull, Milk Protein Quantity and Quality in Low Birth Weight Infants. I. Metabolic Responses and Effects on Growth, Pediatr. 57:659 (1976).
14. B. Reichman, P. Chessex, G. Verellen, G. Putet, J.M. Smith, T. Heim and P.R. Swyer, Dietary Composition and Macronutrient Storage in Preterm Infants, Pediatr. 72:322 (1983).
15. S.B. Roberts and A. Lucas, The Effects of Two Extremes of Dietary Intake on Protein Accretion in Preterm Infants, Early Hum. Dev. 12:301 (1985).
16. R.J. Schanler, C. Garza and B.L. Nichols, Fortified Mothers' Milk For Very Low Birth Weight Infants: Results of Growth and Nutrient Balance Studies, J. Pediatr. 107:437 (1985).
17. P.J. Shenai, J.W. Reynolds and S.G. Babson, Nutritional Balance Studies in Very Low Birth Weight Infants: Enhanced Nutrient Retention Rates by an Experimental Formula, Pediatr. 66:233 (1980).
18. N.W. Svenningsen, M. Lindroth and B. Lindquist, Growth in Relation to Protein Intake of Low Birth Weight Infants, Early Hum. Dev. 6:47 (1982).
19. J.E. Tyson, R.E. Kasky, C.E. Mize, C.J. Richards, N. Blair-Smith, R. Whyte and A.E. Beer, Growth Metabolic Response and Development in Very Low Birth Weight Infants Fed Banked Human Milk or Enriched Formula: I. Neonatal Findings, J. Pediatr. 103:95 (1983).
20. R.K. Whyte, R. Haslam, C. Vlainic, S. Shannon, K. Samulski, D. Campbell, HS. Bayley and J.C. Sinclair, Energy Balance and Nitrogen Balance in Growing Low Birth weight Infants Fed Human Milk or Formula, Pediatr. Res. 17:891 (1983).
21. S. Kashyap, M. Forsyth, C. Zucker, R. Ramakrishnan, R.B. Dell and W.C. Heird, Effects of Varying Protein and Energy Intakes on Growth and Metabolic Response in Low Birth Weight Infants, J. Pediatr. 108:955 (1986).
22. S. Kashyap, K.F. Schulze, M. Forsyth, C. Zucker, R. Ramakrishnan, R.B. Dell and W.C. Heird, Effects of Varying Protein and Energy Intakes on Growth, Nutrient Retention and Metabolic Response of Low Birth Weight Infants, J. Pediatr. (Submitted, 1986).
23. K.F. Schulze, M. Stefanski, J. Masterson, R. Spinnazola, R. Ramakrishnan, R.B. Dell and W.C. Heird, Energy Expenditure, Energy Balance and Composition of Weight Gain in Low Birth Weight Infants Fed Diets of Different Protein and Energy Content, J. Pediatr. In press (1987).
24. E. Danforth, Diet and Obesity, Am.J.Clin.Nutr. 41:1132 (1985).
25. E.E. Ziegler, A.M. O'Donnel, S.E. Nelson and S.J. Fomon, Body Composition of the Reference Fetus, Growth. 40:329 (1976).
26. S. Kashyap, M. Forsyth, C. Zucker, R.B. Dell and W.C. Heird, Relationship between Nitrogen(N) Retention and Retention of Electrolytes and Minerals in Low Birth Weight(LWB) Infants, Pediatr. Res. 20:413A (1986)

27. J.W.T. Dickerson, Changes in Composition of the Human Femur During Growth, Biochem. J. 82:56 (1962).
28. D. Rudman, W.J. Millikan, T.J. Richardson, T.J. Bixler, W.J. Stockhouse and W.C. McGarrity, Elemental Balances During Intravenous Hyperalimentation of Underweight Adult Subjects, J. Clin. Invest. 55:94 (1975).
29. W.C. Heird, Unpublished Data.
30. S. Kashyap, K.F. Schulze, M. Forsyth, C. Zucker, R. Ramakrishnan, R.B. Dell and W.C. Heird, Growth, Nutrient Retention and Metabolic Response of Low Birth Weight Infants Fed Different Preparations of Human Milk, J. Pediatr. (Submitted, 1986).

NUTRIENT UTILIZATION IN TERM INFANTS

Mary Frances Picciano

School of Human Resources and Family Studies
Division of Nutritional Sciences
University of Illinois
Urbana, Illinois 61801

INTRODUCTION

The provision of optimal nutrition during infancy is paramount considering the rapidity with which infants grow during the postnatal period. Human growth rate is the highest in infancy, with the exception of that experienced in utero. Median growth rate in infancy is about 10 g/Kg/d at 2 to 4 weeks, 3.5 g/Kg/d at 12 to 16 weeks, and 1 g/Kg/day at the end of the first year of life (1). During the most intensive growth period of early infancy, human milk or an appropriate substitute is the principal, if not the only source of required nutrients for acquisition and maintenance of bodily tissues. Marked compositional differences exist among human milk and infant formula preparations and, in large part, the physiological significance of these differences has not been evaluated.

Any alteration in the supply of nutrients to infants can have devastating immediate and long-term consequences. Breast-feeding is the ideal method for furnishing all essential nutrients in required amounts during this vulnerable period of life. When breast-feeding is not practiced, for whatever reason, substitutes for human milk are provided. The definition of infant nutritional requirements is essential to enable evaluation of adequacy of the particular milk fed and to determine when and which nutrients are to be furnished by other foods. However, our knowledge of infant nutritional requirements is far from complete. Traditionally, anthropometric measures have served as the main criteria of nutritional adequacy. There is now ample evidence that indices of growth do not provide the necessary degree of sensitivity. This point is well illustrated in the case of mild iron deficiency. Oski and associates recently reported that normally-growing infants without anemia but with biochemical evidence of iron deficiency display abnormal mental development indices which improve within one week of iron therapy (2). Thus, iron deficiency mild enough to be without hematologic and growth manifestations can affect the behavioral and intellectual attainment of infants. This example stresses the need for studies which employ sensitive and specific measures of nutrient utilization and functional consequences in term infants displaying apparently normal growth.

In this selective review, methodology for the assessment of nutritional requirements of infants and factors that influence the selection of appropriate criteria of evaluation will be discussed. The period of infancy is characterized by many unique features of metabolism which impact on nutrient utilization and requirements. Developmental aspects of trace element and nitrogen metabolism will be briefly discussed to illustrate how improved assessment methods have and should continue to provide an understanding of the prime role of early nutrition in the acquisition of full biological potential.

ASSESSMENT OF NUTRIENT NEEDS OF INFANTS

There are basically four general approaches to the assessment of infant nutrient requirements. These are studies of human milk composition, bodily accretion, nutrient balance, and metabolic indices of nutrient utilization. When other data are unavailable, the composition of human milk from well nourished women and levels of intake of apparently healthy breastfed infants often serve as the basis of human infant requirements. This approach is a useful starting point, and there is general agreement that infant formula preparations should contain at least the quantity of essential nutrients contained in human milk. However, differences in bioavailability of nutrients from human milk and infant formulas are well documented, particularly for trace elements (3-6), serving to illustrate the drawbacks of relying on chemical analyses of milk as indicators of nutritional adequacy. In studies which have determined body accretion of nutrients, measurements of the accumulation of the nutrient under investigation in bodily tissues are made temporally. Data for fetuses and newborns from direct whole body chemical analyses are available (7). Such data have been employed to develop estimates of nutrient requirements for growing premature and term infants.

Balance studies which involve measurement of nutrient input (intake) and output (excretion) under well-defined circumstances are useful probes in adulthood when steady-state conditions can be reasonably assured. However, their usefulness in defining infant nutrient requirements is less certain because of the continuous growth experienced during infancy (9). Balance studies are useful, however, in assessing nutrient retention due to the type of infant feeding regimen. Early investigators employed radioisotopes to study tissue accretion *in vivo,* but the use of radioisotopes is no longer feasible because of the hazards of radiation. With advancements in detection methods for stable isotopes, this approach will undoubtedly find increasing application in infant nutrition studies (8).

Finally, as knowledge of nutrient metabolism becomes more refined, the study of bodily metabolites is useful in assessing infant nutritional requirements. This approach involves specific measures of nutrient utilization and is the focus of this discussion. For example, the measurement of serum ferritin can provide useful information on relative iron accretion (10), information which in the early 1970's required a bone marrow aspirate. The soundness of this approach, of course, is dependent upon the use of appropriate indices of nutrient utilization, that is measures which are sensitive, selective, and specific for the nutrient under investigation.

TRACE ELEMENT NUTRITION

There are several very important influencing factors which must be considered when evaluating trace element nutrition of infants. Such factors include fetal endowment, postnatal age, milk content and bioavailability, and introduction of solid foods.

Nutrient requirements for trace elements are strongly influenced by fetal stores. In other words, a milk diet may be adequate for a specific trace element when an infant is born with his/her full fetal endowment of the element but inadequate when fetal endowment is compromised. Liver stores accumulated primarily during the last trimester of pregnancy exert a strong effect on infant copper, iron, and zinc status (11). Term infants are estimated to contain 15 to 17 mg of copper, which is principally accumulated during the last trimester (12). Body iron concentration at birth averages about 75 mg/Kg, of which approximately 75% is found in circulating hemoglobin. Tissue iron stores probably account for 10 to 20 mg of iron. A roughly linear relationship between body iron and body weight occurs throughout fetal development. It is for this reason that premature infants show a greater disposition toward iron deficiency than term infants (13). Body zinc concentration at birth is approximately 38 mg/Kg, 20% of which is in the liver. Only about 2% of total body zinc is in the liver of adults (14). In contrast to copper, iron, and zinc, the human fetal liver does not accumulate manganese (15). Thus, if low intakes were provided, infants would be at great risk for developing manganese deficiency.

Establishing normal values for age-related metabolic indices of utilization is paramount to the assessment of trace element status of infants in nutritional studies. Plasma or serum concentration of copper, 94% of which is bound to circulating ceruloplasmin, is low at birth and steadily rises during the postnatal period. This rise reflects apoceruloplasmin synthesis and is not altered by dietary copper concentration (16). Iron utilization during the first months of life has been well characterized. The increased oxygenation of blood through the pulmonary versus the placental route is associated with decreases in erythrocyte production and hemoglobin concentration postnatally. Infant erythrocytes have a shorter life span than those of adults, and hemoglobin catabolism is accelerated. Since little iron is excreted, these events result in a redistribution of bodily iron from blood to storage sites in early infancy. Almost all of the iron in the serum is bound to the iron-binding protein, transferrin. Serum iron concentrations are low at birth and steadily rise during the first two to three months of life, in concert with synthetic capacity for transferrin in the liver (13). As was the case for copper, dietary iron has no influence on these features of iron metabolism. In direct contrast, plasma zinc concentration does not exhibit a developmental pattern. The finding of low plasma zinc values for cow's milk formula-fed infants in the U.S. led to studies demonstrating low bioavailability of zinc from cow's milk and infant formulas and the need for supplementing zinc at levels above those in human milk (17). The fall in serum selenium concentration during the first 3 months of life first reported by German investigators (18) is now recognized to be due, at least in part, to low selenium intakes from formula preparation (19). Recent evidence from Japan suggests that bioavailability of selenium is also higher from human milk than from cow milk-based formula (20).

As indicated above, there is a large body of evidence to indicate that trace element bioavailability is generally greater from human milk than from cow milk or infant formulas (21). Superior bioavailability of human milk iron and zinc has been clearly demonstrated; however, bioavailability of other trace elements from infant foods is less certain. The mechanisms of iron and zinc absorption from human milk are not completely understood. Iron from human milk is absorbed five times as efficiently as a similar amount from cow milk (4). The profound inhibitory effect of solid foods on iron absorption from human milk also has been demonstrated (10). The interference of solid foods on iron absorption from human milk may provide an explanation for apparently contradictory reports of iron sufficiency in exclusively breastfed infants up to 18 months (22) and of low values for hemoglobin, serum iron, and transferrin saturation in breastfed infants between 6 and 9 months (23). Casey and co-workers (24) reported that consumption of 25 mg of zinc with human milk resulted in a significantly higher plasma Zn response than with either cow milk or infant formula. These findings suggest that human milk may enhance the bioavailability of zinc from solid foods when ingested together, a situation which is directly opposite to that for iron.

AMINO ACID REQUIREMENTS

Present estimates of essential amino acid requirements for infants are derived from studies of formula-fed infants (25). Holt and Snyderman fed semi-synthetic formulas containing mixtures of L-amino acids to infants between the ages of 2 to 6 months and estimated requirements from quantities necessary to achieve maximal nitrogen retention and growth response (26). Fomon and associates (27) fed infant formulas with intact proteins of known amino acid composition and calculated minimum quantities consumed to achieve adequate growth in the 0.5- to 4-month-old infants studied. The Food and Agricultural Organization/World Health Organization (FAO/WHO) estimations of amino acid requirements (25) are composites of data derived from these two lines of inquiry using the lowest values which were adequate for all infants under investigation.

We recently compared amino acid intakes of human milk-fed infants to FAO/WHO recommendations (Table 1). Oral intakes by ten infants were calculated from measurements of milk ingested over a 72-hour period and from direct analyses of human milk samples collected at 2, 4 and 8 weeks of lactation (28). While

Table 1. Amino acid intakes for infants fed human milk at 2, 4, and 8 weeks of age compared to current estimated requirements

Amino Acid	Estimated Requirement[2] (umol/kg/day)	Amino Acid Intakes[1] 2 wk		4 wk		8 wk	
				umol/Kg/day			
phenylalanine	363	461 ± 33[3]	(20%)[4]	363 ± 19	(50%)	296 ± 20	(90%)
tyrosine	359	412 ± 29	(30%)	326 ± 16	(80%)	271 ± 18	(100%)
methionine	182	201 ± 15	(40%)	157 ± 8	(90%)	129 ± 8	(90%)
threonine	731	829 ± 52	(30%)	659 ± 31	(80%)	545 ± 33	(100%)
valine	795	869 ± 62	(50%)	669 ± 26	(90%)	569 ± 36	(100%)
lysine	706	993 ± 71	(10%)	803 ± 35	(10%)	649 ± 40	(80%)
leucine	1229	1765 ± 112	(0%)	1484 ± 80	(0%)	1167 ± 60	(70%)
histidine	180	331 ± 26	(0%)	251 ± 12	(0%)	216 ± 16	(30%)
isoleucine	534	1361 ± 164	(0%)	986 ± 78	(0%)	769 ± 79	(20%)
arginine	-	440 ± 31		337 ± 19		282 ± 21	

[1]Significant temporal variation was apparent for all amino acids; week 2 > week 4 and 8 at $p < 0.01$-0.001. However, the decline from 4 to 8 weeks was statistically significant for only leucine and isoleucine ($p < 0.001$).
[2]FAO/WHO estimates of amino acid requirements (25).
[3]At each time frame, values for each amino acid are mean ± S.E.M. and range of individual intakes.
[4]Values in parentheses indicate the percentage of infants at each time frame consuming less than estimated requirement.

the majority of infants received quantities of essential amino acids that exceeded estimated requirements at 2 weeks, the overwhelming number of infants received less than these estimates for 7 of 9 amino acids studied at 8 weeks. Differences between data at 2 and 8 weeks were primarily due to a 20% decline in human milk concentration of amino acids. It follows that current FAO/WHO recommendations overestimate requirements for phenylalanine, tyrosine, methionine, threonine, valine, lysine and leucine. Amino acid requirements for infants are most likely to be less than or near minimal values, not maximal values, observed in this study.

PROTEIN QUALITY ASSESSMENTS IN HUMAN INFANTS

Results of a number of studies (29-36), initially in preterm infants and more recently in fullterm infants fed cow milk-based formulas, illustrate the metabolic consequences of differences in protein quality and quantity. The protein in these cow's milk formulas is comprised of 18% whey and 82% casein in contrast to human milk, which is believed to be comprised of 60% whey and 40% casein. In these studies, both preterm and fullterm infants fed casein predominant formulas exhibited elevated plasma concentrations of valine, threonine, phenylalanine, and methionine, depressed plasma concentrations of taurine, and elevated serum concentrations of urea nitrogen compared to infants fed human milk.

As shown in Table 2, preterm and fullterm infants do not display similar patterns of plasma amino acids when fed a casein-predominant formula. For example, elevated plasma arginine concentrations were observed only in preterm infants, while elevated plasma isoleucine concentrations were observed only in fullterm infants. The different plasma amino acid concentrations of fullterm and preterm infants are likely related to the degree of developmental immaturity for protein biosynthesis and degradation. The sustained alterations in plasma amino acid concentrations observed in both preterm and fullterm infants are viewed as abnormal (34). Moreover, the possible long-term consequences of such alterations are unknown. There is evidence that hypertyrosinemia for even a short time after birth is associated with impaired perceptual and psycholinguistic abilities which only manifest themselves at age 5 to 8 years (37, 38). Thus, it is appropriate to strive for indices of protein metabolism with formula feeding that are similar to those obtained with human milk feeding.

EFFECTS OF MODIFIED PROTEIN COMPOSITION OF COW'S MILK FORMULA

One approach taken to achieve such indices of protein metabolism was to develop infant formulas with modified protein compositions. The addition of bovine whey to increase the percent of protein as whey from 18 to 60 is now a common practice. The resultant whey predominant formulas are intended to contain the same whey-to-casein ratio as human milk. Feeding a whey predominant formula to preterm infants resulted in plasma aminograms which appeared to be more similar to those of infants fed human milk than those fed casein predominant formula (31, 33). Plasma concentrations of tyrosine were significantly lowered in preterm infants fed whey predominant

Table 2. Plasma Indicators of Protein Metabolism in Infants Fed Casein and Whey-Predominant Formulas Compared to those of Infants Fed Human Milk[1]

	%whey:%casein	VAL	ILE	LEU	THR	PHE	TYR	LYS	MET	TAU	ARG	BUN
Preterm Infants												
Rassin, D.K. *et al*., 1977; Gaull, G.E. *et al*., 1977	18:82	↑[2]	—[3]	—	↑	↑	↑↑ a[4]	↑	↑	↓	↑	↑
	60:40	↑	—	—	↑	↑	↑ b	↑	↑	↓	↑	↑
Fullterm Infants												
Järvenpää, A.-L. *et al*., 1982	18:82	↑	↑	↑	↑ a	↑ a	↑ a	NR[5]	—	↓	NR	↑
	60:40	↑	↑	↑	↑↑ b	— b	— b	NR	—	↓	NR	↑
Mignone, F. *et al*., 1982	18:82	↑	↑	—	—	↑	—	—	—	NR	—	↑
Voltz, V.R. *et al*., 1983	60:40	—	—	—	—	—	—	—	—	—	—	—
Janas, L.M. *et al*., 1985	18:82	↑	— a	—	— a	↑	—	— a	↑	↓	—	↑
	60:40	↑	↑ b	↑	↑ b	↑	—	↑ b	↑	↓	—	↑

[1]VAL = valine, ILE = isoleucine, LEU = leucine, THR = threonine, PHE = phenylalanine, TYR = tyrosine, LYS = lysine, MET = methionine, TAU = taurine, ARG = arginine, BUN = blood urea nitrogen.
[2]Arrows indicate significantly elevated (↑) or depressed (↓) plasma concentrations compared to infants fed human milk.
[3] —— indicates no difference compared to infants fed human milk.
[4]Between formula fed groups in a single study, amino acids designated by different superscripts indicate statistical significance.
[5]NR-plasma concentrations not reported.

formula compared to those fed casein predominant formula, but plasma concentrations of other amino acids were not consistently lowered.

The feeding of whey predominant formula has also been evaluated in fullterm infants (33,34,36). Fasting plasma amino acid concentrations were higher in fullterm than in preterm infants regardless of the type of milk fed. Fullterm infants receiving either a casein or whey predominant formula generally displayed higher plasma concentrations of serum urea nitrogen, total essential amino acids and molar ratios of total essential-to-total amino acids (Figure 1) than those receiving human milk. Plasma values were also elevated for valine and depressed for taurine compared to infants fed human milk (Table 2). At the same time, certain differences between the modes of formula feeding were apparent. Although only tyrosine concentration was lowered in preterm infants fed a whey predominant formula, in fullterm infants plasma concentration of phenylalanine was also lowered in one study. Fullterm infants fed whey predominant formula also displayed a dramatic elevation in threonine concentration not evident for infants fed casein predominant formula.

Data from earlier studies (39) and our laboratory (36) provide a firm basis for another approach to achieving indices of protein metabolism similar to those of human milk-fed infants. From an assessment of dietary intake data and direct analysis of milks consumed by infants, we observed strong correlations between ingested quantities of amino acids and total nitrogen with plasma concentrations of amino acids. Plasma threonine, valine, leucine, isoleucine, methionine, phenylalanine, histidine, and tyrosine levels are directly correlated with dietary intakes of these amino acids. Therefore, these results should allow one to predict the amino acid composition of formula which will result in plasma amino acid concentrations that approximate those of infants fed human milk. This approach was supported by results from studies with infants receiving total parenteral nutrition. Correlations remarkably similar to ours have been reported between levels of infused amino acids and plasma concentrations (39) even though initial metabolic events in the intestine and liver were by-passed.

We also observed positive correlations between plasma concentrations of the branch chain amino acids (valine, leucine, and isoleucine) and lysine with total nitrogen intake (36). In our studies, daily total nitrogen intakes are higher in both groups of formula-fed infants than in human milk-fed infants at 4 and 8 weeks of age. The increased intakes of nitrogen were due to the higher nitrogen content of the formulas compared to human milk (2.5 mg/ml and 2.0 mg/ml, respectively). These findings indicate that plasma amino acid concentrations typical of those in human milk fed infants will be achieved only if total nitrogen and consequently total protein (amino acid) content of formula is reduced, and if the amino acid pattern is altered to more closely approximate that of human milk.

We recently evaluated indices of amino acid utilization in infants fed formulas containing 18% less protein (1.3 mg/ml) than those currently marketed (1.5 mg/ml) and with varying

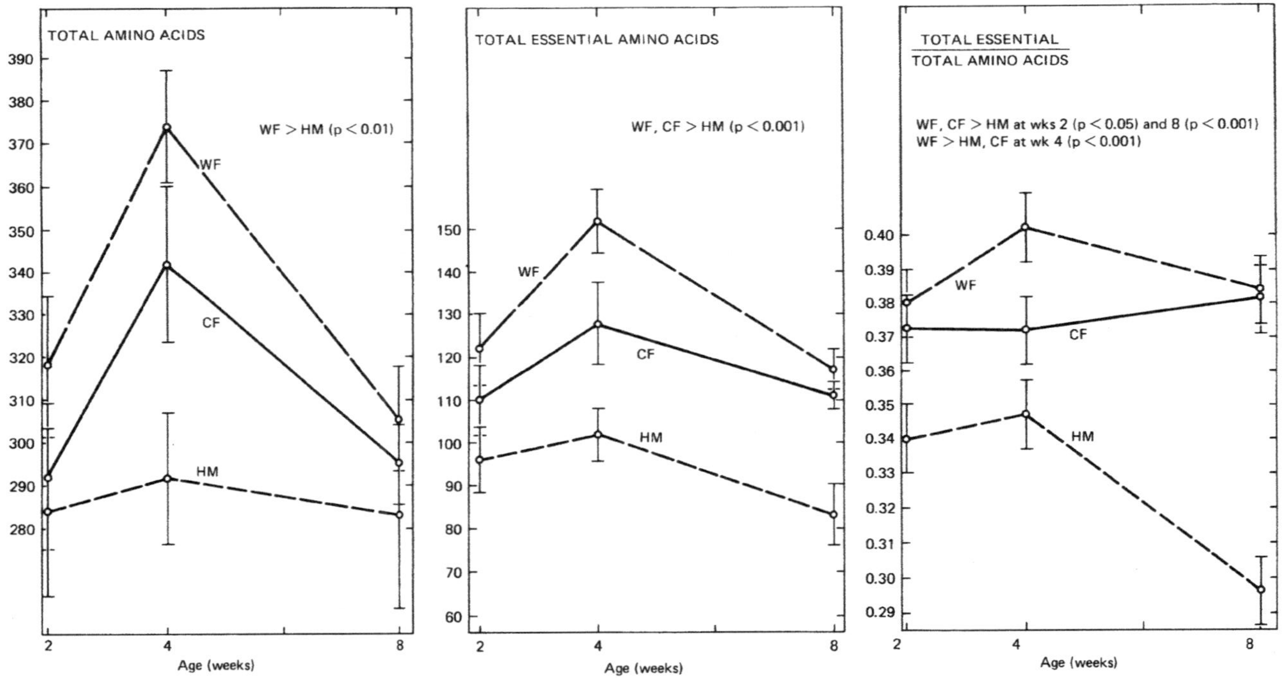

Fig. 1. Mean concentrations of total amino acids (TAA), total essential amino acids (TEAA), and ratios of TEAA/TAA of infants fed various milks. HM designates human milk; WF, whey predominant formula; and CF, casein predominant formula. Only statistically significant group differences are indicated. When a group designation is absent, it did not differ from the others. For all groups, significant temporal variation was apparent for plasma concentrations of TAA and TEAA (wk 4 > wk 2, 8) at $p < 0.001$ and $p < 0.01$, respectively. TEAA included valine, leucine, isoleucine, lysine, threonine, tyrosine, phenylalanine, methionine, cystine and histidine. TAA included TEAA, arginine, citrulline, ornithine, serine, glycine, alanine, proline, aspartate, glutamate, and glutamine. (Reprinted with permission from Pediatrics 75: 775, 1985).

whey-to-casein ratios (40). The three whey-to-casein ratios studied were 50:50, 34:66, and 18:82. Results indicated that dietary protein concentration was the most important determinant of plasma amino acid response. Feeding formulas with reduced protein content resulted in a large number of plasma indices of protein metabolism characteristic of human milk feeding regardless of the specific whey-to-casein ratio. Specifically, mean plasma concentration of total essential and total amino acids and their ratio was similar. Additionally, mean plasma concentrations of valine, lysine, arginine, tyrosine, histidine, threonine, free and total cyst(e)ine and half cystine were similar. However, despite these similarities, formula feeding resulted in alterations relative to human milk feeding, including elevated plasma concentrations of phenylalanine, methionine, isoleucine, and also citrulline and depressed plasma concentrations of tryptophan and taurine. These alterations were common to all three formula groups and, therefore, not a result of differences in specific amino acid pattern of the formula.

The low plasma tryptophan values exhibited by formula-fed infants cannot be explained strictly by intakes, since quantities consumed were similar to or higher than those of human milk-fed infants. Whether low values are a consequence of reduced dietary protein concentration is unclear, since no previous study has evaluated tryptophan. Other possible explanations include reduced bioavailability from heat-treated milk and low digestibility of the whey proteins with highest tryptophan content, alpha-lactalbumin and lactoferrin. Alternately, different hormonal responses of infants fed human milk and formula may be responsible. For example, in 6-day-old term infants, differences have been reported for plasma profiles of a number of hormones such as insulin, cortisol, prolactin, gastric inhibitory peptide, motilin and neurotensin (41-43). These hormonal differences are not evident at 9 months of age when infants are on a mixed diet (44). While insulin administration increases plasma tryptophan concentrations (45), increased C-peptide excretion reported for infants fed formula compared to those fed human milk (46) suggests that the low plasma tryptophan concentrations we observed are not mediated by insulin. Thus, to achieve indices of protein metabolism characteristic of human milk feeding, further modifications of the absolute and relative amounts of amino acids in infant formula preparation are necessary. Thus as predicted, total nitrogen and specific amino acid intakes were important but not exclusive determinants of plasma amino acid responses. As our knowledge of infant protein metabolism advances, identification of other dietary and physiological factors and evaluation of those indicated above may be useful in achieving desired metabolic indices of protein utilization characteristic of human milk feeding with improved formula preparations.

SUMMARY

From the foregoing discussion it becomes obvious that our level of knowledge which serves as the basis for dietary recommendations to ensure optimal nutrition of infants is far from complete. Infant performance in terms of growth measures are too often the only criteria used for evaluating adequacy of infant diets. Human milk feeding is the preferred feeding method but when it is contraindicated for any reason, modern and reliable data on human milk constituents and their physiological significance to the infant should serve as the basis for formula preparations, unless there is good evidence to the contrary. The adequacy of various feeding methods cannot always be predicted from chemical analyses and must be evaluated with specific nutritional metabolic indices as well as traditional anthropometric measures.

REFERENCES

1. Fomon, S.J. "Infant Nutrition," 2nd edition, W.B. Saunders Company, Philadelphia, 1974.
2. Oski, F.A., Honig, A.S., Helu, B. and Howanitz, P. Effect of iron therapy on behavioral performance in non-anemic, iron deficient infants. Pediatrics 71: 877, 1983.
3. McMillan, J.A., Oski, F.A., Lourie, G., Tomarelli, R.M. and Landaw, S.A. Iron absorption from human milk, simulated human milk and proprietary formulas. Pediatrics 60: 896, 1977.
4. Saarinen, U.M., Siimes, M.A. and Dallman, P.A. Iron absorption in infants: High bioavailability of breast milk iron as indicated by the extrinsic tag method of iron absorption and by the concentration of serum ferritin. J. Pediatr. 91: 36, 1977.
5. Hambidge, K.M., Walravens, P.A., Casey, C.E., Brown, R.M. and Bender, C. Plasma zinc concentrations of breast-fed infants. J. Pediatr. 94: 607, 1979.
6. Sandstrom, B., Keen, C.L. and Lonnerdal, B. An experimental model for studies of zinc bioavailability from milk and infant formula using extrinsic labeling. Amer. J. Clin. Nutr. 38: 420, 1983.
7. Widdowson, E.M. Growth and composition of the fetus and newborn, in: "Biology of Gestation, Volume 2," N.S. Assali, ed., Academic Press, New York, 1968.
8. Matthews, D.E. and Beir, D.M. Stable isotope methods for nutritional investigation. Ann. Rev. Nutr. 3: 309, 1983.
9. Fomon, S.J. and Owen, G.M. Comment on metabolic studies as a method of estimating body composition of infants. Pediatrics 29: 495, 1962.
10. Saarinen, U.M. and Siimes, M.A. Iron absorption from breast milk, cow's milk and iron-supplemented formula: An opportunistic use of changes in total body iron determined by haemoglobin, ferritin and body weight in 132 infants. Pediatric Res. 13: 143, 1979.
11. Widdowson, E.M., Dauncey, J. and Shaw, J.C.L. Trace elements in fetal and early postnatal development. Proc. Nutr. Soc. 33 : 275, 1974.
12. Shaw, J.C.L. Trace elements in the fetus and young infant. II. Copper, manganese, selenium and chromium. Am. J. Dis. Child. 134: 74, 1980.
13. Stekel, A. "Iron Nutrition in Infancy," Raven Press, New York, 1984.
14. Hambidge, K.M. Zinc deficiency in man: its origins and effects. Phil. Trans. R. Soc. Lond. B 294: 129, 1981.
15. Widdowson, E.M., Chan, H., Harrison, G.E., and Milner, R.B.G. Accumulation of copper, zinc, manganese,

chromium and cobalt in the human liver before birth. Biol. Neonate 20: 360, 1972.
16. Salmenpera, L., Perheentupa, J., Pakarinen, P. and Siimes, M. Copper nutrition in infants during prolonged exclusive breast-feeding: Low intake but rising serum concentration of Cu and ceruloplasmin. Am. J. Clin. Nutr. 43: 251, 1986.
17. Hambidge, K.M. Trace elements in pediatric nutrition. Adv. Pediatr. 24: 191, 1977.
18. Lombeck, I., Kasperek, K., Karbirsch, H.D., Feinendegen, I.E. and Bremer, H.J. The selenium states of healthy children. I. Serum selenium concentration at different ages; selenium content of food of infants. Eur. J. Pediatr. 125: 81, 1977.
19. Smith, A.M., Picciano, M.F. and Milner, J.A. Selenium intakes and status of human milk and formula fed infants. Am. J. Clin. Nutr. 35: 521, 1982.
20. Hatano, S., Aihara, K., Nishi, Y. and Usui, T. Trace elements (copper, zinc, manganese and selenium) in plasma and erythrocytes in relation to dietary intake during infancy. J. Ped. Gastr. and Nutr. 4: 87, 1985.
21. Lonnerdal, B. Dietary factors affecting trace element bioavailability from human milk, cow's milk and infant formulas. Progress in Food and Nutrition Science 9: 35, 1985.
22. McMillan, J.A., Lanaw, S.A. and Oski, F.A. Iron sufficiency in breast-fed infants and the availability of iron from human milk. Pediatrics 58: 686, 1976.
23. Saarinen, U.M. Need for iron supplementation in infants on prolonged breast feeding. J. Pediatr. 93: 177, 1978.
24. Casey, C.E., Walravens, P.A., and Hambidge, K.M. Availability of zinc: loading test with human milk, cow's milk and infant formulas. Pediatrics 68: 394, 1981.
25. World Health Organization. "Energy and Protein Requirements," Technical Report Series No. 724, FAO/WHO, Geneva, 1985.
26. Holt, L.E., and Snyderman, S.E. The amino acid requirements of children, in: "Amino Acid Metabolism and Genetic Variation," W.L. Nyhan, ed., McGraw-Hill, New York, 1967.
27. Fomon, S.J., Thomas, L.N., Filer, L.J., et al. Requirements for protein and essential amino acids in early infancy. Acta Paediatr. Scand. 62: 33, 1973.
28. Janas, L.M., and Picciano, M.F. Quantities of amino acids ingested by human milk-fed infants. J. Pediatr. 109: 802, 1986.
29. Raiha, N.C.R., Heinonen, K., Rassin, D.K., and Gaull, G.E. Milk protein quantity and quality in low-birth-weight infants. I. Metabolic responses and effects on growth. Pediatrics 57: 659, 1976.
30. Rassin, D.K., Gaull, G.E., Heinonen, K., and Raiha, N.C.R. Milk protein quantity and quality in low-birth-weight infants. II. Effects on selected aliphatic amino acids in plasma and urine. Pediatrics 59: 407, 1977.
31. Rassin, D.K., Gaull, G.E., Raiha, N.C.R., and Heinonen, K. Milk protein quantity and quality in low-birth-weight infants. IV. Effects on tyrosine and phenylalanine in plasma and urine. J. Pediatr. 90: 356, 1977.

32. Gaull, G.E., Rassin, D.K., Raiha, N.C.R., and Heinonen, K. Milk protein quantity and quality in low-birth-weight infants. III. Effects on sulfur amino acids in plasma and urine. J. Pediatr. 90: 348, 1977.
33. Jarvenpaa, A.L., Raiha, N.C.R., Rassin, D.K., and Gaull, G.E. Milk protein quantity and quality in the term infant. I. Metabolic responses and effects on growth. Pediatrics 70: 214, 1982.
34. Jarvenpaa, A.L., Rassin, D.K., Raiha, N.C.R., and Gaull, G.E. Milk protein quantity and quality in the term infant. II. Effects on acidic and neutral amino acids. Pediatrics 70: 221, 1982.
35. Mignone, F., Oggero, R., Galvagno, G., and Bonetti, G. Plasma amino acids during the first five months of life. Minerva Pediatrica 34: 337, 1982.
36. Janas, L.M., Picciano, M.F., and Hatch, T.F. Indices of protein metabolism in term infants fed human milk, whey-predominant formula, or cow's milk formula. Pediatrics 75: 775, 1985.
37. Mamunes, P., Prince, P.E., Thornton, N.H., Hunt, P.S., and Hitchcock, E.S. Intellectual deficits aftr transient tyrosinemia in the term neonate. Pediatrics 57: 675, 1976.
38. Sternowsky, H., and Heigl, K. Tyrosine and its metabolites in urine and serum of premature and mature newborns: Increased values during formula versus breast feeding. Eur. J. Pediatr. 132: 179, 1979.
39. Winters, R.W., Heird, W.C., Dell, R.B., and Nicholson, J.F. Plasma amino acids in infants receiving parenteral nutrition, in: Amino Acid. Part III. Delivery of Nitrogen. "Clinical Nutrition Update," 1977.
40. Janas, L.M., Picciano, M.F., and Hatch, T.F. Indices of protein metabolism in term infants fed formula with reduced protein concentrations and altered whey to casein ratios. J. Pediatr. (in press).
41. Lucas, A., Adrian-Sarson, D.L., Blackburn, A.M., et al. Breast vs. bottle: Endocrine responses are different with formula feeding. Lancet 2: 1267, 1980.
42. Lucas, A., Boyes, B., Bloom, S.R., et al. Metabolic and endocrine responses to a milk feed in 6 day old term infants: Differences between breast and cow's milk formula feeding. Acta Paediatr. Scand. 70: 195, 1981.
43. Lucas, A., Sarson, D.L., Bloom, S.R., et al. Developmental aspects of gastric inhibitory poly peptide (GIP) and its possible role in the enteroinsular axis in neonates. Acta Paediatr. Scand. 69: 321, 1980.
44. Aynsley-Green, A. Metabolic and endocrine interrelations in the human fetus and neonate. Am. J. Clin. Nutr. 41: 399, 1985.
45. Fernstrom, J.D., and Wurtman, R.J. Brain serotonin content: Physiological dependence on plasma tryptophan levels. Science 173: 149, 1971.
46. Ginsburg, B.E., Lindblad, B.S., Persson, B., et al. Plasma valine and urinary C-peptide in infants. The effect of substituting breast-feeding with formula or formula with human milk. Acta Paediatr. Scand. 74: 615, 1985.

DOES NUTRITION IN EARLY LIFE HAVE LONG TERM METABOLIC EFFECTS?

CAN ANIMAL MODELS BE USED TO PREDICT THESE EFFECTS IN THE HUMAN?

Margit Hamosh and Paul Hamosh

Department of Pediatrics, and Department of Physiology and Biophysics
Georgetown University Medical Center
Washington, D.C. 20007

INTRODUCTION

The mode of infant feeding is thought to affect the metabolism of the adult. It is the aim of my presentation to evaluate whether infant feeding practice affects lipid metabolism in later life. Specifically, I would like to address the following questions:

1. Does breast feeding alter adipose tissue development in a fashion that protects against obesity in later life?

2. Does early exposure to dietary cholesterol (present in human milk at much higher level than in formula) result in more efficient cholesterol catabolism in the adult and thus in lower incidence of hypercholesterolemia and atherosclerosis in later life?

3. Does the unique fatty acid composition of human milk, which provides adequate amounts of docosahexaenoic acid, affect brain composition and function, especially in premature infants?

4. Is membrane structure and function affected by the differences in composition of the fat in human milk and infant formula?

The following is a short review of the current status of information on the relationship between early feeding practices and health in later life. The literature review is by no means exhaustive, rather, the studies have been cited with the aim to illustrate trends and hypotheses. The part dealing with early nutrition, brain function and membrane composition is rather speculative summarizing mainly information on full term and preterm (i.e. produced by mothers of full term or premature infants) milk and the possible effect of different fatty acid composition on cell and organ function.

INFANT FEEDING PRACTICES, WEIGHT GAIN AND OBESITY

The wide spread occurence of obesity and atherosclerosis in developed societies and the difficulty to find efficient treatment modalities have increased the emphasis on prevention of these disorders. One question is whether breast feeding in infancy might be associated with lower incidence of obesity and atherosclerosis in adult life.

Follow-up studies that compare the effect of breast feeding and of formula feeding on growth, weight, fat content, and incidence of obesity have been carried out during the last 20 years. While the earlier studies clearly showed that breast-fed babies had a tendency to be leaner than the bottle fed group, recent studies fail to show this effect. It seems thus, that in the late sixties and early seventies, bottle fed infants were often overfed, while more recently, with the greater awareness of good dietary habits, bottle fed infants receive volumes similar to those of infants fed human milk; furthermore, formula composition is also closer to that of human milk.

Follow-up studies on the effect of early nutrition on growth and obesity can be divided into 3 groups: short term studies (6 weeks to 6 months after birth), slightly longer term follow-up (1 year to 5 years after birth) and long term studies that evaluate the status of older children, adolescents and adults. The earlier studies have been reviewed by Weil (1) and Taitz (2) in 1977.

It is difficult to reach clear cut conclusions from the studies conducted in this field. The marked differences in experimental design, techniques of assessing weight gain and especially adiposity and furthermore, the poorly defined breast feeding variable (i.e., duration and exclusiveness of breast feeding) are major problems in the critical evaluation of this large number of studies.

Early Effects: Infants and Toddlers

In a study conducted in Sheffield in 1970 only 8% of 261 infants studied were breast fed (3). The author stresses the excessive weight gain during the first six weeks of life, which was much greater than in earlier decades, when most infants were breast fed. Taitz (3) also questions the wisdom of over-feeding formula and judging infant well-being solely by weight gain. Oakley (4) on the other hand found greater skinfold thickness at 6 weeks of age in breast fed than in bottle fed infants. Infants who were fed formula plus cereal supplements, had, however, the highest skinfold thickness.

More recently (5), when bottle fed infants were allowed to self-regulate food intake, they consumed almost identical amounts of kcal/kg/day and only slightly higher volumes than a group of breast fed infants, during the first 3 months after birth. The marked change in infant feeding habits during the period 1962-1982 is also evident in additional studies. Thus while Dubois (6), and Ferris (7) report similar weight gains in breast fed and formula fed infants in 1982 and 1980, respectively, Hooper, in studies published in 1965 (8) and 1971 (9) reports a much higher incidence of obesity ("Michelin tire baby" appearance) in bottle fed than in breast fed infants and higher morbidity in the former infants (8). The decrease in tendency to over-feed bottle fed infants (10) is evident also in recent reports of significantly later birth weight doubling (7,11) and tripling times (11), than reported in the mid seventies (12). The more rapid tripling times in formula fed infants, as compared to breast fed infants suggest however, that there still remains a risk of overfeeding the formula fed infant. Furthermore, Jung and Czajka-Narins show race and sex differences in infant growth and suggest that local standards may be more useful to monitor growth of certain populations (10). Indeed, two studies from Scandinavia report similar growth rate and adiposity in proprietary formula fed and breast fed infants in the mid and late seventies (13, 14). The Finnish study (14) shows differences in the type of artificial feeding: thus, while there was no difference between infants fed proprietary milk or breast milk, in infants fed cow's milk; weight for age and skinfold thickness were greater. The authors conclude therefore, that, since volumes were similar and the energy content of the cow's milk was actually lower, the differences between the three groups of infants might be due to differences in the composition of the feeds (especially of protein and fat). Rational feeding is attributed in the Swedish study (13) to the low incidence of obesity (as compared to earlier British studies (10)) in both breast and bottle fed infants. The authors suggest that genetics may be an important factor in the development of obesity. An 18 month follow-up of 316 infants led Yeung et al to conclude that fatness during infancy is not determined by type of feeding (breast milk or formula) or the time of introduction of solids, that in formula fed infants weight gain is determined by the amount of energy consumed and most important, fat infants do not necessarily remain fat (15).

A different view is expressed in the recent study of Kramer et al (16). At 12 months of age, maternal and paternal adiposity do not appear to affect the infant's adiposity. This study also shows that birth weight, sex, age at introduction of solids and duration of breast feeding are all significant predictors of weight at 12 months (r^2= 0.196, $p < 0.0001$). The study concludes that "each week of exclusive breast-feeding resulted in a 16.5 gm reduction in weight at 12 months".

Preschool Children: ages 3-5

In a study published in 1983, Vobecky et al (17) show no difference in weight between breast fed and bottle fed infants at 3 years of age. It is not known how many infants were in each group and how long the breast feeding lasted. This study raises the important point that the infants in the highest weight group did not eat more than those with normal or low weight. Similar findings have been reported by several other investigators (6,13,16,18,19). Thus, the question of energy expenditure, i.e. activity, has to be introduced. Weight at 6 months has only minimal predictive value for weight at 2-5 years (16,17,20,21).

As far as the long-term preventive effect of breast feeding on incidence of obesity at preschool age of 4-5 years, several reports fail to show such an effect (20-22). Furthermore, length of breast feeding and delayed introduction of solids into the diet had no relationship to the incidence of obesity at age 4 (22).

Thus, while the young infant and toddler might, to a certain extent, be protected from overweight by breast feeding, the effect seems to be lost during the following 2-3 years.

School age Children, Adolescents, and Adults

A relationship between height, weight and adiposity in infancy (6 months to one year) and childhood obesity has been reported in several earlier studies (19,24,25). In 1970 Eid, confirmed these observations and showed that rapid weight gain during the first six months of life is positively correlated to excessive weight at age 6-8 years. (26) There was no statistical difference between breast fed and bottle fed infants, although the breast fed infants tended to be leaner. Contrary to these findings in Britan, a Swedish study carried out during the same time concluded that weight gain during the first year of life is not a strong predictor of obesity at age 7-10 years, except for boys in the highest

Table 1: RELATIONSHIP BETWEEN MODE OF FEEDING AND WEIGHT GAIN DURING THE FIRST TWO YEARS OF LIFE

STUDY	YEAR	COUNTRY	REF #	FOLLOW-UP PERIOD	BREAST FED (%)	WEIGHT GAIN	ADIPOSE TISSUE	OTHER
Taitz	1971	England	3	6 Weeks	8	HM < F		
Oakley	1977	England	4	6 Weeks	33	HM = F	HM > F	
Holfvander	1982	Sweden	5	1-3 months	50			HM Consumption $\leq$ F
Dubois	1979	Canada	6	4-9 months	7			HM=F in Incidence of Obesity
Ferris	1980	USA	7	6 months	80	HM = F		
Hooper	1965	England	8	12-18 months	24	HM < F	HM < F	Morbidity: Obese > lean
Hooper	1971	England	9	12-18 months	15	HM < F	HM < F	Morbidity: Obese = lean
Neuman	1976	USA	12	6-12 months	39	HM < F		DT: HM - 124d; F - 113d
Jung	1985	USA	11	6-24 months	37	HM = F		DT: HM - 149d; F - 136d
Saarinen	1979	Finland	14	1-12 months	42	HM = F	HM = F	
Kramer	1985	Canada	16	6-12 months	58	HM < F	HM < F	Length of breast feeding important

HM - human milk; F - formula; DT - birth weight doubling times weight percentiles.(27-29) Similar conclusions were reached in a Swiss study in which no relationship was found between skinfold thickness during infancy at 6 and 9 months, and at age 15 years (except for the subscapular skinfolds in girls) (30). In contrast to these studies, several other investigators report a strong correlation between increased weight (and adiposity) in infancy and childhood and obesity in the adult (31,32).

Only very few studies have addressed the question of long term effects of breast feeding on the prevention of obesity. Kramer (33) finds a significant protective effect of breast feeding against obesity in adolescents and young adults. The magnitude of the protective effect appears to rise slightly with the duration of breast feeding. Marmot et al, on the other hand, report higher mean weight and skin fold thickness in men who have been breast fed (34).

From the studies reported so far it is difficult to assess whether breast feeding indeed might result in a lower incidence of obesity.

Table 2: RELATIONSHIP BETWEEN MODE OF FEEDING IN INFANCY AND INCIDENCE OF OBESITY IN CHILDHOOD AND ADULT LIFE

STUDY	YEAR	COUNTRY	REF #	FOLLOW-UP PERIOD	BREAST FED (%)	PREDICTIVE VALUE OF INFANT WEIGHT	OBESITY INCIDENCE
PreSchool, age 3-5 years							
Vobecky	1983	Canada	17	3 yrs	?	Weak	HM = F
Wolman	1984	USA	22	4 yrs	28	None	HM = F
Dine	1979	USA	20	5 yrs	15	Weak	HM = F
Poskitt	1978	England	21	5 yrs	32	None	HM = F
School Age and Adults							
Fomon	1984	USA	79	8 yrs	37	None	
Eid	1970	England	26	6-8 yrs	40	Strong	HM ≤ F
Mellbin	1976	Sweden	28,29	7-10 yrs	?	Weak, only m	
Hernesniemi	1974	Switzerland	30	15 yrs	?	Weak, only f	
Kramer	1981	Canada	29	12-18 yrs	18	Strong	HM < F
Marmot	1980	England	30	32 yrs	72		HM > F only m
Stark	1981	England	31	26 yrs	?	Strong	
Charney	1976	USA	32	20-30 yrs	?	Strong	

HM - human milk; F - formula; m - males, f - females

When evaluating the relationship of infant nutrition to the development of obesity, many factors in the data gathering process, have to be taken into consideration. Some of these are: expression of growth and ponderosity (23) (weight gain, weight gain normalized per age, weight/height ratio index, weight/height^2 (quetelet index), weight/height$^{1/3}$ (ponderal index), length of breast feeding, type of artificial feeding (proprietary formula vs. cow's milk); has fat accretion been measured (skin fold thickness, densitometry, ^{40}K), length of follow-up period: are the infant data based on recorded information or parental recall? Studies using similar techniques might provide a clear cut answer to this important question. An excellent discussion of the problems encountered in epidemiologic studies and possible ways for improvement are presented in this book in Dr. Kramer's chapter (34A).

RATIONALE FOR AN EFFECT OF BREAST FEEDING ON WEIGHT GAIN AND OBESITY IN LATER LIFE

In 1975 Barbara Hall proposed that changes in composition of human milk during each nursing act as a cue for the duration of nursing (35); the marked increase in fat content during the feed might be associated with changes in the taste and texture of the milk. Thus, the infant might stop sucking before milk fat concentration reaches a high level of 4-6 g/dl. Since a considerable volume of milk remains in the breast after the infant stops nursing (35,36), the changes in composition of human milk during the feed might be associated with development of an appetite control mechanism in breast fed infants. The bottle fed infant would not develop such a control mechanism because of the uniform composition of formula and also because the amount consumed might be strongly affected by the mother's desire to "empty the bottle" at each feed. Although Hall's hypothesis was not supported by actual cross feeding experiments of human milk containing different amounts of fat (37), the studies of Lucas et al show marked differences in the pattern of milk intake between breast fed and bottle fed infants (38,39).

Differences in the Pattern of Milk Intake, Gastric Emptying and Endocrine Responses between Breast Fed and Bottle Fed Infants

Using a nipple shield sampling system that permits measurement of milk composition on small sequential milk samples obtained during a breast feed (40) and careful infant weighing (on an electronic balance accurate to 1 g) the following patterns were found for breast and bottle fed infants: Breast fed infants - milk flow had a definite pattern, 50% of the feed from each breast being taken in during the first two minutes of sucking and a mean of 80-90% being ingested by four minutes (38). Thus, in a ten minute feed from one breast only the initial phase is nutritive and the later phase is non-nutritive. Nutritive sucking starts again immediately after the baby is tranferred to the second breast. There seems to be a major physiological change in a feed at about four minutes that may relate to the change in milk composition. Indeed milk fat concentration rises markedly only after the initial four minutes of nursing (40). Since 80-90% of the feed from each breast is consumed within 4 minutes, the high-fat energy rich hind milk (40,41) is consumed only in very small amounts. Bottle fed infants: In contrast to the biphasic pattern of milk intake seen in breast fed infants (Fig. 1), bottle fed infants had a linear pattern of milk intake during the first 10 minutes of feeding, at which time 81% of the feed has been consumed (39). Thus, although breast fed and bottle fed infants consumed similar volumes per feed (67±2 ml versus 75±6 ml, respectively), at 10-15 minutes after the start of the feed the bottle fed infants had consumed significantly more ($p < 0.001$) milk than the breast fed group (60-70 ml vs 37 ml, respectively). The different patterns of milk intake could be the reason for the physiological differences betwen breast and bottle fed infants (42).

There are marked differences in the pattern and rate of gastric emptying of human milk as compared to infant formula (43,44) (Table 3). Gastric emptying of human milk follows a biphasic emptying pattern, with an initial fast phase whereas gastric emptying of formula follows a

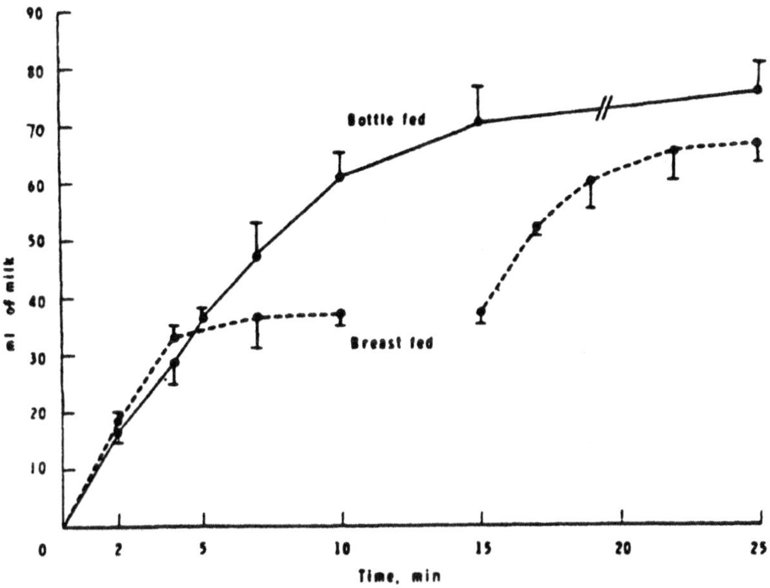

Figure 1: Pattern of milk intake (Mean ± SEM) during a breast feed or bottle feed on the sixth postnatal day. In the breast fed group, the feed from each breast lasted 10 minutes with a 5 minute pause in between. Twenty five minutes was the mean time for completion of the feed. Data from ref. # 39 (Lucas et al (1981), Early Human Dev, 5, 195-199).

linear pattern in both full term and preterm infants. Gastric emptying rates are significantly slower ($p < 0.01$) in formula fed than in milk fed infants, irrespective of gestational age. Gastric emptying is affected by the osmolality and composition of the meal. Since the human milk and the infant formula, used in these studies, had identical osmolality, Cavell attributes the difference in rates of gastric emptying to differences in the fat and protein composition of the feeds (43,44).

The more rapid gastric filling (39) and delayed gastric emptying (43,44) in formula fed infants probably lead to marked differences in gastric distension between breast fed and formula fed infants. This could be the stimulus for the release of hormones triggered by gastric distension.

The passage from the intrauterine milieu, characterized by constant nutrient infusion through the placenta, to extrauterine life, i.e. intermittent oral feeding, is accompanied by marked functional and structural changes in the gastrointestinal tract. These changes are initiated and regulated by the secretion of gastrointestinal hormones released in response to food ingestion (45). Most of the studies in this field have been carried out in full term and preterm infants during the first month after birth - 1st feed, and at 6,13, and 24 days (46,47). Thus, from the data at hand, we can only evaluate the early developmental period. Comparison of hormonal responses to feeding of human milk or formula shows marked differences associated with these types of feeding (42,48). While the functional significance of the endocrine and metabolic differences reported in bottle fed and breast fed infants require further study, it has been suggested that these

Table 3: GASTRIC EMPTYING OF HUMAN MILK AND INFANT FORMULA

	Full Term Infants		Preterm Infants	
	Human Milk	Formula	Human Milk	Formula
Gastric half-emptying time (min)	48±15	78±14**	25.1±11.5	51.9±9.8**
Gastric emptying* rate (ml/0.1m²)	29.5±4.6	22.7±4.2**	19.4±4.5	13.8±2.8**

* Data are expressed as ml/0.1m² of body surface; data are mean ± SEM.
** Statistically significant difference between human milk and formula ($p < 0.011$)
Meal volume av. 32 ml/kg body weight. 17 full term infants and 11 preterm infants were studied. The non absorbable marker used was PEG Macrogoll 3000.
Data from refs. # 43 and 44 (Cavell (1979), Acta Paediatr Scand 68, 725-730 and (1981) 70, 639-641.

differences might cause early (and possibly persistent) differences in fat depositon between the two groups of infants (45). Among the postprandial changes in circulating hormones, the differences in the levels of insulin, motilin, enteroglucagon, pancreatic polypeptide and neurotensin between breast fed and bottle fed infants might be associated with the suggested lower incidence of obesity in breast fed infants (8, 9,14,16,26,29). Basal levels of vasointestinal peptide (VIP), gastric inhibitory peptide (GIP), neurotensin and motilin are higher in bottle fed than in breast fed infants. Furthermore, postprandial levels of plasma insulin, neurotensin, pancreatic polypeptide (PP) and motilin are also higher in bottle fed than in breast fed infants (Fig. 2). Although the exact function of these hormones in the newborn has not been evaluated, pancreatic polypeptide is thought to inhibit pancreatic exocrine secretion (42) and neurotensin may modulate insulin and glucagon release (42); enteroglucagon is trophic to the gut mucosa, motilin probably contributes to the development of gastric and intes-tinal motility (and the differences in circulating hormone levels might explain the differences in stool frequency between breast fed and bottle fed infants); GIP may be the principal effector of the enterovascular axis (49), whereby a greater insulin release occurs after oral rather than intravenous glucose administration. Therefore, the differences between breast fed and bottle fed infants in release of these hormones indicate differences in GI function, as well as in pancreatic exocrine and endocrine function in the immediate postnatal period. Whether these differences are sustained throughout the entire nursing period remains to be established.

Infant Nutrition and the Metabolic-Cellular Basis of Obesity

The greater insulin release in bottle fed infants is of particular interest. Since insulin stimulates fat deposition, it might affect the early development of adipose tissue and thus have a relationship to cell number as well as cellular fat content of adipose tissue (50,51). While there is still no absolutely convincing evidence that breast fed children are leaner than bottle fed ones, or that the obese child becomes an obese adult, there is convincing evidence from animal studies (50) and some evidence from human studies (51,52) that overfeeding in infancy leads to an increase in adipocyte number and fat content, two of the characteristics of obesity. Whether the higher insulin levels in bottle

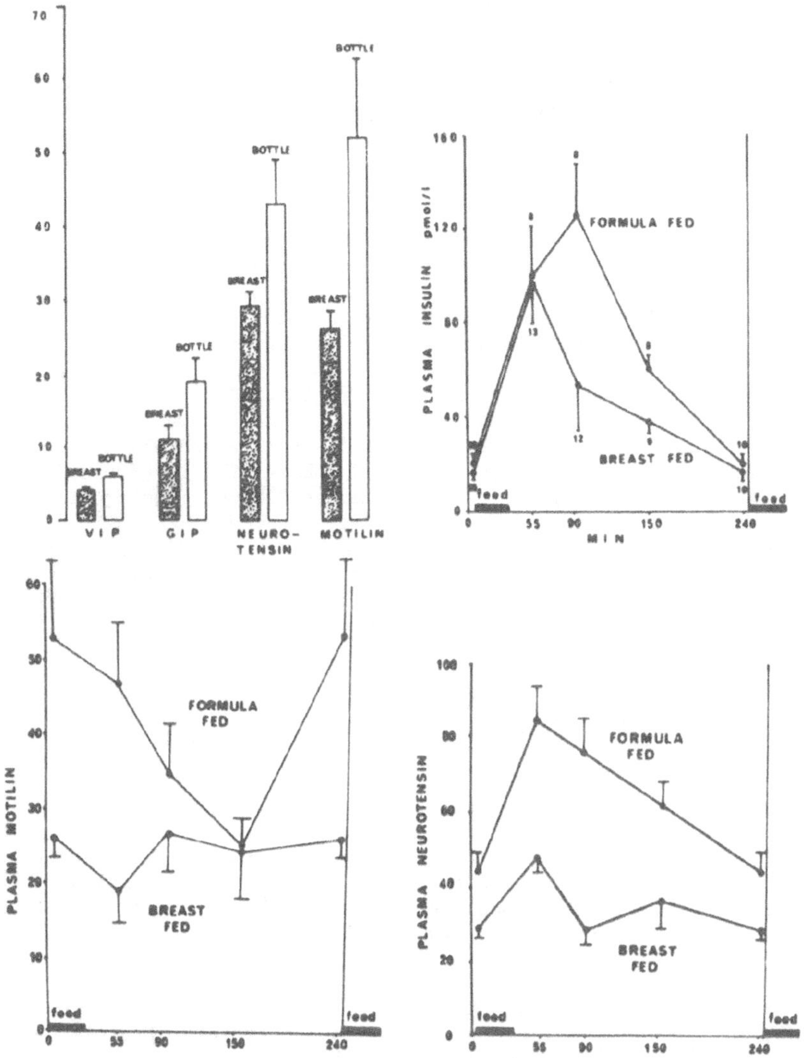

Figure 2: Effect of human milk or infant formula on basal levels of plasma VIP, GIP, neurotensin and motilin, (upper, left). Effect of a single feed on plasma insulin, motilin and neurotensin levels, (upper, right and lower, left and right). Data, in pmol/L, mean ± SEM, from refs. # 42 and 48 (Lucas et al (1980), Lancet 1, 1267-1268 and Acta Paediatr Scand (1981) 70, 195-200.

fed infants induce higher adipose tissue lipoprotein lipase (53) and thereby more rapid lipid filling of fat cells is unknown at present. Animal studies show that lipoprotein lipase activity is present in preadipocytes and has a regulatory role in the lipid filling of this tissue during the early postnatal period (54). Developing adipose tissue in the human (fetus and infant) also contains small adipocytes (25 um) (55) suggesting that preadipocytes (i.e. smaller cells in the early stages of fat accumulation) are a major component of human adipose tissue during pre- and postnatal development. Two critical periods for the proliferation of adipose tissue have been identified by Knittle et al (52): one is before the age of two and the other during the adolescent growth spurt. Whereas the first period might be directly related to infant nutrition, the adolescent adipocyte proliferation may be the result of excessive fat content of adipocytes, since coupling between fat cell size and adipocyte multiplication has been suggested in the human (56) and demonstrated in the rat (54). Rats undernourished during the early weeks of life have a permanent reduction in the number and size of fat cells (50). Fat cell size, can however, change in the adult rat as a result of overfeeding (58). There is at present no information on fat cell size and number in breast fed as compared to bottle fed infants and present methods to evaluate body fat cannot answer this question. Furthermore, because of the marked difference in metabolic activity of human adipose tissues at various anatomical locations (59), the standard practice of studying only subcutaneous adipose specimens might not be sufficient. One might, however, assume that childhood obesity, even in breast fed infants, might be due to overfeeding after weaning.

After three months of age, breast fed infants do not take in sufficient nutrients to support maximal growth. A more detailed discussion of this concept is given in the chapter of Neville and Oliva-Rasbach (60A) in this book. The composition and volume of milk provided to the human infant is probably the result of an evolutionary compromise that aims to keep the diad of mother and infant in reasonably good health, i.e. nutrients might be somewhat lower in concentration than needed to achieve maximal potential growth in order to avoid depriving the mother of her nutritional requirements (61); thus, by being slightly "underfed" the breast fed infant might be protected from obesity, provided there is no overfeeding after weaning.

IS THERE A RELATIONSHIP BETWEEN MODE OF FEEDING DURING EARLY INFANCY AND SUBSEQUENT CONCENTRATION OF SERUM CHOLESTEROL?

In 1972, based on studies in pigs and rats, Reiser and Sidelman proposed the intriguing hypothesis that newborns exposed to high levels of dietary cholesterol are better able to cope with dietary cholesterol as adults (62). The absence of hypercholesterolemia in the "cholesterol challenged" adult would be due to changes in the activity rates of enzymes involved in cholesterol metabolism (either synthesis or catabolism). These changes would be "imprinted" during the early postnatal period.

Because human milk contains about 15 mg/dl cholesterol whereas proprietary infant formulas contain only trace amounts of cholesterol, if the initial animal studies apply to the human, the breast fed individual would be better prepared to handle dietary cholesterol than the adult who was bottle fed in infancy. Because of its possible clinical relevance in the prevention of atherosclerosis, the "cholesterol challenge" hypothesis has received much attention and has been tested in further animal and human studies. The animal studies have produced conflicting results which suggest marked species differences in the early "conditioning" of cholesterol metabolism. Most of the studies were carried out in rats, but several studies were also carried out in guinea pigs, rabbits, and baboons. Some of the rat studies support the "cholesterol challenge theory" whereas others refute it. The supportive studies show that:

1. Rats prematurely weaned to a low fat diet developed higher serum cholesterol levels than rats prematurely weaned to a high fat diet (63).

2. High levels of cholesterol in artificial diets fed to newborn rats did not protect the adult from hypercholesterolemia after challenge; suckled rats had, however, lower serum cholesterol after the choles-

terol challenge (75±2 vs 106±6 mg/dl, respectively), suggesting that cholesterol metabolism systems are affected in early life (64).

3. Exposure to high cholesterol during gestation and lactation markedly reduces the activity of hepatic 3 hydroxy-3 methyl-glutaryl CoA reductase (cholesterol synthesis) and increases cholesterol 7 hydroxylase (cholesterol catabolism) in the weanling and adult rat (65). Specific temporal differences in dietary modulation of these enzymes indicate that HMG CoA reductase is more sensitive to dietary changes during gestation than during lactation, whereas 7 hydroxylase is affected by high dietary cholesterol only during lactation (66).

Studies that do not support the "cholesterol challenge hypothesis" show that:

4. High cholesterol intake in early life did not prevent the increase in serum cholesterol after challenge in the adult (67).

5. Furthermore, animals originally fed high fat diets retain the capacity to increase serum cholesterol levels in response to a fat challenge in adult life (68). Indeed, the data of Coates et al suggest that exposure to low fat diet in infancy protects against hypercholesterolemia (68).

Studies in other species have not been able to substantiate the "cholesterol challenge hypothesis".

<u>In the guinea pig</u> stimulation of cholesterol catabolism (by feeding cholestiramine after birth) rather than cholesterol feeding, or early weaning, can influence the response to dietary cholesterol in the adult (69). The efficient handling of dietary cholesterol in the cholestiramine fed guinea pig is directly associated with markedly higher levels of cholesterol 7 -hydroxylase, the rate limiting enzyme in cholesterol catabolism (70).

<u>In the rabbit</u>, very high cholesterol diets of the lactating doe, resulted in only transient hypercholesterolemia in the suckling pups. Thus, increased cholesterol intake early in life (by feeding high cholesterol diets to the lactating doe) led to a rise in serum cholesterol at weaning, followed by a return to normal serum cholesterol levels at 11 weeks of age (71).

<u>In the baboon</u>, animals which were breast fed or formula fed as infants had significant differences in the response of HDL cholesterol concentration and ApoA I levels to the type of fat fed in later life. However, breast feeding or feeding formulas containing various levels of cholesterol for 3 months during infancy did not result in statistically significant differences in total serum cholesterol, VLDL + LDL cholesterol and Apo B concentrations (72).

Some of the animal studies suggest, therefore, that early nutritional management affects subsequent cholesterol homeostasis. A hypocholesterolemic effect of milk (bovine) in the adult rat has also been reported (73), and it has been suggested that it might be mediated by the hydroxymethyl glutaric acid or orotic acid present in milk (74).

<u>Human studies</u>

Human studies have failed to demonstrate an effect of early nutrition on subsequent serum cholesterol concentrations (Table 4). In general, cord blood cholesterol levels are low (50-70 mg/dl) and increase immediately after birth. In a recent study, Lane and McConathy found little differences in serum lipid classes (including cholesterol) and apoprotein levels in breast fed and bottle fed infants during the first month of life (at 3, 14, and 24 days after birth) (75). Follow-up studies from infancy to childhood and adolescence show that, while serum cholesterol concentrations are generally higher during breast feeding than formula feeding, these differences do not persist beyond this time. The Bogalusa Heart Study (76) examined the effect of cow milk, cow milk formula and soy formula on total and lipoprotein cholesterol at six and 12 months after birth. Although the infants fed animal fat had higher levels of cholesterol and lipoprotein, than the infants fed vegetable fat at 6 months of age, by age one year these differences had disappeared. In an earlier study, Glueck et al (77) reported that alternating "moderate" cholesterol containing formula with "low" cholesterol containing formula during the first 12 months of life (each formula being fed for 6 months) had no effect on serum cholesterol levels in normal cholesterolemic infants irrespective of the order of feeding.

Table 4: RELATIONSHIP BETWEEN INFANT FEEDING AND CONCURRENT/SUBSEQUENT CHOLESTEROL LEVELS

Study	Year	Country	Ref #	Follow-up years	Serum Cholesterol during study	Concentration at follow-up
Glueck	1972	USA	77	1	-	HC = LC
Farris	1982	USA	76	1	HC > LC*	HC = LC
Ward	1980	USA	91	2.5	-	HC > LC
Nestel	1979	Australia	92	0.4-1.3	HC > LC	-
Huttunen	1983	Finland	78	5	HC > LC	HC = LC
Fomon	1984	USA	79	8	HC > LC	HC = LC
Hodgson	1976	USA	80	7-12	-	HC = LC
Friedman	1975	USA	81	15-19	HC > LC	HC = LC
Marmot	1980	England	34	33	-	HC = LC male HC < LC female

* HC = high cholesterol diet, human milk, cow's milk, cow's milk based formula;
LC - low cholesterol diet = proprietary formula

Longer term follow-up into childhood shows similar results: Huttunen et al followed infants fed human milk, cow's milk formula, and proprietary formula from birth to age 5 years (78). Higher serum cholesterol levels in breast and cow milk formula fed infants were detected only during the first 6 months after birth; all three groups of infants had however, identical serum cholesterol at 9 months, 3, 4 and 5 years. Almost identical data were reported by Fomon et al (79), who studied a greater number of breast or formula fed infants (469 vs 84 in the Finnish study (78)) repeatedly in infancy and again at 8 years of age (79). Although the breast fed infants had higher serum cholesterol levels during the nursing period, at 8 years of age there was no difference in serum cholesterol level between the breast and formula fed groups. The significant correlation between serum cholesterol concentration in infancy and at age 8 years, in formula fed males only, suggests, moreover, that low dietary cholesterol intake in infancy leads to lower serum cholesterol in childhood. In a study of 97 school children tested between 7 - 12 years, Hodgson et al (80) reported that feeding a low cholesterol diet in infancy (proprietary formula) resulted in significantly lower serum cholesterol levels at school age than feeding cholesterol containing human or cow's milk formula (157 mg/dl vs 172 and 174 mg/dl, respectively).

Two longer follow-up studies, one into adolescence and the other to age 32 years, have been reported. The first found no difference in serum cholesterol levels at age 15-19 years between breast and formula fed infants, although the former groups had higher serum cholesterol levels at 4 and 12 months of age (81). This study was severely criticized by Reiser on the grounds that: a) the same infants were not tested at each time point and b) no adequate dietary record was kept between infancy and adolescence (82). The second report (34) on a small number of subjects found no difference in serum cholesterol levels in adult breast or bottle fed males, although breast fed females had slightly lower serum cholesterol (209 mg/dl) than bottle fed ones (229 mg/dl).

We agree with the conclusion of Fomon et al (79) and Reiser (82) that childhood and adolescence are perhaps too early an age to detect the possible beneficial effect of breast feeding on cholesterol homeostasis in later life. It is also possible that quantitation of serum cholesterol is too insensitive to detect early changes and that quantitation of lipoprotein classes and apoprotein concentrations might be necessary in order to assess whether there are indeed differences in cholesterol handling by adults who were breast or bottle fed in infancy.

It seems, however, that based on our present status of knowledge, animal studies seem to be inadequate to address this question. Furthermore, the complexity of possible causative factors of human obesity and atherosclerosis indicate that while limited answers might be obtained in animal studies, the major questions will have to be solved by well controlled, carefully executed human studies.

ESSENTIAL FATTY ACIDS, BRAIN DEVELOPMENT, MEMBRANE STRUCTURE AND FUNCTION AND NEONATAL DIET

Brain Development

Essential fatty acids are important in brain development, cell proliferation, myelination (83) and retinal function (84). It was thought until recently that the structural lipids (which contain most of the essential fatty acids) are much more resistant to change (in fatty acid composition) than the storage fats (which are mainly adipose tissue triglycerides). Recent studies, in primates maintained on diets deficient in n-3 essential fatty acids, show that brain with an abnormal fatty acid composition (C22:5, n-3) has a remarkable capacity to change its fatty acid content and composition (to C22:6, n-3 docosahexaenoic) after ingestion of fish oil, implying a greater lability of the fatty acids of brain phospholipids than previously assumed (85). Phosphoglycerides of brain contain high levels of docosahexaenoic acid (22:6, n-3). Recent studies show that human brain lipids change during development, from very little 22:6 n-3 at 6-12 weeks post conception to 4% and 18% of total fatty acids in phosphatidylethanolamine at 17-18 weeks and later during the second trimester, respectively. Similar increases were observed in the other brain phospholipid classes (86).

The increasing survival of very low birth weight infants has led recently to the study of the developmental accretion of essential fatty acids during the last trimester of fetal development through the early weeks of life. These studies have shown that, during the last trimester, brain levels of linoleic acid were low and that substantial accretion of long chain polyunsaturated essential fatty acids (22:4n6, 22:5n3 and 22:6n3) occurred (86). Postnatal brain development is characterized by an increase in brain linoleic acid content. Increase in chain elongation -desaturation products did not occur for several weeks postnatally. These results suggest that placental transfer of polyenoic fatty acids is of primary importance in accretion of these fatty acids in the fetus. Accretion of polyenoic fatty acids in the fetal brain during the last trimester of pregnancy occurs at a rate of 43 mg n-6 and 22 mg n-3 polyenoic fatty acids per week (86). We and others have shown that polyenoic fatty acid are present in the milk of women who deliver prematurely or at term. Furthermore, the level of these fatty acids is significantly higher in colostrum and milk of mothers of premature infants than of mothers of full term infants (87). Thus, human milk is able to meet the needs for these fatty acids in preterm infants and to efficiently substitute for their placental supply to the fetus (86,87). Commercial formula, however, contains only little polyenoic fatty acids. The obvious question is: will brain development and later structure and function be affected by the mode of feeding of very low birth weight infants?

Membrane structure and function

Erythrocyte membrane phospholipids contain higher levels of polyenoic acids (C20 and C22) in breast fed than in formula fed infants (88), and preliminary data suggest differences in the thermodynamic characteristics (energy of activation) of erythrocyte-membrane enzymes (in this case, acetylcholine esterase) of breast fed and butter-fat formula-fed infants (89). Since these are short term studies, conducted only concurrent and not subsequent to infant feeding, it is not known how long these differences will last. Dietary induced changes in membrane structure and function in many organs occur rapidly in the adult (90). The question to be answered is: even if the early feeding mode has only a transient effect on membrane structure and function, will these changes, during a critical period of development, have long lasting effects?

We have much work ahead of us, and hopefully it will not be too long before we get the answers!

SUMMARY

Nutrition in infancy could have long lasting effects that might affect the metabolic activity of the adult. It is thought that breast feeding (as compared to bottle feeding) might affect: 1) body weight and fat content, and possibly protect against obesity; 2) blood cholesterol levels, leading to lower incidence of atherosclerosis; 3) membrane composition, thus membrane fluidity and function; 4) brain composition, especially that of premature infants. Careful evaluation of the first two parameters shows however, that at present there is great variation in the results of studies aimed to answer these questions and that additional carefully planned studies are needed before definite conclusions can be reached. With respect to the remaining two parameters, there are as yet no long term studies to permit an evaluation of possible differences on membrane structure and function or on brain composition and function.

As to the second question raised in my presentation, "can animal studies predict these effects in the human?", studies on adipose tissue development suggest similarities between several species, including the human, in the effect of nutrition on cell number and cell size and their relationship to obesity. Because of the multifactorial aspect of atherosclerosis it is much more difficult to compare studies on the effect of early nutrition on cholesterol metabolism in experimental animals and man.

The long term effects (if indeed they exist) of breast versus bottle feeding on membrane function and brain development cannot be evaluated at present.

Acknolwedgements

The authors' studies are supported by NIH grant HD-20833. We are grateful to our daughter, Tamar Hamosh for her assistance in library research and to Barbara Runner for expert secretarial help.

REFERENCES

1. Weil, WB (1977): Current controversies in childhood obesity. J Pediatr, 91, 175-187.
2. Taitz LS (1977): Obesity in pediatric practice: infantile obesity. Pediatr Clin N Am, 24, 107-115.
3. Oakley JR (1977): Differences in subcutaneous fat in breast and formula-fed infants. Arch Dis Child, 52, 79-80.
4. Taitz LS (1971): Infantile overnutrition among artificially fed infants in the Sheffield region. Br Med J, 1, 315-316.
5. Hofvander Y, Hagman U, Hillervik C, Sjolin S (1982): The amount of milk consumed by 1-3 months old breast or bottle fed infants. Acta Paediatr Scand, 71:953-958.
6. Dubois S, Hill DE, Beaton GH (1979): An examination of factors believed to be associated with infantile obesity. Am J Clin Nutr, 31, 1997-2007.
7. Ferris AG, Laus, MJ, Hosmer DW, Beal VA (1980): The effect of diet on weight gain in infancy. Am J Clin Nutr, 33, 2635-2642.
8. Hopper PD (1965): Infant feeding and its relationship to weight gain and illness. Practitioner, 194, 391-395.
9. Hooper PD, Alexander EL (1971): Infant morbidity and obesity. Practitioner, 207, 221-227.
10. Shukla A, Forsythe HA, Anderson CM, Marwok SM (1972): Infantile over-nutrition in the first year of life: A field study in Dudley, Worcestershire, Br Med J, 4, 507-515.
11. Jung E, Czajka-Narins DM (1985): Birth weight doubling and tripling times: an updated look at the effects of birth weight, sex, race and type of feeding. Am J Clin Nutr, 42, 182-189.
12. Neumann CG, Alpaugh M (1976): Birth weight doubling time: a fresh look. Pediatrics, 57, 469-473.
13. Sveger T, Lindberg T, Weilbull B, Olsson VL (1975): Nutrition, over-nutrition, and obesity in the first year of life in Malmo, Sweden. Acta Paediatr Scand. 64, 635-640.
14. Saarinen UM, Siimes MA (1979): Role of prolonged breast feeding in infant growth. Acta Paediatr Scand, 68, 245-250.
15. Yeung DL, Pennell MD, Leung M, Hall J (1981): Infant fatness and feeding practices: a longitudinal assessment. J Am Diet Assoc, 79, 531-535.
16. Kramer MS, Barr RG, Leduc DG, Boisjoly C, McVey-White L, Pless IB (1985): Determinants of weight and adiposity in the first year of life. J Pediatr, 106, 10-14.
17. Vobecky JS, Vobecky J, Shapcott D, Demers PP (1983): Nutrient intake patterns and nutritional status with regard to relative weight in early infancy. Am J Clin Nutr, 38, 730-739.
18. Sveger T (1978): Does overnutrition or obesity during the first year affect weight at age four? Acta Paediatr Scand, 67, 465-467.

19. Huenemann RL (1974): Obesity in six months old children, Environmental factors associated with preschool obesity. J Am Diet Assoc, 64, 480-487.
20. Dine MS, Gartside PS, Glueck CJ, Rheines L, Greene G, Khoury P (1979): Where do the heaviest children come from? A prospective study of white children from birth to 5 years of age. Pediatrics, 63, 1-7.
21. Poskitt EME, Cole TJ (1978): Nature, nurture, and childhood overweight. Br Med J, 1, 603-605.
22. Wolman PG (1984): Feeding practices in infancy and prevalence of obesity in preschool children. J Am Diet Assoc, 84, 436-438.
23. Fisch RO, Bilek MK, Ulstrom R (1975): Obesity and leanness at birth and their relationship to body habitus in later childhood. Pediatrics, 56, 521-526.
24. Asher PC (1965): Fat babies and fat children: The prognosis of obesity in the very young. Arch Dis Child, 41, 672-673.
25. Heald FP, Hollander RJ (1965): The relationship between obesity in adolescence and early growth. J Pediatr, 67, 35- 40.
26. Eid EE (1970): Follow-up study of physical growth of children who had excessive weight gain in first six months of life. Br Med J, 2, 74-76.
27. Mellbin T, Vuille JC (1973): Physical development at 7 years of age in relationship to velocity of weight gain in infancy with special reference to incidence of overweight. Br J Prev Soc Med 27, 225-235.
28. Mellbin T, Vuille JC (1978): Weight gain in infancy and physical development between 7 and 10 1/2 years of age. Br J Prev Soc Med, 30, 233-238.
29. Mellbin T, Vuille JC (1978): Relationship of weight gain in infancy to subcutaneous fat and relative weight at 10 1/2 years of age. Br J Prev Soc Med, 30, 239-243.
30. Hernesniemi I, Zackman M, Prader A (1974): Skinfold thickness in infancy and adolescence. A longitudinal correlation study in normal children. Helv Padiat Acta 29, 523-530.
31. Stark O, Atkins E, Wolff OH, Douglas JWB (1981): Longitudinal study of obesity in the National Survey of Health and Development. Br Med J 283, 13-15.
32. Charney E, Goodman HC, McBride M, Lyon B, Pratt R (1976): Childhood antecedents of adult obesity. Do chubby infants become obese adults? N Engl J Med, 295, 6-9.
33. Kramer MC (1981): Do breast-feeding and delayed introduction of solid foods protect against subsequent obesity? J Pediatr, 98, 883-887.
34. Marmot MG, Page CM, Atkins E, Douglas JBW (1980): Effect of breast feeding on plasma cholesterol and weight in young adults. J Epidemiol Comm Health, 34, 164-167.
34A Kramer MS (1987): Breast feeding and child health - methodologic issues in epidemiologic research. In Human Lactation, Vol. 3: The Effects of Human Milk upon the Recipient Infant. A. Goldman, SA Atkinson and LA Hanson, Eds, Plenum Press, In Press.
35. Hall B, (1975): Changing composition of human milk and early development of an apetite control. Lancet, 1, 779-781.
36. Neville MC, Personal communication
37. Woolridge MW, Baum JD, Drewett RF (1980): Does a change in the composition of human milk affect sucking patterns and milk intake? Lancet, 2,1292-1294.
38. Lucas A, Lucas PJ, Baum JD (1979):Pattern of milk flow in breast fed infants. Lancet, 2, 57-58.
39. Lucas A, Lucas PJ, Baum JD (1981): Differences in the pattern of milk intake between breast and bottle fed infants. Early Human Dev, 5, 195-199.

40. Lucas A, Lucas PJ, Baum JD (1980): The nipple shield system: a device for measuring the dietary intake of breast fed infants. Early Human Dev, 4, 365-372.
41. Hytten FE (1954): Clinical and chemical studies in human lactation. II Variation in major constituents during a feed. Br Med J 1, 176-179.
42. Lucas A, Blackburn AM, Aynsley-Green A, Adrian TE, Sarson DL, Bloom SR (1980): Breast vs bottle: Endocrine responses are different with formula feeding. Lancet, 1, 1267-1269.
43. Cavell B (1979): Gastric emptying in preterm infants. Acta. Paediatr Scand, 68, 725-730.
44. Cavell B (1981): Gastric emptying in infants fed human milk or infant formula. Acta Paediatr Scand, 70, 639-641.
45. Aynsley-Green A (1983): Plasma hormone concentrations during enteral and parenteral nutrition in the human newborn. J Ped Gastroenterol Nutr, 2 (Suppl), S108-S112.
46. Lucas A, Bloom SR, Aynsley-Green A (1980): The development of gut hormone responses to feeding in neonates. Arch Dis Child, 55, 678-682.
47. Lucas A, Bloom SR, Aynsley-Green A (1982): Postnatal surges in plasma gut hormones in term and preterm infants. Biol. Neonate. 41, 63-67
48. Lucas A, Boyles S, Bloom SR, Aynsley-Green A (1981): Metabolic and endocrine responses to a milk feed in six day old term infants: differences between breast and cow's milk formula feeding. Acta Paediatr Scand, 70, 195-200.
49. Dupre J, Ross SA, Watson D, Brown JC (1973): Stimulation of insulin secretion by gastric inhibitory polypeptide in man. J. Clin. Endocrine Metab. 37, 826-828.
50. Knittle JL, Hirsch J (1968): Effect of early nutrition on the development of rat epididymal fat pads: cellularity and metabolism. J Clin Invest, 47, 2091-2098.
51. Brook CGD, Lloyd JR, Wolff OH (1972): Relation between age of onset of obesity and size and number of adipose cells. Br Med J, 1, 25-27.
52. Knittle JK, Timmers K, Ginsberg-Fellner F, Brown RE, Katz DP (1979): The growth of adipose tissue in children and adolescents. J Clin Invest. 63, 239-246.
53. Hamosh M, Hamosh P (1983): Lipoprotein lipase; its physiological and clinical significance. Mol Aspects Med, 6, 199-289.
54. Hietanen E, Greenwood MRC (1977): A comparison of lipoprotein lipase activity and adipocyte differentiation in growing male rats. J Lipid Res, 18, 480-490.
55. Dunlop M, Court JM, Hobbs JB, Boulton TSC (1978): Identification ofsmall cells in fetal and infant adipose tissue. Pediatr Res, 12, 905-907.
56. Hager A, Sjostrom L, Arvidsson B, Bjorntorp P, Smith U (1977): Body fat and adipose cellularity in infants: a longitudinal study. Metabolism, 26, 607-614.
57. Faust IM, Johnson PR, Stern JS, Hirsch J (1978): Diet induced adipocyte number increase in adult rats: a new model of obesity. Am J Physiol. 235, E279-E286.
58. Foust IM, Johnson PR, Hirsch J (1980): Longterm affects of early nutritional experience on the development of obesity in the rat. J Nutr 110, 2027-34.
59. Hamosh M, Hamosh P, Bar-Maor A, Cohen H (1963): Fatty acid metabolism by human adipose tissue. J Clin Invest, 42, 1649-1652.
60. Dugdale AE (1986): Evolution and breast feeding. Lancet 1, 670-673.
60A Neville MC, Oliva-Rasbach J (1987): Is maternal milk production limiting for infant growth during the first year of life in breast-fed infants? In Human Lactation, Vol. 3: The Effects of Human Milk Upon the Recipient Infant. A Goldman, SA Atkinson and LA Hanson, Eds. Plenum Press, In Press.

61. Hamosh M (1987): Fat needs for term and preterm infants. In "Nutrition from conception to weaning" Tsang RC, Nicols B (eds), CV Mosby Co., St. Louis, In Press.
62. Reiser R, Sidelman J, (1972): Control of cholesterol homeostasis by cholesterol in the milk of the suckling rat. J Nutr. 102, 1009-1016.
63. Hahn P, Kirby L (1973): Immediate and late effects of premature weaning and of feeding a high fat diet or high carbohydrate diet to weanling rats. J Nutr, 103, 690-696.
64. Kris-Etherton PM, Layman DK, York PV, Frantz T, (1979): The influence of early nutrition on the serum cholesterol of the adult rat. J Nutr, 109, 1244-1257
65. Naseem SY,Khan MA, Heald FP, Nair PO (1980): The influence of cholesterol and fat in maternal diet of rats on the development of hepatic cholesterol metabolism in the offspring. Atherosclerosis. 36, 1-5.
66. Naseem SY, Khan MA, Jacobson MS, Nair PO, Heald FP (1980): The influence of dietary cholesterol and fat on the homeostasis of cholesterol metabolism in the rat. J Nutr, 111, 276-286.
67. Green MH, Dohner EL & Green JB (1981): Influence of dietary fat and cholesterol on milk lipids and on cholesterol metabolism in the rat. J Nutr, 111, 276-286.
68. Coates PM, Brown SA, Sonawane BR, Koldovksy O (1983): Effect of early nutrition on serum cholesterol levels in adult rats challenged with high fat diet. J Nutr, 113, 1046-1050.
69. Li RU, Bale LK, Subbiah MTR (1979): Effect of enhancement of cholesterol degradation during neonatal life of guinea pig on its subsequent response to dietary cholesterol. Atherosclerosis, 32, 93-98.
70. Li JR, Bale LK, Kottke BA (1980): Effect of neonatal modulation of cholesterol homeostasis on subsequent response to cholesterol challenge in adult guinea pig. J Clin Invest, 65, 1060-1068.
71. Whatley BJ, Green JB, Green MH (1981): Effect of dietary fat and cholesterol on milk composition, milk intake and cholesterol metabolism in the rabbit. J Nutr, 111, 432-441.
72. Mott GE, McMahan CA, Kelley JL, Farley CM, McGill HC Jr (1982): Influence of infant and juvenile diets on serum cholesterol, lipoprotein cholesterol, and apolipoprotein concentrations in juvenile baboons. Atherosclerosis, 45, 191-202.
73. Kritchevsky D, Tepper SA, Morrissey RB, Czarnecki SK, Klurfeld DM (1979): Influence of whole or skim milk on cholesterol metabolism in rats. Am J Clin Nutr, 32, 597-600.
74. Richardson T (1978): The hypocholesterolemic effect of milk. A review. J Food Protect, 41, 226-233.
75. Lane DM, McConathy WJ (1986): Changes in serum lipids and apolipoproteins in the first four weeks of life. Pediatr Res, 20, 332-337.
76. Farris RP, Frank CG, Webber LS, Srinivasan SR, Berenson GS (1982): Influence of milk source on serum lipids and lipoproteins during the first year of life, Bogalusa Heart Study. Am J Clin Nutr, 35, 42-49.
77. Glueck CJ, Tsang R, Balisteri W, Fallat R (1972): Plasma and dietary cholesterol in infanty: Effects of early low or moderate dietary cholesterol intake on subsequent response to increased dietary cholesterol. Metabolism, 21, 1181-1192.
78. Huttunen JK, Saarinen UM, Kostianen E, Siimes MA (1983): Fat composition of the infant diet does not influence subsequent serum lipid levels in man. Atherosclerosis, 46, 87-94.
79. Fomon SJ, Rogers RR, Ziegler ER, Nelson SE, Thomas LN (1984): Indices of fatness and serum cholesterol at age 8 years in relation to feeding and growth during early infancy. Pediatr Res, 18, 1233-1238.

80. Hodgson PA, Ellefson RD, Elveback LR, Harris LE, Nelson RA, Weidman WH (1976): Comparison of serum cholesterol in children fed high, moderate, or low cholesterol milk diets during neonatal period. Metabolism, 25,739-745.
81. Friedman G, Goldberg SJ (1975): Concurrent and subsequent serum cholesterol in breast- and formula-fed infants. Am J Clin Nutr, 28, 42-45.
82. Reiser R (1975): Experimentation with human subjects. Am J Clin Nutr, 28,2
83. Crawford MA, Hassam AB, Rivers JPW (1978): Essential fatty acids requirements in infancy. Am J Clin Nutr, 31, 2181-2190.
84. Neuriger M, Connor WE, Van Petten C, Barstad L (1984): Dietary omega-3 fatty acid deficiency and visual loss in infant rhesus monkeys. J Clin Invest, 73, 272-276.
85. Connor WE, Neuriger M, Lin D, Neuwelt E (1985): The incorporation of docosahexaenoic acid into the brain of monkeys deficient in w-3 essential fatty acids. XIII Internat. Cong Proc Abstr, 104.
86. Clandinin MT, Chappell JE, Heim T, Swyer PR, Chance GW (1981): Fatty acid utilization in perinatal de novo synthesis of tissues. Early Human Dev, 5, 355-366.
87. Bitman J, Wood DL, Hamosh M, Hamosh P, Mehta NR (1983): Comparison of the lipid composition of breast milk from mothers of term and preterm infants. Am J Clin Nutr, 38, 300-313.
88. Putnam JC, Carlson SE, DeVoe PW, Barness LW (1982): The effect of variations in dietary fatty acids on the fatty acid composition of erythrocyte phosphatidylcholine and phosphatidylethanolamine in human infants. Am J Clin Nutr, 36, 106-114.
89. Hall B, Muller DPR, Aggett PJ, Butcher PD (1985): The effect of milk lipids on the activity of acetylcholine esterase in red blood cells. In "Composition and Physiological Properties of Human Milk", J Schaub (ed) Elsevier Amsterdam, 179-188.
90. Clandinin MT, Foot M, Robson L (1983): Plasma membrane: can its structure and function be modulated by dietary fat? Comp Biochem Physiol, 76B, 335-339.
91. Ward SD, Melin JR, Lloyd FP, Norton JA, Christian JC (1980): Determinants of plasma cholesterol in childrem, a family study, Am J Clin Nutr, 33, 63-70.
92. Nestel PJ, Poyser A, Boulton TJC (1979): Changes in cholesterol metabolism in infants in response to dietary cholesterol and fat. Am J Clin Nutr, 32, 2177-2182.

NUTRIENT UTILIZATION: WORKSHOP SUMMARY

Stephanie A. Atkinson, and Bo Lönnerdal

Department of Pediatrics, McMaster University Medical Centre, Hamilton, Ontario, Canada and Department of Nutrition, University of California, Davis, California

The central theme of this workshop was the utilization of specific minerals and vitamins from human milk and an evaluation of the adequacy of these nutrients in milk to support growth and development of normal term and premature infants. The trace elements iron, zinc, copper, manganese and selenium; the minerals calcium and phosphorus; and a water soluble vitamin, folic acid were discussed.

The trace element requirements of infants are poorly understood: of particular relevance is the issue of bioavailability of trace minerals from various infant diets. One of the important factors affecting bioavailability is the presence of trace mineral binding proteins in milk as was discussed by Bo Lonnerdal from the USA. For iron, it appears that the high bioavailability demonstrated from human milk is due, at least in part, to the presence of a specific iron-binding protein lactoferrin. In an infant Rhesus monkey model, Lonnerdal's research group has identified a possible mucosal receptor for lactoferrin which facilitates iron uptake and is expressed through life, although the number of receptors is highest in infancy. Thus, there may be a specific mechanism for assuring iron accrual in early life. The fate of the lactoferrin molecule was discussed by Christian Barth from the FRG who reported on the use of lysine-labelled lactoferrin to study absorption of this protein in the pig. A higher degree of disappearance of the labelled protein was found in the adult than in the young pig, indicating higher degradation and/or absorption.

With respect to zinc and copper, Lonnerdal described three binding ligands in human milk: citrate, serum albumin and casein. Other suggested ligands such as picolinate and lactoferrin have not been proven to bind zinc. The difference in bioavailability of zinc and copper from human milk and cow's milk appears to be more due to an absence of factors limiting trace element bioavailability in human milk than to the presence of factors stimulating uptake. There is insufficient knowledge regarding manganese deficiency or toxicity. From limited data, it appears that manganese is absorbed and retained to a large extent in the young, possibly because of low bile output in early life. The body load of manganese from various diets is highly variable, with human milk providing low amounts, cow's milk formula intermediate amounts and soy formula large amounts. Animal experiments showed that the retained manganese becomes redistributed into the brain, the major site of manganese toxicity. Therefore, there is a potential risk for manganese toxicity. Erika Sievers from the FRG emphasized the difficulties with the balance technique as such and the micronutrient manganese in particular for studying manganase balance in term and preterm infants. The manganese content of the breast milk was quite low, about 6 ng/ml, while the cow's milk formula used contained a much higher concentration of manganese. Manganese intake was immediately reflected in the amount of manganese retained, indicating lack of homeostatic regulation. Most balances in preterm infants appeared to be positive and therefore further manganese supplementation of such infants was not found to be warranted.

Selenium content of human milk, in contrast to other trace elements, parallels the levels of this element in the environment and maternal diet. Ann Smith from the USA presented animal data to suggest that lactation itself invokes a relative depletion of body stores of selenium, but the functional significance of such depletion is unknown. For the infant, selenium status is dependent on the form of feeding since some formulas such as soy-based ones contain considerably less of this element than human milk. Factors affecting the bioavailability may include selenium-binding compounds in human milk and the predominance of casein as opposed to whey proteins in bovine milk.

Precise requirements for calcium and phosphorus for growing premature infants is highly controversial and many factors (such as dietary vitamin D, Ca/P ratio, metabolic capacity of the infant, mineral mineral interactions, heat processing of milk and rate of infant growth) affecting utilization of these minerals. Issues related to the optimal dietary Ca/P ratio, nitrogen/P ratio, and the amount of vitamin D necessary to optimize the retention of calcium and phosphorus and bone mineralization were discussed by Jacques Senterre from Belgium. Stephanie Atkinson from Canada and Frank Pohlandt from FRG agreed with Senterre that human milk whether banked or from the infant's mother does not provide sufficient calcium and phosphorus to meet the needs of growing premature infants; consequently a phosphate-depletion state develops. However, these scientists presented diverse views on the practical approach to supplementation of human milk with those minerals. The prescription of Senterre is provided in his chapter. Pohlandt prefers to supplement human milk with calcium and phosphorus until hypercalciuria is produced and bone mineral density parallels intrauterine accretion. A word of caution to this approach was given by Atkinson who presented data on the association of oral calcium supplementation with increased fatty acid malabsorption and the potential for hypercalciuria and nephrocalcinosis particularly when prolonged diuretic therapy must be maintained in low birthweight infants.

The relationship between folate status of the mother, milk intake of folate and folate status of the infant was discussed by Johan Ek from Norway. It is clear that the fetus has mechanisms of folate accrual, making both plasma and red cell folate higher in the fetus and the newborn infant than in the mother. Breast milk folate levels reflect maternal folate status but generally decrease during lactation while infant serum and RBC folate levels are maintained. Since folate levels in infants fed formula with similar folate concentration as human milk are lower, it appears that either there is a lower biological activity of folate from formula than from human milk, or higher metabolic demands for folate in formula-fed than in breast-fed infants. These data suggest that formulas should contain at least 40% more folate than in human milk in order to assure proper folate nutrition of infants. Most current proprietary infant formulas are above this suggested level.

The WORKSHOP ON NUTRIENT UTILIZATION provided new information which demonstrated that minerals and vitamins in human milk are highly bioavailable to the infant as a result of specific facilitory factors and absorptive mechanisms that are not completely understood. Despite the excellent nutritional value of human milk for the normal term infant, the premature infants' needs for some nutrients such as calcium and phosphorus may not be met. Clearly, future research must address the optimal amount and ratios of minerals and vitamins provided for infants to maximize nutrient utilization from all forms of feedings.

TRACE ELEMENT BINDING LIGANDS IN HUMAN MILK: FUNCTION IN TRACE ELEMENT UTILIZATION

Bo Lönnerdal

Department of Nutrition
University of California
Davis, CA

INTRODUCTION

Several essential trace elements are present in human milk in much lower concentrations than in milk from other species or in other infant diets.[1] However, recent research has shown that the bioavailability of these trace elements (iron, zinc, copper and manganese) is very high from human milk.[2] As a consequence, trace element deficiency in breast-fed infants is rare.

In order to overcome a lower bioavailability of trace elements from cow's milk and soy, the common protein sources used when manufacturing infant formulas, trace elements are added to supplement the original content of the processed raw materials. Our knowledge about the appropriate level and chemical form of these supplements, however, is limited. Increased understanding of the factors binding trace elements in human milk and how they facilitate trace element uptake can aid in deciding on optimum levels and modes of supplementation.

It should be recognized that ligands which bind trace elements in the diet will be subject to digestion before reaching the enterocytes of the small intestine. However, limitations in the digestive capacity of the newborn as well as the resistance against proteolysis of some components may result in a distribution of trace elements among ligands in partially digested human milk in the intestine which is not fundamentally different from that in the milk originally ingested. Thus, studies on binding ligands in the milk must be followed by direct studies on the effect of individual components on mucosal uptake of specific elements as well as whole body retention.

Iron

The ligands binding iron in human milk are described in Table 1. Of these ligands, lactoferrin has received by far the most attention. This protein, originally reported by Johansson in 1960,[6] can bind two ferric (+III) ions with the concomitant binding of two anions, carbonate or bicarbonate.[7] The amino acid sequence of human lactoferrin was recently determined[8] and the carbohydrate structures of the two glycans linked to the single polypeptide have also been elucidated. As can be seen in Table 1, lactoferrin binds only 30% of the iron in human milk. This means that lactoferrin is present in a largely unsaturated form; only 1-5% of its iron-binding capacity is utilized.[3] There may be several reasons why lactoferrin does not bind a larger proportion of iron in human milk: 1) Although the binding constant of lactoferrin for iron is very high,[7] $K_{diss} \cong 10^{30}$, other ligands with lower binding constants such as citrate and casein are present in higher concentrations and complicated equilibria among these ligands, other cations and iron will determine iron distribution; 2) Iron bound to xanthine oxidase,[9] localized within a prosthetic flavin group which is found in the hydrophobic milk fat globule membrane, may be inaccessible for other ligands. However, as mentioned earlier, it is possible that with digestion and pH changes, iron may become re-distributed in the stomach and intestine as compared to the milk. A likely candidate for accumulating iron released by this re-distribution would be lactoferrin with its high binding constant, provided that the molecule survives passage through the upper part of the gastrointestinal tract. An example of such re-distribution would be that iron bound to citrate would be released at pH 5 while lactoferrin would bind iron at this pH in the presence of bicarbonate.

Table 1. Iron Binding Ligands in Human Milk[1,2]

Ligand	Concentration (μM)	MW	% of Iron in Human Milk
Lactoferrin	25	80,000	26
Citrate	3000	210	32
Casein	70	25-35,000	9
Xanthine oxidase	0.02	300,000	33

[1]From refs. 3-5.

[2]Concentrations and % of iron in human milk represent average values.

Stability of lactoferrin against proteolytic degradation *in vitro* has been demonstrated by Brock et al.[10] This was especially pronounced for the iron-saturated form. Spik et al.[11] demonstrated that larger fragments of lactoferrin produced *in vitro* maintain the ability to bind iron. We have collected feces from exclusively breast-fed infants and found significant quantities of immunologically detectable lactoferrin in the extracts from both term[12] and premature[13] infants fed their own mothers' milk. The proportion of lactoferrin remaining intact was larger in young term infants (less than 2 months) than in older infants (2-6 months).[12] Schanler et al.[14] have demonstrated significantly higher levels of lactoferrin in fecal samples from very low birth weight infants fed fortified human milk as compared to cow's milk formula. Spik et al.[11] also found lactoferrin in fecal samples of term infants when lactoferrin was added to infant formula.

High iron absorption, as measured by the use of radioisotopes, has been documented from human milk in both human infants[15] and adults.[16] Attempts to evaluate the specific role of lactoferrin in iron absorption have been inconclusive.[16-18] We have utilized the infant Rhesus monkey as an animal model to investigate this question. Rhesus monkey milk is similar to human milk in gross nutrient composition and also contains a high concentration of lactoferrin as compared to most other species.[19] Iron absorption from human milk as well as monkey milk was found to be 40-60%.[20] We subsequently demonstrated the presence of a specific brush border membrane receptor that facilitates iron uptake.[21,22] It appears that the carbohydrate structure of the side chain is crucial for the binding of lactoferrin to the receptor and therefore lactoferrin from species other than the human and monkey that have different glycan structures do not bind to the receptor. Cox et al.[18] have earlier suggested the presence of a receptor for lactoferrin in adult humans. Recently, Mazurier et al.[23] visualized a receptor for lactoferrin in the brush border from guinea pigs. Our preliminary results on human infant intestinal samples indicate that there is a similar specific mechanism for iron accrual via lactoferrin receptors in the breast-fed infant.

Zinc and Copper

The ligands binding zinc and copper in human milk appear to be similar (Table 2). Citrate was reported in 1980 to be the low molecular weight ligand binding zinc in human milk[24]; subsequently serum albumin and casein were identified as zinc-binding proteins in human milk.[25] These ligands were also found to be copper-binding.[25,26] The amount of copper in human milk is low and therefore the small fraction of copper bound to low

Table 2. Zinc- and Copper-Binding Ligands in Human Milk[1]

Ligand	Concentration (μM)	MW	% Zn bound	% Cu bound
Serum albumin	7	68,000	28	39
Casein	70	25-30,000	14	28
Citrate	3000	210,000	29	24
Alkaline Phosphatase	--	155,000	18	0[2]
Amino Acids	--	< 200	--	--[3]

[1]From refs. 4,9,24-26.

[2]Fat fraction: 9%

[3]Included in citrate fraction.

molecular weight ligands was not detected.[27] However, by adding copper to human milk, Martin et al.[26] were able to identify citrate and some free amino acids as potential copper-binding ligands. Zinc and copper in human milk are also found in the milk fat globule membrane.[9] Preliminary studies indicate that zinc in part is bound to alkaline phosphatase in the membrane,[9] while no copper-binding protein in membranes has yet been identified.

Isolation and identification of metal binding ligands is methodologically complicated and therefore requires appropriate control experiments. Several mistakes with regard to method precautions such as using appropriate pH, ionic strength, trace element concentration of buffer, and avoiding nonspecific binding of gels have led to erroneous identifications of metal binding ligands in human milk. One such "identified" ligand, picolinic acid,[28] cannot be found in human milk at the concentrations claimed and, more importantly, if picolinic acid is chromatographed either with zinc or when added to human milk, the elution position for zinc-picolinate is different than that for zinc in human milk.[29] While it theoretically can be argued that redistribution of zinc among ligands could occur during chromatography, a minimum prerequisite for claiming a "zinc binding ligand" would be that the added ligand behaves as the in vivo ligand. The fact that picolinic acid is a good chelator of zinc in vitro should not overshadow the fact that it does not bind zinc in human milk. Another suggested candidate for zinc-binding is lactoferrin.[30,31] This iron-binding protein can be shown in vitro to also bind zinc, but with a lower affinity than for iron.[32] When lactoferrin was tentatively identified as a zinc-binding protein, however, this protein merely co-chromatographed with zinc. With regard to the study of Ainscough et al.,[30] it was later shown that the zinc could be attributed to the cation-binding capacity of the ion-exchanger used (CM-Sephadex) and that running the column without a sample would yield a zinc peak at the very same position as the one described for lactoferrin.[25] Furthermore, it has been shown that when ^{65}Zn is added to human milk and applied to anti-lactoferrin-Sepharose, an affinity absorbent binding lactoferrin, this protein but not ^{65}Zn is completely removed.

Recently, Eckhert[34] reported a "zinc absorption enhancing protein" in human milk. The protein was not purified, merely chromatographed by gel filtration and dialyzed. In fact, in this area of the chromatogram several known milk proteins would be eluted, such as α-lactalbumin, lysozyme and ribonuclease. Dialysis against distilled water removed most zinc from this protein fraction (only 1:30 molar ratio of zinc to protein), illustrating that it is not a zinc binding protein. A positive effect on zinc absorption from this fraction as compared to a similar fraction from cow's milk was claimed based on observations on plasma zinc uptake in one human subject. Because of high intraindividual variation in zinc absorption in man, such a claim is unlikely to be supported by proper experimental studies.

The bioavailability of zinc from human milk has been assessed by several methods. Indirect evidence for a high bioavailability was obtained from the study by Hambidge et al.[36] in which it was found that infants fed formula based on cow's milk with a level of zinc similar to that of human milk had lower plasma zinc levels than breast-fed infants. In addition, linear growth of male infants fed formula with this level of zinc was slower than that of breast-fed infants, which is an indication of zinc deficiency. Zinc supplementation of the formula to three times its original value overcame the negative effect on growth and normalized plasma zinc values. Subsequently, Casey et al.[37] reported higher plasma uptake of zinc from human milk than from cow's milk or formula in adults. However, plasma uptake of zinc does not necessarily reflect a higher bioavailability of zinc, i.e., higher utilization and retention of zinc.[38] We utilized a radioisotope of zinc, ^{65}Zn, in human adults, suckling rat pups and infant Rhesus monkeys to assess zinc bioavailability.[39-41] In all these studies, zinc retention from human milk was higher than from other milks or formulas. The magnitude of the difference in zinc retention among the diets was considerably higher in the infant animal models, emphasizing that when the digestive capacity is low and the absorptive capacity is high, the effect of the dietary components on zinc bioavailability is amplified.[2]

The components responsible for the high bioavailability of zinc from human milk have received much attention. After the initial identification of citrate as the major low molecular weight ligand binding zinc in human milk, several investigators wanted to ascribe a unique role for this ligand in zinc absorption. We suggested that zinc bound to citrate represents an accessible pool of zinc for absorption while other dietary components such as cow's milk casein would inhibit zinc absorption.[24] This was based on our finding that both human and cow's milk contain large amounts of citrate, but the amount of zinc bound to citrate is higher in human milk. In a subsequent study, we were able to show that addition of citrate beyond what is originally in milk did not significantly affect zinc absorption.[39] This is likely due to the fact that there is already a roughly 200:1 molar excess of citrate to zinc in both human and bovine milk as well as in many formulas. The negative effect of cow's milk casein was demonstrated in our rat pup model.[42] Intubation with isolated fractions from human and cow's milk demonstrated that zinc absorption was high from whey, fat and the low molecular weight fraction from both milks but lower from cow's milk casein than from human milk casein. The considerably higher concentration of casein in cow's milk as compared to human milk would further compound this effect. Thus, the high absorption of zinc from human milk may well be due to its relative lack of factors having an inhibitory effect on zinc absorption.

The bioavailability of copper from infant diets has received far less attention than zinc, most likely because of the lack of effective methods to study copper absorption. We used the rat pup model developed for zinc absorption to study the uptake of copper from milks and formulas.[43] Similar to zinc and iron, copper absorption was highest from human milk. Copper absorption from cow's milk formula was not found to be significantly different from human milk, while that of cow's milk was. This is consistent with the fact that when copper deficiency has been observed in human infants, cow's milk feedings (often in combination with premature birth) was used.[44] Thus, processing of cow's milk into formula or the addition of some nutrient may have a beneficial effect on copper absorption. The individual components of human and bovine milk which influence copper absorption still remain to be investigated.

Table 3. Manganese Binding Ligands in Human Milk[1]

Ligand	Concentration (μM)	MW	% Mn bound
Lactoferrin	25	80,000	67
Casein	70	25-35,000	11
Low molecular weight ligands	--	< 200	4
Milk fat globule membrane protein	--	∿ 60,000	18

[1]From ref. 33.

Manganese

The ligands binding manganese in human milk are shown in Table 3. As can be seen, lactoferrin is the major manganese-binding ligand in human milk.[33] It should be emphasized, however, that the amount of manganese in human milk is very low, 4-8 ng/ml (i.e., the proportion of iron to manganese in human milk is about 2000:1). Very low molecular weight compounds in human milk bind manganese but their physical-chemical structures remain to be characterized. Similarly, the protein(s) in the milk fat globule membrane that bind manganese has not yet been identified.

We have assessed manganese uptake from milks and formulas in our rat pup model.[45] In young animals, manganese retention was very high from all infant diets studied, indicating that absorption is high and/or excretion low. We have subsequently shown that there is little if any control of uptake at the mucosal level,[46] and that retained manganese is re-distributed within the body and not excreted. Miller et al.[47] have shown that this early absence of excretion is related to low bile output. Thus, in young animals, the body load of manganese is directly correlated to manganese intake from the diet. Since some infant formulas contain considerably higher concentrations of manganese than human milk,[48] the possibility of manganese toxicity cannot be excluded. The finding of redistribution of manganese from other body compartments to the brain, the major target tissue for manganese toxicosis,[49] emphasizes the need for scrutinizing manganese intake of formula-fed infants.

SUMMARY

The trace elements iron, zinc, copper and manganese are bound to various ligands in human milk. Concentrations of ligands and their affinity constants toward different cations will determine the distribution of trace elements among ligands. In human milk, lactoferrin has been shown to bind iron and manganese, but not zinc and copper. The latter two ions are bound to casein, serum albumin and citrate. Proteins in the milk fat globule membrane also contain significant proportions of these trace elements. During passage through the stomach and the upper small intestine, pH changes as well as digestion will take place and re-distribution of trace elements will occur. Lactoferrin has been shown to be resistant to low pH and proteolysis and can be found undigested in feces of breast-fed infants. The presence of a receptor for lactoferrin in the brush border membrane of the small intestine could explain the exceptionally high absorption of iron and manganese in newborns. Human casein and serum albumin are easily digested and a re-distribution of zinc and copper to citrate and possibly free amino acids is likely to occur. This pool of zinc and copper would be easily accessible for absorption and may explain the high bioavailability of zinc and copper in newborns.

REFERENCES

1. B. Lönnerdal, C. L. Keen, and L. S. Hurley, Iron, copper, zinc and manganese in milk, Ann. Rev. Nutr. 1:149 (1981).
2. B. Lönnerdal, Dietary factors affecting trace element bioavailability from breast-milk, cow's milk and infant formula, in: "Progress in Food and Nutrition Science," Vol. 9, R. K. Chandra, ed., Pergamon Press, Inc., Elmsford, NY, pp. 35-62 (1985).
3. G.-B. Fransson and B. Lönnerdal, Iron in human milk, J. Pediatr. 96:380 (1980).
4. G.-B. Fransson and B. Lönnerdal, Distribution of trace elements and minerals in human and cow's milk, Pediatr. Res. 17:912 (1983).
5. B. Lönnerdal, Iron in breast milk, in: "Iron Nutrition in Infancy and Childhood," A. Stekel, ed., Nestle, Vevey/Raven Press, New York, pp. 95-118 (1984).
6. B. G. Johansson, Isolation of an iron-containing red protein from human milk, Acta Chem. Scand. 14:510 (1960).
7. P. L. Masson, "La lactoferrine," Editions Arscia SA, Brussels (1970).
8. M. H. Metz-Boutique, J. Jollés, J. Mazurier, F. Schoentger, D. Legrand, G. Spik, J. Montreuil, and P. Jollés, Human Lactoferrin: amino acid sequence and structural comparisons with other transferrins, Eur. J. Biochem. 145:659 (1984).
9. G.-B. Fransson and B. Lönnerdal, Iron, copper, zinc, calcium and magnesium in human milk fat, Am. J. Clin. Nutr. 39:185 (1984).
10. J. H. Brock, F. Arzabe, F. Lampreave, and A. Pineiro, The effect of trypsin on bovine transferrin and lactoferrin, Biochim. Biophys. Acta 446:214 (1976).

11. G. Spik, B. Brunet, C. Mazurier-Dehaine, G. Fontaine, and J. Montreuil, Characterization and properties of the human and bovine lactoferrins extracted from the feces of newborn infants, Acta Paediatr. Scand. 71:974 (1982).
12. L. A. Davidson and B. Lönnerdal, Lactoferrin and secretory IgA in feces of exclusively breast-fed infants, Am. J. Clin. Nutr. 41:852 (1985).
13. S. M. Donovan, S. A. Atkinson, and B. Lönnerdal, Whey proteins in feces of preterm infants receiving preterm milk and infant formula, in this volume.
14. R. J. Schanler, R. M. Goldblum, C. Garza, and A. S. Goldman, Enhanced fecal excretion of selected immune factors in very low birth weight infants fed fortified human milk, Pediatr. Res. 20:711 (1986).
15. U. M. Saarinen, M. A. Siimes, and P. R. Dallman, Iron absorption in infants: high bioavailability of breast milk iron as indicated by the extrinsic tag method of iron absorption and by the concentration of serum ferritin, J. Pediatr. 91:36 (1977).
16. J. A. McMillan, S. A. Landaw, and F. A. Oski, Iron sufficiency in breast-fed infants and the availability of iron from human milk, Pediatrics 58:686 (1976).
17. J. A. McMillan, F. A. Oski, G. Lourie, R. M. Tomarelli, and S. A. Landaw, Iron absorption from human milk, simulated human milk, and proprietary formulas, Pediatrics 55:686 (1975).
18. T. M. Cox, J. Mazurier, G. Spik, J. Montreuil, and T. J. Peters, Iron binding proteins and influx of iron across the duodenal brush border. Evidence for specific lactotransferrin receptors in the human intestine, Biochim. Biophys. Acta 588:120 (1979).
19. L. A. Davidson and B. Lönnerdal, Isolation and characterization of Rhesus monkey milk lactoferrin, Pediatr. Res. 20:197 (1986).
20. B. Lönnerdal, L. Davidson, and C. L. Keen, Development of a Rhesus monkey model for the study of iron and manganese from infant diets, Fed. Proc. 44:1850 (1985).
21. L. A. Davidson and B. Lönnerdal, Specific binding of monkey milk lactoferrin to its brush border receptor. Fed. Proc. 44:1673 (1985).
22. L. A. Davidson and B. Lönnerdal, The intestinal lactoferrin receptor: presence and specificity during development, Fed. Proc. 45:588 (1986).
23. J. Mazurier, J. Montreuil, and G. Spik, Visualization of lactotransferrin brush-border receptors by ligand-blotting, Biochim. Biophys. Acta 821:435 (1985).
24. B. Lönnerdal, A. G. Stanislowski, and L. S. Hurley, Isolation of a low molecular weight zinc binding ligand from human milk, J. Inorg. Biochem. 12:71 (1980).
25. B. Lönnerdal, B. Hoffman, and L. S. Hurley, Zinc and copper binding proteins in human milk, Am. J. Clin. Nutr. 36:1170 (1982).
26. M. T. Martin, K. F. Licklider, J. G. Brushmiller, and F. A. Jacobs, Detection of low molecular weight copper (II) and zinc (II) binding ligands in ultrafiltered milks -- the citrate connection, J. Inorg. Biochem. 15:55 (1981).
27. B. Lönnerdal, C. L. Keen, B. Hoffman, and L. S. Hurley, Copper ligands in human milk: a vehicle for copper supplementation in the treatment of Menkes' disease?, Am. J. Dis. Child. 134:802 (1980).
28. G. W. Evans and P. E. Johnson, Characterization and quantitation of a zinc-binding ligand in human milk, Pediatr. Res. 14:876 (1980).
29. L. S. Hurley and B. Lönnerdal, Zinc binding in human milk: citrate versus picolinate, Nutr. Rev. 40:65 (1982).
30. E. W. Ainscough, A. M. Brodie, and J. E. Plowman, Zinc transport by lactoferrin in human milk, Am. J. Clin. Nutr. 33:1314 (1980).
31. P. Blakeborough, D. N. Salter, and M. I. Gurr, Zinc binding in cow's milk and human milk, Biochem. J. 209:505 (1983).

32. E. W. Ainscough, A. M. Brodie, and J. E. Plowman, The chromium, manganese, cobalt and copper complexes of human lactoferrin, Inorg. Chim. Acta 33:149 (1979).
33. B. Lönnerdal, C. L. Keen, and L. S. Hurley, Manganese binding proteins in human and cow's milk, Am. J. Clin. Nutr. 41:550 (1985).
34. C. D. Eckhert, Isolation of a protein from human milk that enhances zinc absorption in humans, Biochem. Biophys. Res. Comm. 130:264 (1985).
35. B. Arvidsson, Å. Cederblad, E. Björn-Rasmussen, and B. Sandström, A radionuclide technique for studies of zinc absorption in man, Int. J. Nucl. Med. Biol. 5:104 (1979).
36. K. M. Hambidge, P. A. Walravens, C. E. Casey, R. M. Brown, and C. Bender, Plasma zinc concentrations of breast-fed infants, J. Pediatr. 94:607 (1979).
37. C. E. Casey, P. A. Walravens, and K. M. Hambidge, Availability of zinc: loading tests with human milk, cow's milk, and infant formulas, Pediatrics 68:394 (1981).
38. L. S. Valberg, P. R. Flanagan, J. Brennan, and M. J. Chamberlain, Does the oral zinc tolerance test measure zinc absorption?, Am. J. Clin. Nutr. 41:37₀(1985).
39. B. Sandström, A. Cederblad, and B. Lönnerdal, Zinc absorption from human, cow's milk and infant formula, Am. J. Dis. Child. 137:726 (1985).
40. B. Sandström, C. L. Keen, B. Lönnerdal, An experimental model for studies of zinc bioavailability from milk and infant formulas using extrinsic labelling, Am. J. Clin. Nutr. 38:420 (1983).
41. B. Lönnerdal, J. G. Bell, A. G. Hendrickx, and C. L. Keen, Improved zinc bioavailability from dephytinized soy formula, Am. J. Clin. Nutr. 43:674 (1986).
42. B. Lönnerdal, C. L. Keen, J. G. Bell, and L. S. Hurley, Zinc uptake and retention from chelates and milk fractions, in "Trace Elements in Man and Animals (TEMA)-5," C. F. Mills, I. Bremner, and J. K. Chester, eds., Commonwealth Agricultural Bureaux, Farnham Royal, UK, pp. 427-430 (1985).
43. B. Lönnerdal, J. G. Bell, and C. L. Keen, Copper absorption from human milk, cow's milk and infant formulas using a suckling rat model, Am. J. Clin. Nutr. 42:836 (1985).
44. A. Cordano, Copper deficiency in clinical medicine, in "Zinc and Copper and Clinical Medicine," K. M. Hambidge and B. L. Nichols, Jr., eds., SP Medical and Scientific Books, New York, pp. 119-126 (1978).
45. C. L. Keen, J. G. Bell, and B. Lönnerdal, The effect of age on manganese status and retention from milk and infant formulas in rats, J. Nutr. 116:395 (1986).
46. J. G. Bell, C. L. Keen, and B. Lönnerdal, Manganese uptake by brush-border membrane vesicles from rat small intestine, Fed. Proc. 45:368 (1986).
47. S. T. Miller, G. C. Cotzias, and H. A. Evert, Control of tissue manganese: initial absence and sudden emergence of excretion in the neonatal mouse, Am. J. Physiol. 229:1980 (1975).
48. B. Lönnerdal, C. L. Keen, M. Ohtake, and T. Tamura, Iron, zinc, copper, and manganese in infant formulas, Am. J. Dis. Child. 137:433 (1983).
49. C. L. Keen, B. Lönnerdal, and L. S. Hurley, Manganese, in "Biochemistry of the Essential Ultratrace Elements," E. Frieden, ed., Plenum Publ. Co., New York, pp. 89-132 (1985).

CALCIUM AND PHOSPHORUS BALANCE IN PRETERM INFANTS FED HUMAN MILK OR HUMAN MILK SUPPLEMENTED WITH VITAMIN D AND MINERALS

Jacques Senterre

Department of Pediatrics, State University of Liège
Hôpital de la Citadelle
Liège, Belgium

INTRODUCTION

The recommended intakes for calcium (Ca), phosphorus (P), and vitamin D in the preterm infant remain an area of ongoing controversy.[1-3] Because of the rapid rate of growth, hypomineralization of bone or overt rickets have been frequently observed in preterm infants fed human milk or standard formulas. The pathogenesis of the skeletal lesions is often multifactorial. Inadequate Ca and P intakes, copper and zinc deficiencies, and poor vitamin D status have all been implicated.[1-3]

Rapid fetal skeletal growth normally takes place <u>in utero</u> during the last trimester, in association with the placental transfer of large amounts of Ca and P. From chemical analyses of human fetuses, it can be calculated that the mean accumulations of Ca and P during the third trimester of gestation are about 130 and 75 mg/kg/day, respectively.[4]

Although the most appropriate goals of nutrition of the premature infant are not definitively known, achieving postnatal growth and nutrient accretion rates that approximate the <u>in utero</u> rates of the normal fetus is a widely accepted approach to the nutritional management of preterm infants. Employing a factorial method to estimate requirements, Ziegler et al.[5] summed the tissue accretion of Ca and P with estimates of urinary and dermal losses and the extent of absorption to derive theoretical mineral requirements for preterm infants of varying gestational ages. They calculated that the advisable intake of Ca and P for 800 to 1200 g preterm infants reaches 210 and 140 mg/kg/day, respectively.

Human milk, particularly the infant's own mothers' milk, is advocated as a nutrient source for preterm infants. Term or preterm human milks, however, contain only about 30 mg Ca and 15 mg P per dl (45 and 22 mg per 100 kcal).[6] It is obvious that the amounts of mineral supplied by breast milk are inadequate to allow preterm infants to accumulate Ca and P at intrauterine rates, even if all the Ca and P are absorbed and retained. However, the theoretical estimates of mineral requirements must be validated by measuring absorption and retention of oral intakes and by measuring bone mineralization in growing infants.

The present report summarizes our results of mineral balance studies carried out in preterm infants fed pooled banked pasteurized human milk unmodified supplemented with vitamin D, phosphate, or calcium and phosphate.[7-12] From these data, we will draw some practical conclusions about the quantities of Ca, P, and vitamin D which are required to prevent rickets and to support adequate mineralization in preterm infants fed human milk.

METHODS

Three-day metabolic balance studies were carried out at 21 ± 5 days of postnatal age in matched groups of growing preterm boys with birthweights under 1500 g. All infants were appropriate for gestational age. Parental consent had been given. The infants were nursed in incubators. During balance period, the baby was loosely restrained on a hammock-shaped metabolic bed. Separate collection of stools and urine were performed as previously described.[13] The infants were tube-fed pooled banked pasteurized expressed human milk every 3 h. Carmine markers were added to the first and last feedings. From aliquots of milk, fecal homogenates and urines, Ca was measured by atomic absorption spectroscopy after ashing at 550° C for 14 h, and P was measured by the method of Briggs after digestion with H_2SO_4.

Results were expressed as mean ± 1SD. Statistical analyses were carried out with Student's t test.

RESULTS

Human milk

The preterm infants fed human milk received on the average 56 mg of Ca/kg/day (Table). There was a net absorption of 50 % of ingested Ca. Mean urinary excretion of Ca reached 9 mg/kg/day. As a result, mean Ca retention was only 19 mg/kg/day. Mean P intake was 28 mg/kg/day. Almost all P ingested was absorbed and retained, but retention reached only 24 mg/kg/day.

Effects of vitamin D supplementation

In the group of preterm infants receiving 30 μg (1200 IU) of supplemental vitamin D from birth, Ca intake was similar but net absorption was significantly improved compared to the group fed human milk without vitamin D supplementation. However, the better absorption of Ca lead to an increase in the urinary excretion of Ca so that Ca retention was not significantly improved. Vitamin D supplementation did not significantly modify the P absorption and retention.

Effects of P supplementation

Ten additional preterm infants receiving 1200 IU of vitamin D from birth were fed human milk supplemented with a molar solution of K_2HPO_4 (3 ml/l) in order to increase the P content of milk by 9 mg/dl. Although the Ca:P ratio in the milk decreased from 2.0 to 1.1, net absorptions of Ca and P were not modified. Urinary excretion of P increased and there was a marked decrease in the urinary excretion of Ca. Mean Ca and P retentions were significantly improved.

Effects of Ca and P supplementation

A fourth group of 8 additional preterm infants receiving 1200 IU of vitamin D from birth were given human milk supplemented with 27 mg/dl of Ca (Ca gluconate) and 25 mg/dl of P (Na_2HPO_4). Calcium intake was about doubled and the Ca:P ratio in the milk was 1.5. Net absorptions of Ca and P were not modified. Urinary excretion of Ca was lower than that of P. Mean Ca and P retentions were significantly greater than in the other groups.

Table. Effects of Vitamin D and Mineral Supplementation on Calcium and Phosphorus Retention in Preterm Infants Fed Human Milk

	HM (n = 10)	HM+D (n = 10)	HM+P (n = 10)	HM+CaP (n = 8)
Calcium				
Intake (mg/kg/day)	56 ± 9	58 ± 11	53 ± 8	90 ± 6*
Feces (mg/kg/day)	28 ± 8	17 ± 10*	14 ± 7*	24 ± 13
Urine (mg/kg/day)	9 ± 5	16 ± 8	3 ± 1*	3 ± 2*
Retention (mg/kg/day)	19 ± 9	25 ± 12	35 ± 7*	63 ± 12*
Net absorption (%)	50 ± 11	71 ± 14*	74 ± 13*	73 ± 13*
Net retention (%)	34 ± 10	43 ± 11	68 ± 12*	70 ± 13*
Phosphorus				
Intake (mg/kg/day)	28 ± 4	30 ± 7	49 ± 11*	62 ± 5*
Feces (mg/kg/day)	3 ± 1	3 ± 1	3 ± 1	4 ± 1
Urine (mg/kg/day)	1 ± 1	Trace	12 ± 7*	5 ± 4*
Retention (mg/kg/day)	24 ± 5	27 ± 6	34 ± 8*	53 ± 4*
Net absorption (%)	89 ± 4	90 ± 4	94 ± 3*	93 ± 2
Net retention (%)	86 ± 5	90 ± 5	69 ± 10*	85 ± 6

* Values significantly different (p<0.01) compared with human milk group
Data from Ref. 11 and 12.

DISCUSSION

Vitamin D requirement

Recent studies[14] have shown that vitamin D activity in human milk is much lower than previously thought.[15] In the current studies, we supplemented human milk with 1200 IU vitamin D per day. This dosage in older children could result in elevated 25-hydroxyvitamin D (25-OHD) levels. However, there is evidence that the vitamin D requirement of preterm infants is higher than that of term infants because of the very high rate of bone growth and possible impairment in vitamin D absorption and/or metabolism.[1-3] A close correlation exists between maternal and cord plasma 25-OHD levels. However, there may be sizeable differences in vitamin D status between mothers in North Europe and mothers in United States and Canada. In some North European countries including Belgium,[17] most newborn infants have low plasma levels of 25-OHD (8 ± 4 ng/ml) while in North America 25-OHD levels of newborn infants are in the range of 12 to 45 ng/ml.[2,3,18,19,31] We have shown that, in case of low 25-OHD level at birth, daily administration of at least 1200 IU of vitamin D is necessary to raise 25-OHD to adequate levels.[10]

In preterm infants supplemented with 400 IU D, Hillman et al.[18] observed that low or normal serum 25-OHD at birth may decline during the first nine weeks of life. They also showed that preterm infants given 800 IU D had higher serum Ca and 25-OHD concentration and less frequent hypomineralization than those given 400 IU D.[19] Recently, Pettifor et al.[20] reported that, despite a daily intake of 750 IU D, a number of preterm infants had persistently low 25-OHD values. On the contrary, Markestad et al.[21], demonstrated that adequate serum 25-OHD levels can be achieved using supplements as low as 500 IU D per day for the first 1-2 months of life. Administration of as little as 80-200 IU 25-OHD/kg/day leads to a rapid elevation of low serum 25-OHD concentrations,[19,22] but is not as effective as 800 IU D in preventing bone hypomineralization.[19]

$1,25(OH)_2D$ is the most potent metabolite of vitamin D in stimulating intestinal Ca absorption. In preterm infants, Glorieux et al.[23] have clearly shown that there is a strong positive correlation between 25-OHD and $1,25(OH)_2D$ levels. The production of $1,25(OH)_2D$, therefore, appears to be substrate dependent. This is in accordance with other observations : in premature infants with rickets, Seino et al.[24] found low $1,25(OH)_2D$ levels associated with low 25-OHD levels, whereas Rowe et al.[25] and Steichen et al.[26] reported elevated $1,25(OH)_2D$ levels associated with adequate 25-OHD levels but inadequate mineral intake. Elevated $1,25(OH)_2D$ concentrations have been observed during periods of increased demand of minerals such as infancy, puberty, pregnancy, and lactation. So the observed high levels in preterm infants may be physiological. Seino et al.[24] performed treatment trials of 1-alpha-OHD versus 600 IU D in very low-birth-weight infants and observed a significant reduction of the incidence of rickets. We did balance studies in formula fed preterm infants first on 1200 IU D and then on 0.5 μg $1,25(OH)_2D$.[9] Calcium absorption and retention increased on $1,25(OH)_2D$, but hypercalcemia and hypercalciuria occured. Since the production of $1,25(OH)_2D$ is tightly regulated and seems operative in preterm infants, and since the major function of $1,25(OH)_2D$ at the bone is resorption of minerals and inhibition of matrix formation, we do not recommend the routine administration of $1,25(OH)_2D$ in preterm infants. It is preferable to adapt vitamin D intake in order to get adequate 25-OHD levels.

Phosphorus depletion syndrome

We have demonstrated that almost all of the P ingested is absorbed and retained in preterm infants fed human milk. Calciuria, however, is very high, particularly after vitamin D supplementation, because of the increased Ca absorption. This wasting of Ca can be explained by the limited dietary supply of P in rapidly growing preterm infants fed human milk. Indeed, contrary to Ca which is essentially deposited in the bone, absorbed P is utilized not only for bone Ca deposition with a Ca:P ratio of 2.2:1.0, but also for soft tissue growth with a N:P ratio of 17:1. In case of relative deficiency of P, phosphate seems to be preferentially used for building up protoplasm rather than for skeletal calcification. In these circumstances, a classical P depletion syndrome sets in. It is characterized by a severe hypophosphatemia despite an increased tubular reabsorption of P, and no urinary excretion of P. Low levels of serum phosphate stimulate the synthesis of $1,25(OH)_2D$ which in turn increases Ca gut absorption and bone resorption of minerals. This results in a tendency to hypercalcemia, symptomatic hypercalciuria, elevated plasma alkaline phosphatase, and rickets. Serum PTH concentrations are normal or low.

Forty years ago, Von Sydow[27] reported a high incidence of hypophosphatemia, elevation of alkaline phosphatase, and rickets in preterm infants fed human milk. With the resurgence of breast feeding, increasing numbers of preterm infants are being fed with their own mothers' milk, and there have been several reports of phosphate deficiency rickets in very small breast-fed premature infants.[20,25,27-30,32-34] The P depletion syndrome may be precipitated by the use of human milk supplemented with protein and/or Ca salts without increasing the P intake.[20,36]

Our balance studies have shown that adding about 10 mg of P per dl of human milk prevents hypophosphatemia and hypercalciuria, and increases significantly Ca and P retentions.[8-12] Similar observations were subsequently reported by others groups.[32-39] Unless phosphate supplementation of human milk is routinely performed, we recommend that serum phosphate and/or urinary concentrations of Ca and P be closely monitrored. In our experience, a serum phosphate level below 5.5 mg/dl (1.8 mmol/l) and/or a Ca:P ratio (in mg) in the urine below 1 has to be taken as signs of P deficiency.

Calcium and phosphorus supplementation

Carey et al.[40] recently questioned whether supplementation of human milk with phosphate alone is desirable since increased P intake resulted in increased Ca utilization, thereby unmasking the concomitant dietary Ca deficiency. This could lead to hypocalcemia and secondary hyperparathyroidism which in turn would result in increased renal excretion of P. Hypocalcemia[25,41] and elevation of PTH[42] have been reported after P supplementation, but not confirmed in a other study.[37] We think that supplementation of human milk with phosphate alone has, at least, the advantage of reducing the risk of renal calcinosis[43] and hypophosphatemic rickets, which is a more severe bone disease than that due to secondary hyperparathyroidism.[44] Of course, the supplementation with P alone should not be expected to overcome deficient Ca ingestion. Both Ca and P need to be supplied if appropriate growth and mineralization of bone are to occur.

We have previously demonstrated by balance studies that premature infants fed human milk supplemented with Ca and P received 90 mg Ca and 62 mg P/kg/day, and retained twice as much Ca and P compared to infants fed human milk or human milk supplemented with P alone.[12] On fortifying human milk with minerals it is important to add phosphate salts first into milk to avoid calcium phosphate precipitation.[12] Those preterm infants receiving human milk fortified with Ca and P had no biochemical evidence of P and Ca insufficiency. Similarly, Schanler et al.[36] found that combined Ca and P supplementation resulted in normal serum Ca, P and alkaline phosphatase levels, and no clinical or radiological signs of rickets in preterm infants fed fortified mothers' milk. Their data and ours suggest that the mineral intakes in those preterm infants were adequate even thought it resulted in a lower Ca and P retention rate than in utero. Thus, fetal retention may not be an appropriate criterion of adequacy. Greer et al.[45] reported a complete resolution of rickets and dramatic increase in bone mineral content in a very-low-birth-weight infant fed human milk supplemented with minerals in order to provide a Ca and P intake of 110 and 50 mg/kg/day, respectively. Greer et al.[46] and Chan et al.[38] found that preterm infants fed high mineral formulas had bone mineral contents approximating intrauterine bone mineralization. They noticed, however, that bone mineral content decreased from birth in one third of the infants. Decreasing density may be a physiological event in preterm as in term infants during the first postnatal months because of bone remodelling. In fact, mimicking in utero accretion rates may not be necessary. In a follow-up study, Helin and coworkers[47] in Sweden measured bone mineral content in 75 preterm infants at the ages of 4 to 16 years. Those infants had had a mean birthweight of 1580 g and were presumably breast-fed in early life. Their bone mineral content was appropriate for height and similar to normative data obtained in normal children born at term.

It has been shown that indiscreminate use of high dose of Ca and P for supplementing human milk is not without risk. Milk osmolality increases substantially.[12] Too high an intake of P may lead to hyperphosphatemia, hypocalcemia, and secondary hyperparathyroidism.[1-3] Too high an intake of Ca may impede fat absorption[8,10,12,48,49] and may induce metabolic acidosis since hydroxyapatite deposition is a source of hydrogen ion.[50]

CONCLUSION

There is increasing evidence that during rapid postnatal growth, defeciency of dietary Ca and P is a major factor in the etiology of rickets in very-low-birth-weight infants. Term as well as preterm breast milks have low Ca and P contents and do not provide enough minerals for adequate bone mineralization in preterm infants. Adequate vitamin D intake is required for maximal Ca absorption. Accordingly Ca and P and vitamin D need to be supplied if satisfactory bone growth and skeletal mineralization are to occur. Mimicking fetal mineral accretion rate may, however, not be necessary. Indiscriminate use of high dose of Ca and P may induce adverse effects.

A phosphorus depletion syndrome characterized by hypophosphatemia, hypercalciuria, elevated serum alkaline phosphatase levels, and signs of rickets is frequently observed in preterm infants fed human milk. Phosphate supplementation of human milk prevents or corrects all signs of P deficiency, and improves slightly Ca and P retention, but it can not overcome deficient Ca ingestion.

In our experience, preterm infants fed human milk supplemented with minerals in order to provide 90-110 mg of Ca and 60-70 mg of P/kg/day and 1200 IU vitamin D/day do not manifest biochemical signs of Ca or P deficiency. This can be achieved either by supplementing human milk with Ca and P salts[12] or with a powdered fortifier containing proteins and minerals,[39] or by mixing human milk with a cow's milk based preterm formula.

REFERENCE

1. Atkinson SA. Calcium and phosphorus requirements of low birth weight infants : a nutritional and endocrinological perspective. Nutr Rev 41:69-78 (1983).
2. Hillman LS. Mineralization and late mineral homeostasis in infants. Role of mineral and vitamin D sufficiency and other factors, in: "Perinatal Calcium and Phosphorus Metabolism," Holick MF, Gray TK, Anast CS, eds, Elsevier, Amsterdam (1983).
3. Greer FR, Tsang RC. Calcium phosphorus, magnesium, and vitamin D requirements for the preterm infant, in: "Vitamin and Mineral Requirement for the Preterm Infants," Tsang RC, ed., Marcel Dekker, New York (1985).
4. Widdowson EM. Changes in body composition during growth, in: "Scientific Foundations of Paediatrics," Davis JA, Dobbing J, eds, William Heinemann Medical Books Ltd, London (1981).
5. Ziegler EE, Biga RL, Fomon SJ. Nutritional requirement of the premature infant, in: "Textbook of Pediatric Nutrition," Suskind RM, ed., Raven Press, New York (1981).
6. Atkinson SA, Radde LC, Chance GW. Macro-mineral content of milk obtained during early lactation from mothers of premature infants. Early Hum Dev 4:5-14 (1980).
7. Senterre J. Endogenous faecal calcium, total digestive juice calcium net and true calcium absorption in premature infants, in: "Perinatal Medicine," Stembera ZK, Polacek K, Sabata V, eds, Georg Thieme, Stuttgart (1976).
8. Senterre J. Calcium and phosphorus retention in preterm infants, in: "Intensive care in the newborn, II," Stern L, Oh W, Friis-Hansen B, eds, Masson, New York (1978).

9. Senterre J, David L, Salle B. Effects of 1,25-dihydroxycholecalciferol on calcium, phosphorus and magnesium balance, and on circulating parathyroid hormone and calcitonin in preterm infants, in: "Intensive care in the newborn III," Stern L, Salle B, eds, Masson, New York (1980).
10. Senterre J, Salle B. Calcium and phosphorus economy of the preterm infant and its interaction with vitamin D and its metabolites. Acta Paediatr Scand 296:85-92 (1982).
11. Senterre J, Putet G, Salle B, Rigo J. Effects of vitamin D and phosphorus supplementation on calcium retention in preterm infants fed banked human milk. J Pediatr 103:305-7 (1983).
12. Salle B, Senterre J, Putet G, Rigo J. Effects of calcium and phosphorus supplementation on calcium retention and fat absorption in preterm infants fed pooled human milk. J Pediatr Gastroenterol Nutr 5:638-42 (1986).
13. Senterre J, Sodoyez-Goffaux F, Lambrechts A. Metabolic balance studies in premature babies. I. Methodology. Acta Paediatr Belg 25:133-42 (1971).
14. Reeve LE, Chesney RW, Deluca HF. Vitamin D of human milk : identification of biologically active forms. Am J Clin Nutr 36:122-6 (1982).
15. Greer FR, Reeve LE, Chesney RW, Deluca HF. Water-soluble vitamin D in human milk : a myth. Pediatrics 69:238 (1982).
16. Delvin EE, Glorieux FH, Salle BL, David L, Varenne JP. Control of vitamin D metabolism in preterm infants : fetomaternal relationships. Arch Dis Child 57:754-7 (1982).
17. Bouillon R, Van Baelen H, De Moor P. 25-hydroxyvitamin D and its binding protein in maternal and cord serum. J Clin Endocrinol Metabol 45:679-84 (1977).
18. Hillman LS, Hoff N, Salmons S, Martin L, McAlister W, Haddad J. Mineral homeostasis in very premature infants : serial evaluation of serum 25-hydroxyvitamin D, serum minerals , and bone mineralization. J Pediatr 106:970-80 (1985).
19. Hillman LS, Hollis B, Salmons S, Martin L, Slatopolsky E, McAlister W, Haddad J. Absoption, dosage, and effect on mineral homeostasis of 25-hydroxycholecalciferol in premature infants : comparison with 400 and 800 IU vitamin D_2 supplementation. J Pediatr 106:981-9 (1985).
20. Pettifor JM, Stein H, Herman A, Ross FP, Blumenfeld T, Moodley GP. Mineral homeostasis in very low birth weight infants fed either own mother's milk or pooled pasteurized preterm milk. J Pediatr Gastroenterol Nutr 5:248-53 (1986).
21. Markestad T, Aksnes L, Finne PH, Aarkog D. Vitamin D nutritional status of premature infants supplemented with 500 IU vitamin D_2 per day. Acta Paediatr Scand 72:517-20 (1983).
22 Salle BL, David L, Glorieux FH, Delvin E, Senterre J, Renaud H. Early oral administration of vitamin D and its metabolites in premature neoanates : effect on mineral homeostasis. Pediatr Res 16:75-8 (1982).
23. Glorieux FH, Salle BL, Delvin EE, David L. Vitamin D metabolism in preterm infants : serum calcitriol levels during the first five days of life. J Pediatr 99:640-3 (1981).
24. Seino Y, Ishii T, Shimotsuji T, Ishida M, Yabuuchi N. Plasma active vitamin D concentration in low birthweight infants with rickets and its response to vitamin D treatment. Arch Dis Child 56:628-32 (1981).
25. Rowe JC, Wood DH, Rowe DW, Raisz LG. Nutritional hypophosphatemic rickets in a premature infant fed breast milk. N Engl J Med 300:293-6 (1979)
26. Steichen JJ, Tsang RC, Greer FR, Ho M, Hug G. Elevated serum 1,25-dihydroxyvitamin D concentrations in rickets of very low-birth-weight infants. J Pediatr 99:293-8 (1981).
27. Von Sydow G. A study of the development of rickets in premature infants. Acta Paediatr Scand suppl.11:1-22 (1946).
28. Callenbach JC, Sheehan MB, Abramson SJ, Hall RT. Etiologic factors in rickets of very low-birth-weight infants. J Pediatr 98:800-5 (1981).

29. Atkinson SA, Radde IC, Anderson GH. Macromineral balances in premature infants fed their own mothers' milk or formula. J Pediatr 102:99-106 (1983).
30. Gross SJ. Growth and biochemical response of preterm infants fed human milk or modified infant formula. N Engl J Med 308:237-41 (1983).
31. Hillman LS, Salmons SJ, Slatopolsky E, McAlister WH. Serial serum 25-hydroxyvitamin D and mineral homeostasis in very premature infants fed preterm human milk. J Pediatr Gastroenterol Nutr 4:762-70 (1985).
32. Lyon AJ, McIntosh N. Calcium and phosphorus balance in extremely low birthweight infants in the first six weeks of life. Arch Dis Child 59:1145-50 (1984).
33. Lyon AJ, McIntosh N, Wheeler K, Brooke OG. Hypercalcaemia in extremely low birthweight infants. Arch Dis Child 59:1141-4 (1984).
34. Rowe J, Rowe D, Horak E, Spackman T, Saltzman R, Robinson S, Philipps A, Raye J. Hypophosphatemia and hypercalciuria in small premature infants fed human milk : Evidence for inadequate dietary phosphorus. J Pediatr 104:112-7 (1984).
35. Sagy M, Birenbaum E, Balin A, Orda S, Barzilay Z, Brish M. Phosphate-depletion syndrome in a premature infant fed human milk. J Pediatr 96:683-5 (1986).
36. Schanler RJ, Garza C, O'Brian Smith E. Fortified mothers' milk for low birth weight infants : Results of macromineral balance studies. J Pediatr 107:767-74 (1985).
37. Sann L, Loras B, David L, Durr F, Simonnet C, Baltassat P, Bethenod M. Effect of phosphate supplementation to breast fed very low birthweight infants on urinary calcium excretion, serum immunoreactive parathyroid hormon and plasma 1.25-dihydro-vitamin D concentration. Acta Paediatr Scand 74:664-8 (1985).
38. Chan GM, Mileur L, Hansen JW. Effects of increased calcium and phosphorus formulas and human milk on bone mineralization in preterm infants. J Pediatr Gastroenterol Nutr 5:444-9 (1986).
39. Modanlou HD, Lim MO, Hansen JW, Sickles V. Growth, biochemical status, and mineral metabolism in very-low-birth-weight infants receiving fortified preterm human milk. J Pediatr Gastroenterol Nutr 5:762-7 (1986).
40. Carey DE, Goetz CA, Horak E, Rowe JC. Phosphorus wasting during phosphorus supplementation of human milk feedings in preterm infants. J Pediatr 107:790-4 (1985).
41. Kovar IZ, Mayne PD, Robbe I. Hypophosphotemic rickets in the preterm infant, hypocalcaemia after calcium and phosphorus supplementation. Arch Dis Child 58:629-31 (1983).
42. Koo WWK, Gupta JM, Nayanar VV, Wilkinson M, Posen S. Continuous nasogastric phosphorus infusion in hypophosphotemic rickets of prematurity. Am J Dis Child 138:172-5 (1984).
43. Goldsmith MA. Renal calcification in premature infants. Pediatrics 71:992 (1983).
44. Harrison JE, Hitchman JW, Hitchman A, Hasany SA, McNeill KG, Tam CS. Differences between the effects of phosphate deficiency and vitamin D deficiency in bone metabolism. Metabolism 29:1225-9 (1980).
45. Greer FR, Steichen JJ, Tsang RC. Calcium and phosphate supplements in breast milk-related rickets. Results in a very-low-birth-weight infant. Am J Dis Child 136:581-3 (1982).
46. Greer FR, Steichen JJ, Tsang RC. Effects of increased calcium, phosphorus, and vitamin D intake on bone mineralization in very-low-birthweight infants fed formulas with polycose and medium-chain triglycerides. J Pediatr 100:951-5 (1982).
47. Helin I, Landin LA, Nilsson BE. Bone mineral content in preterm infants at age 4 to 16. Acta Paediatr Scand 74:264-7 (1985).
48. Katz L, Hamilton JR. Fat absorption in infants of birth weight less than 1,300 gm. J Pediatr 85:608-14 (1974).

49. Chappell JE, Clandinin MT, Kearney-Volpe C, Reichman B, Swyer PW. Fatty acid balance studies in premature infants fed human milk or formula : effect of calcium supplementation. J Pediatr 108:439-47 (1986).
50. Kildeberg P, Engel K, Winters RW. Balance of net acid in growing infants. Endogenous and transintestinal aspects. Acta Paediatr Scand 58:321-9 (1969).

SELENIUM NUTRITION DURING LACTATION AND EARLY INFANCY

Anne M. Smith and Mary Frances Picciano

University of Utah, Salt Lake City, Utah 84112

University of Illinois, Urbana, Illinois 61801

INTRODUCTION

The nutritional importance of selenium has only recently been discovered. Selenium was originally considered only to be a toxic element after it was found to cause the disabling Alkali Disease and the Blind Staggers in livestock grazing in parts of the western United States during the early 1900's. These areas were and still are sites of selenium-accumulating plants. About 50 years later the beneficial aspects of selenium began to be recognized when Schwarz and Foltz (1) found that selenium prevented liver necrosis in rats.

A metabolic role for selenium was not demonstrated until 1973 when the element was shown to be an integral component of glutathione peroxidase (GPx), an enzyme involved in the destruction of hydroperoxides (2). More recently, the importance and possible essentiality of selenium in human metabolism has been indicated. Several patients on long-term total parenteral nutrition therapy with solutions low in selenium have experienced severe muscle pain which was corrected with selenium supplementation (3, 4). Additional evidence for the essentiality of selenium is its ability to maximize growth of mammalian cells in culture (5). Other functions, independent of its role in GPx, cannot be ruled out, especially since the loss of GPx activity is not enough to explain the connection between selenium deficiency and many diseases with which it is associated.

SELENIUM NUTRITION DURING LACTATION AND MILK SELENIUM CONTENT

Historically, interest in selenium nutrition during lactation stemmed from a 1975 report by Shearer and Hadjimarkos which indicated a geographic variation in the selenium content of human milk in the United States (6). These investigators demonstrated that the selenium content of milk from women living in areas producing selenium adequate crops (28 ng/ml) was greater than that of milk from women in low-selenium areas (13 ng/ml). Since this early report, geographic differences in the selenium content of human milk have been closely linked to variations in dietary selenium intake. Low values, 10 ng/ml or less, have recently been reported for mature human milk from New Zealand (7), Finland (8), and Belgium (9), countries known to have low dietary selenium intakes for adults. In contrast, higher mean values, 16-28 ng/ml, have been reported for mature human milk from Germany (10), the United States (11), and Japan (12).

The lowest human milk selenium concentration (5.8 ng/ml) was reported for Finnish mothers at 6 months of lactation and was probably related to their low dietary selenium intake (8). Their mean selenium intake was 33 ug/day, which is below the safe and adequate intake of 50 ug/day recommended by the U.S. National Academy of Sciences (13). Between 1976 and 1980 the importation and use of wheat higher in selenium increased the maternal dietary selenium intake and milk selenium concentrations in Finland (8).

Recent data from our laboratory indicate a direct relationship between maternal selenium status during lactation and milk selenium content (14). A significant positive correlation ($r = 0.61$, $p = 0.003$) was found between maternal plasma selenium concentrations at 4 and 8 weeks of lactation and their milk selenium concentrations. Data from this study also suggest that more selenium is required during lactation to maintain maternal selenium status. Lactating women were shown to have significantly lower plasma and erythrocyte selenium concentrations and plasma GPx activities compared to non-lactating women. A significant relationship ($r = 0.53$, $p = 0.009$) was also found between plasma selenium and GPx activity for lactating but not for control subjects. These differences between groups may be because maximal plasma GPx activity was not achieved by lactating subjects and the increased amount of selenium required for milk production was not provided by the diet.

Using the rat as an animal model we have shown that more selenium is required in the diet during lactation to maintain maternal tissue selenium concentrations and GPx activities similar to non-reproducing animals fed the National Research Council recommendation (0.1 mg selenium/kg diet) (15, 16). An increase in dietary selenium concentration to at least 0.2 mg/kg, if provided as selenite, is also necessary to maintain milk selenium concentrations that result in maximal activity of GPx in the tissues of the nursing pups (15). Tissue GPx activity is considered a better index of selenium status than tissue selenium concentrations because, whereas selenium concentrations will increase with each increase in dietary selenium intake, GPx activity will plateau at a maximal level.

MATERNAL SELENIUM SUPPLEMENTATION

Because of the marginal selenium status of lactating women in Finland, and its apparent effect on milk selenium concentration, selenium supplementation of Finnish lactating women has been a recent area of interest. Kumpulainen et al. (17) have shown that maternal selenium supplementation with 100 micrograms of selenite or Se-yeast daily, increased maternal serum selenium, milk selenium concentrations, and the serum selenium concentration of the recipient infants. The Se-yeast supplement, however, was more bioavailable than the selenite, resulting in mean milk selenium concentrations and maternal and infant serum selenium concentrations similar to those we have observed in the United States (11, 14). In the Finnish population, however, GPx activities of maternal or infant serum were not assessed, and, therefore, the response of this functional Se parameter to maternal selenium supplementation is still unclear. The possibility exists that the greater increase in selenium concentrations with organic Se-yeast may be a result of the non-specific incorporation of selenoamino acids into general proteins rather than into functional selenoproteins such as GPx.

Recently we have used the rat model to study the bioavailability of selenomethionine and Se-yeast, relative to that of selenite, during lactation (18). Using both tissue selenium concentrations and GPx activities as indices, we found the bioavailability of selenomethionine and Se-yeast to be greater than that of selenite in both the lactating dam and her nursing pups. Specifically, bioavailability estimates for milk selenium concentrations and pup erythrocyte selenium and GPx activities were greater for selenomethionine than for Se-yeast and less for selenite. The greater availability of organic selenium to pup tissues may be a function of the greater concentration as well as the form of selenium in milk from dams fed organic selenium.

SELENIUM NUTRITION OF INFANTS

The provision of adequate selenium during early infancy is important in terms of the well established role of selenium as a component of GPx. GPx functions as part of an antioxidant system within the cell and its function is complimentary to that of vitamin E. GPx is found primarily in the cytosol and reduces lipid hydroperoxides to prevent free radical generation. Based on the role of GPx in the metabolism of hydroperoxides, low selenium status may increase the risk of oxidative damage. Newborn infants, especially those born prematurely, need an adequate supply of selenium to maintain erythrocyte GPx and prevent hemolytic anemia. Gross (19) studied vitamin E sufficient premature infants from 1 to 8 weeks of age and found that declining erythrocyte selenium concentrations and GPx activities were associated with an increase in hemolysis. Supplemental selenium was found to increase GPx and prevent the hemolysis. Until this report in 1976, vitamin E deficiency was the only known cause of hemolytic anemia of prematurity. Adequate selenium may also be necessary for proper drug metabolism by the neonate, since selenium appears to play a role in maintaining cytochrome P-450 levels either by decreasing heme catabolism or increasing heme utilization (20).

Early studies of selenium nutrition of infants focused primarily on the depressed selenium status observed in several diseases of infancy and childhood. Low blood selenium concentrations have been observed in children with protein calorie malnutrition (21) and in children fed synthetic diets for inborn errors of metabolism (22). In addition, over half of the reported cases of selenium deficiency in patients on total parenteral nutrition have been in children (23, 24). Selenium deficiency is also closely related to the occurrence of Keshan disease, a potentially fatal cardiomyopathy that affects post-weaning infants, children, and women of child-bearing age in the low-selenium Keshan province of China (25).

One of the first reports of the selenium status of healthy infants was made by Lombeck et al. (26) who studied serum selenium concentrations in infants, children, and adults in Germany. These investigators found that, not only are infants born with serum selenium concentrations much lower than those of adults, but that serum selenium concentrations decline during the first six months of life. The German infants studied, however, were all fed cow milk formula and their selenium status may have been a reflection of the low selenium content of those formulas. We studied the selenium content of infant foods in the United States and found that commercial infant formulas marketed in the United States are low in selenium relative to that of human milk (11). Up to this time, selenium is not added per se to infant formulas and what is present is derived from other ingredients, primarily protein.

In human milk-fed and formula-fed infants observed at 3 months, we found that, although formula-fed infants consumed significantly larger quantities of milk, their selenium intakes (7.2 ug/day) were significantly lower than those of human milk fed infants (10.0 ug/day). The selenium intakes of both groups of infants were also low compared to the 10 to 40 ug/day recommended by the Food and Nutrition Board for infants 0 to 6 months old (13). Sixty percent of the human milk-fed and 90 percent of formula-fed infants studied had intakes below the range.

Based on the above observations, the possibility exists that the recommended intake (13), which is extrapolated from animal data, overestimates the human need for selenium. Consequently, we examined selenium intakes and biochemical status in order to determine whether infants may be receiving less than optimal levels of selenium. We found that the serum selenium concentrations of human milk-fed and formula-fed infants at 3 months of age were directly correlated ($r = 0.42$; $p < 0.05$) with their selenium intakes, with human milk-fed infants having significantly greater serum selenium levels (78.2 $\pm$ 19.2 ng/ml) than formula-fed infants (55.0 $\pm$ 16.8 ng/ml) (18).

In a subsequent study we determined 1) how early in infancy the type of milk consumed influenced serum and erythrocyte selenium concentrations and GPx activities, and 2) how altering the protein composition of infant formula altered the bioavailability of selenium. Three groups of healthy full-term infants were studied at 2, 4 and 8 weeks of age. From birth, group I was fed human milk, group II, a whey-predominant formula and group III, a cow-milk based formula. All formula was supplied by Ross Laboratories (Columbus, Ohio) in ready-to-use (32 fl. oz.) cans. Milk was the sole nutritional source of all infants during the 8 weeks of the study. The total selenium content of the formulas was similar (5.8 ng/ml) but lower than that of human milk (16 ng/ml). Milk plasma and erythrocyte selenium concentrations were determined using a gas chromatograph equipped with an electron capture detector (27). Plasma and erythrocyte GPx activity was determined using the coupled assay of Paglia and Valentine (28).

Plasma and erythrocyte selenium and GPx activities are shown in Table 1. Human milk-fed infants maintained plasma selenium concentrations significantly higher than both formula groups and this difference was evident as early as 2 weeks after birth. By 8 weeks, infants fed the whey-predominant formula had plasma selenium concentrations significantly lower than the other two groups. Infants fed the whey-predominant formula also had significantly lower plasma GPx than the human milk fed group at 2, 4 and 8 weeks. No significant differences were found among the 3 groups for either RBC selenium or GPx activities, blood measures indicative of more long-term selenium status. These parameters would not be expected to reflect recent dietary selenium intake as quickly as plasma selenium or GPx. The results of this study suggest that the bioavailability of selenium from whey-protein is lower than that from casein or human milk proteins.

SUMMARY

Evidence is available indicating that additional selenium is required during lactation, especially in low-selenium countries such as Finland. This increase in selenium intake appears necessary to ensure adequate milk selenium concentrations and adequate selenium status in the recipient infant, while at the same time maintaining maternal selenium status. Availability of selenium to the infant appears to be governed by two factors, the total selenium concentration of milk and the molecular form of the milk selenium. In order to better understand the utilization of selenium from human milk and to help optimize the design of infant formula with respect to selenium bioavailability further studies are necessary to characterize the nature of the selenium-binding compounds in human milk.

TABLE 1. Plasma and RBC selenium and GPx activities of infants 2, 4, and 8 weeks of age fed either human milk (Group I), casein-predominant formula (Group II) or whey-predominant formula (Group III)

Biochemical Measurement	Feeding Group	n	Weeks 2	Weeks 4	Weeks 8
Plasma Selenium (ng/ml)	I	10	82.8 ± 4.0[a]	85.3 ± 4.5[a]	83.3 ± 6.0[a]
	II	14	52.1 ± 2.1[b]	48.7 ± 3.7[b]	40.8 ± 3.5[c]
	III	12	53.5 ± 2.3[b]	49.4 ± 2.5[b]	49.4 ± 2.6[b]
Plasma GPx (units/g protein)	I	10	2.1 ± 0.3[a]	2.1 ± 0.2[a]	1.8 ± 0.1[a]
	II	14	1.6 ± 0.1[b]	1.5 ± 0.2[b]	1.5 ± 0.1[b]
	III	12	1.8 ± 0.2[a,b]	1.9 ± 0.3[a,b]	1.6 ± 0.1[a,b]
RBC Selenium (ng/ml)	I	10	255 ± 31	213 ± 18	231 ± 19
	II	14	198 ± 13	204 ± 25	179 ± 23
	III	12	254 ± 27	229 ± 19	204 ± 25
RBC GPx (units/g Hb)	I	10	9.7 ± 0.8	8.7 ± 0.7	10.2 ± 0.9
	II	14	10.4 ± 1.5	9.8 ± 1.6	8.9 ± 1.2
	III	12	9.4 ± 0.9	8.3 ± 1.0	9.9 ± 1.4

[1]Values are means ± SE. For each parameter, values with unlike superscripts are significantly different at $p < 0.05$ or less.

REFERENCES

1. Schwarz, K. and Foltz, C.M. Selenium as an integral part of factor 3 against dietary liver degeneration. J. Am. Chem. Soc. 79, 3292. (1957).
2. Rotruck, J.T., Pope, A.L. Ganther, M.E., Swanson, A.B., Hafeman, D.G. and Hoekstra, W.G. Selenium: biochemical role as a component of glutathione peroxidase. Science 179, 588. (1973).
3. van Rij, A., Thomson, C.D., McKenzie, J.M. and Robinson, M.F. Selenium deficiency in total parenteral nutrition. Am. J. Clin. Nutr. 32, 2076. (1979).
4. Baker, S.S., Lerman, R.H., Krey, S.H., Crocker, K.S., Hirsh, E.F. and Cohen, H. Selenium deficiency with total parenteral nutrition: reversal of biochemical and functional abnormalities by selenium supplementation: A case report. Am. J. Clin. Nutr. 38, 769. (1983).
5. McKeehan, W.L., Hamilton, W.G., and Ham, R.G. Selenium is an essential trace nutrient for growth of WI-38 diploid human fibroblasts. Proc. Natl. Acad. Sci. 73, 2023. (1976).
6. Shearer, T.R. and Hadjimarkos, D.M. Geographic distribution of selenium in human milk. Arch. Environ. Health 30, 230. (1975).
7. Williams, M.M.F. Selenium and glutathione peroxidase in mature human milk. Proc. Univ. Otago. Med. Sch. 61, 20. (1983).
8. Kumpulainen, J., Vuori, E., Kuitonen, P., Makinen, S. and Kara, R. Longitudinal study on the dietary selenium intake of exclusively breast-fed infants and their mothers in Finland. Intl. J. Vit. Nutr. Res. 53, 420. (1983).

9. Robberecht, H., Roekens, E., Deelstra, H., Van Caillie-Bertrand, M. and Clara, R. Longitudinal study of the selenium content in human breast milk in Belgium. Acta Paed. Scand. 74, 254. (1985).
10. Lombeck, I., Kasperek, K., Bonnermann, B., Feinendegen, L.E. and Bremer, H.J. Selenium content of human milk, cow's milk and cow's milk infant formulas. Eur J. Pediatr 129, 139. (1978).
11. Smith, A.M., Picciano, M.F. and Milner, J.A. Selenium intakes and status of human milk and formula fed infants. Am. J. Clin. Nutr. 35, 521. (1982).
12. Higashi, A., Tamari, H., Kuroki, Y. and Matsuda, I. Longitudinal changes in selenium content of breast milk. Acta Paediatr. Scand. 72, 433. (1983).
13. National Academy of Sciences - National Research Council. Recommended dietary allowances. 9th ed. Washington, D.C.: National Academy of Sciences. (1980).
14. Mannan, S. and Picciano, M.F. Influence of maternal selenium status on human milk selenium concentration and glutathione peroxidase activity. Am. J. Clin. Nutr. in press. (1986).
15. Smith, A.M. and Picciano, M.F. Evidence for increased selenium requirement for the rat during pregnancy and lactation. J. Nutr. 116, 1068. (1986).
16. National Research Council. Nutrition Requirements of Laboratory Animals, No. 10, 3rd rev. ed. National Academy of Sciences, Washington, D.C. (1978).
17. Kumpulainen, J., Salmenpera L., Siimes, M.A., Koivistoinen, P. and Perheentupa, J. Selenium status of exclusively breast-fed infants as influenced by maternal organic or inorganic selenium supplementation. Am. J. Clin. Nutri. 42, 829. (1985).
18. Smith, A.M. and Picciano, M.F. Relative bioavailability of seleno-compounds in the lactating rat. J. Nutr. in press. (1986).
19. Gross, S. Hemolytic anemia in premature infants: Relationships to vitamin E, selenium, glutathione peroxidase, and erythrocyte lipids. Sem. Hematol. 13, 187. (1976).
20. Correia, M.A. and Burk, R.F. Defective utilization of haem for selenium - deficient rat liver. Biochem. J. 214, 53. (1983).
21. Burk, R.F. Jr., Pearson, W.N., Wood, R.P. and Viteri, F. Blood selenium levels and in vitro red blood cell uptake of Se-75 in kwashiorkor. Am. J. Clin. Nutr. 20, 723. (1967).
22. Lombeck, I., Kasperek, K., Harbisch, H.D. et al. The selenium state of children. II Selenium content of serum, whole blood, hair and the activity of erythrocyte glutathione peroxidase in dietetically treated patients with phenylketonuria and maple syrup-urine disease. Eur. J. Pediatr. 128, 213. (1978).
23. Collipp, P.J. and Chen, S.Y. Cardiomyopathy and selenium deficiency in a two-year old girl. Correspondence. N. Eng. J. Med. 304, 1304. (1981).
24. Kien, C.L. and Ganther, H.E. Manifestations of chronic selenium deficiency in a child receiving total parenteral nutrition. Am. J. Clin. Nutr. 37, 319. (1983).
25. Chen, X., Yang, G., Chen, J., Chen, X., Wen, Z. and Ge, K. Studies on the relations of selenium and Keshan disease. Bio. Tr. Ele. Res. 2, 91. (1980).
26. Lombeck, I., Kasperek, K., Harbisch, H.D., Feinendegen, L.E. and Bremer, H.J. The selenium state of healthy children. I. Serum selenium concentration at different ages; activity of glutathione peroxidase of erythrocytes of different ages; selenium content of food of infants. Eur. J. Pediatr. 125, 81. (1977).

27. McCarthy, T.P., Brodie, B., Milner, J.A. and Bevill, R.F. Improved method for selenium determination in biological samples by gas chromatography. J. Chromatog. 225, 9. (1981).
28. Paglia, D.E. and Valentine, W.N. Studies on the quantitative and qualitative characterization of erythrocyte glutathione peroxidase. J. Lab Clin. Med. 70, 158. (1967).

MANGANESE UTILIZATION IN BREAST-FED AND FORMULA-FED INFANTS

Klaus Dörner, Erika Sievers, and Stefan Dziadzka

University Children's Hospital
Schwanenweg 20
D-2300 Kiel 1, FRG

INTRODUCTION

Manganese balance data from newborns or infants are very scarce. There seem to be several reasons for this.

(1) The clinical importance of manganese appears to be less important than copper or zinc, as a symptomatic deficiency state of this trace element has not been defined.
(2) Balance techniques in young infants are laborious.
(3) Determinations of manganese in biological materials are difficult and render conflicting results even in well recognized laboratories.
(4) Since manganese has only one stable isotope, kinetic balance studies are not possible unless radioactive isotopes are administered. This is of course not feasible in healthy infants.

As a consequence, recommendations on "safe and adequate intakes" are based on intake measurements, preferably from breast-milk. But a broad range of manganese concentrations in breast-milk has been published during the last twenty years all over the world (table 1, mainly based on data from Iyengar, 1982). Geographic dissimilarities exist (Iyengar and Parr, 1985), but those differences are probably due to analytical problems rather than to biological variations.

Recommended Dietary Allowances (RDA 1980) are oriented to these high breast-milk values and reflect a "safety factor" added to account for the assumed lesser bioavailability of manganese from cow's milk formulae.

Three approaches have been followed to adress the many questions on manganese requirement, on bioavailability and on balances. The first used the rat pup model by Lönnerdal and his group: Keen et al. (1986) added ^{54}Mn as an extrinsic marker to different types of milk, fed the mixture and studied the dependency of resorption and retention on age of the pup, time course of feeding and the body distribution. One important finding was that the bioavailability of manganese in human milk was about 80% until the 15th day of life.

Table 1: Representative manganese concentrations in breast-milk (µg/l, range or SD) measured by different authors during the past twenty years (based on Iyengar 1982)

USSR	1968	19 - 29
USSR	1969	40 ± 8
USA	1971	120 ± 70
New Zealand	1972	15 (12 - 20)
FRG	1975	20 ± 4.4
USA (var.)	1977	14 - 25
Finland	1979	appr. 4 (2 - 6)
Australia	1983	4 ± 2
USA	1984	4.9
USA	1984	3.5 - 6.6
Our results	1985	6.2 (4.2-9.0)

A second one was that no significant differences in manganese bioavailability were found between human milk, cow's milk formula and cow's milk.

Keen's consistent results found in rat pups up to 18 days old contrasted with findings in newborns at day 10 published by Mena (1981). The latter found by radiotracer studies a total body retention of 8 ± 2% and 16 ± 3% for term and the preterm respectively. In adults an intestinal absorption of $^{54}MnCl_2$ of 3 ± 0.5% was found, an amount which is the same in manganese exposed miners suggesting a constant absorption not influenced by high tissue levels. On the other hand, individuals with increased iron absorption showed an increased manganese absorption.

It remains to be clarified whether the suckling rat pup model is limited to the early newborn period and is not transferable to older infants, or whether Mena's experimental approach renders too low resorptions. Because ethical problems rule out the use of radiotracers in healthy infants, the answer has to be found by classical balance studies, which first and solely were performed by Widdowson (1969) in one week old children. She found strongly negative balances (11 $\mu g \cdot kg^{-1} \cdot day^{-1}$).

The studies presented in this paper were done with infants during the first five months of life and are based on the classical balance technique. The limitations of this approach (Beisel 1979) include:
(1) the need of steady-state conditions;
(2) reliable measurement of intake;
(3) reliable timing of sample collection;
(4) accurate collecting of feces and urine;
(5) precise and accurate analytical methods.

We tried to fulfill these five preconditions in our balance studies with term and preterm infants and one infant with phenylketonuria.

MATERIALS AND METHODS

Six groups of infants were studied: I = 11 breast-fed terms, II = 4 terms and III = 3 preterms fed with an unsupplemented adapted formula, IV = 5 terms and V = 3 preterms fed with a trace element (not manganese!) fortified formula, and VI = 1 term infant with phenylketonuria fed with the adequate special diet to maintain the phenylalanine blood level at 2-4 mg/dl.

Balances were regularly performed five times at mean age of 2.5, 5.5, 8.5, 12.3 and 15.9 weeks. These ages refer to the breast-fed group, the others may differ slightly. Infants of group III were triplets with a gestational age of 34 weeks, group V were also triplets. They had a gestational age of 36 weeks. The mean weights of the five groups in the first and the last collection periods are given together with the total of balances and the number of infants in each group in Table 2. Collecting periods of 72 hours began with a carmine (Nacarat) marker; the end of period was marked in the same way. Since carmine used contained less than 1% of the average manganese intake of breast-fed infants, this quantity was ignored in the intake calculations.

The two cow's milk formulae fed groups had similar manganese concentrations (98.6 and 77.0 µg/l, respectively). Their iron contents differed considerably (1.12 and 10.1 mg/l, respectively). Furthermore, three infants in group II and IV had occasionally received additional iron medication.

Breast-milk intakes were measured by test-weighing. In each suckling procedure, milk samples were collected before and after feeding. Urines were collected with special urine bags, and feces on trace element free napkin inlayers.

Table 2: Mean-weights and SD during the first and the fifth collection period of the infants groups studied, number of balances evaluated and number of children (the child with phenylketonuria is not included); * group III was studied only in the first two collecting periods. For explanation of the groups see the text.

Diet groups	I	II	III*	IV	V
Weight - kg (mean ±1SD)	4.1±0.4 - 6.7±0.7	3.4±0.7 - 6.2±0.8	1.8±0.3 - 2.2±0.5	3.4 - 6.6±0.8	2.3±0.2 - 5.2±0.3
Number of balances	45	20	6	19	13
Number of infants	11	4	3	5	3

Milk and fecal samples were lyophilized, ashed under pressure with nitric acid and analysed by atomic absorption spectrometry (AAS) with graphite furnace and partly by flame-AAS under standard conditions. Urines were measured directly without ashing.

Precision and accuracy were determined by serial and day-to-day measurements of available milks, urine and fecal samples, reference materials and addition assays. The results have been published elsewhere (Dörner et al. 1985).

RESULTS AND DISCUSSION

In this study manganese concentrations were measured in 2339 breast-milk samples. The median was found to be 6.2 µg/l. The concentrations increased from the beginning to the end of suckling very slightly but the differences were statistically significant. Clinically significant changes during lactation were not found.

The main excretion path of manganese is through the biliary tract. Since urine excretion is minor, retention is largely a function of intake and fecal excretion.

The results of manganese balances in breast-fed infants are given in the left and the upper part of Fig. 1 where intake (x-axis) and retention (y-axis) are compared. Intakes generally are below 5 $\mu g \cdot kg^{-1} \cdot 72h^{-1}$ and the retention is positive in the majority of the 45 balances. There is no dependency of the retention on the age of the infants. The mean relative retention is 43%.

The main problem with the classical balance scheme is the impossibility of deciding between apparent retention and true absorption. Statements on the bioavailability of manganese from breast-milk are therefore difficult. Manganese circulates enterohepatically. It may be accepted therefore, that bioavailability from breast-milk is more - perhaps considerably more - than 40%. This finding is close to those of Keen et al. (1986) in rat pups older than 15 days, but higher than those of Mena (1981) who found a 10% retention and quite far from the negative balances in 1 week old newborns reported by Widdowson (1969).

The relatively broad scatter of the data (Fig. 2) is mainly caused by the difficulty in collecting true 72-hour stool samples. Carmine is useful, but when it appears in a stool sample, it is not possible (and not allowed by definition!) to decide, whether it is from the end or the beginning of the stool sample. In infants with low stool frequencies, serious over- and underestimations of fecal manganese excretion are possible.

In the right part of Fig. 1 the results from formula-fed term infants are shown. Two adapted formulas were fed which practically did not differ in composition and in manganese

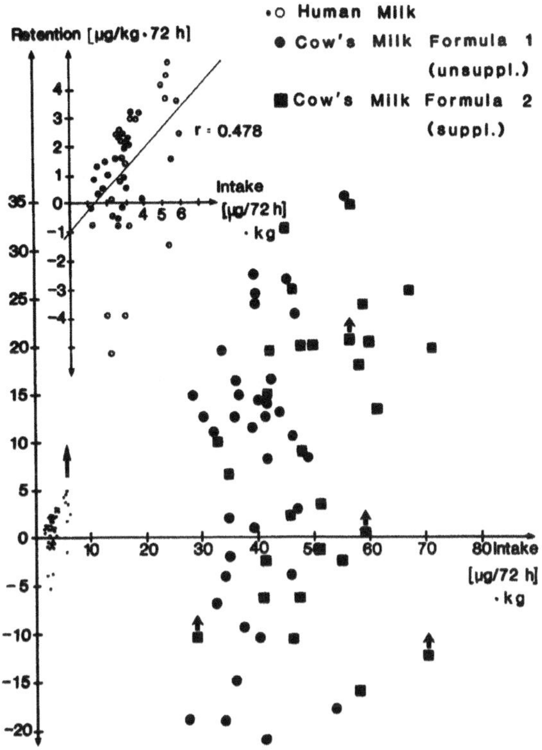

Fig. 1: Intake and retention in breast-fed infants (small dots on the left side and in the upper part) and formula-fed term infants; short arrows indicate additional (unmeasured) tea feeding (adapted from Dörner et al. 1985).

contents, but in the contents of the trace elements iron, copper, zinc and iodine. Interferences in intestinal absorption of trace metals are well known for iron, copper and zinc. The data concerning manganese are ambigious. Stastny et al. (1984) found no influence of formula supplementation with iron on serum manganese, yet the tissue manganese in suckling mice was found by Keen et al. (1984) to be markedly lower in mice fed with a supplemented diet containing 5 mg/l iron and extremely high manganese concentration of 13 mg/l. In our studies the iron supplementation either in the formula or as an additional medication rendered no obvious deviations. Manganese retention was 3.0±4.9, 2.4±5.0 and 2.2±4.9 $\mu g \cdot kg^{-1} \cdot day^{-1}$ in the unsupplemented formula group, the supplemented formula group and the infants with additional iron medication respectively. We therefore combined the two groups with formula 1 and 2. The mean intake of this group is more than ten times higher than in the breast-fed group, the mean retention about 6 times higher. Mean and standard deviations are shown in Fig. 2. The mean relative retention is 20%, half of that found in breast-fed infants. Because of the wide scattering and the higher content in formula milk we hesitate to declare a better bioavailability of manganese from breast milk although the relative retention from breast milk is double.

The fetal liver accumulates copper. This copper supply is needed to compensate for the large physiological copper losses in early infancy. Contrary to copper the fetal liver accumulates no manganese. One could argue that manganese deficiency may occur more easily in preterm infants, as they also have negative balances in early infancy, as published by Widdowson (1969). Indeed, in our first collection period, 2

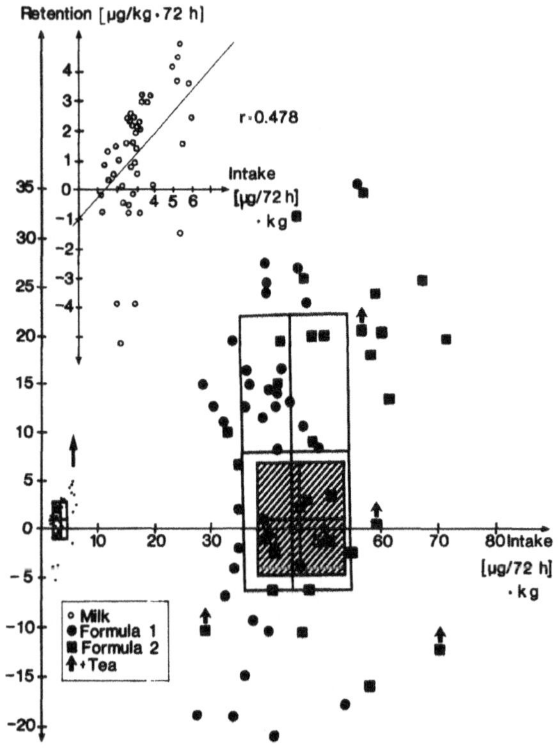

Fig. 2: Means ± 1 standard deviation of manganese intake and retention added to fig. 1; the hatched area presents the data of preterms (dots are not given).

of 4 balances were negative and in the second collection period 3 of 6 were negative. We can not decide whether this is caused by an increased biliary excretion, a diminished reabsorption of biliary manganese or a diminished absorption of manganese from milk. The mean intake and the mean retention, which are shown in Fig. 2, however, do not depart materially from the group of term infants. The arithmetic mean of relative retention in the preterm is 18% and thus in the range of the formula-fed term infants. Our data do not support a higher manganese supplementation of formula milk for preterms than for term infants.

Trace element intake of patients with special diets, such as in phenylketonuria, are often based upon the guidelines of RDA rather than experimental data. We had the opportunity to study trace element balances in an infant with early detected phenylketonuria. The manganese results are given in Fig. 3. The most impressive fact is the extremely high intake, which was first believed to be erroneous. The manganese intake in this child is up to a hundred times higher than in breast-fed infants, the absolute retention 30 times higher and the mean relative retention with 12% lower than in any other group described here. It should be mentioned that the high manganese concentration was not derived from natural sources or as an impurity introduced by the manufacturing process, but was added to the special diet by the producer according to RDA-recommendations.

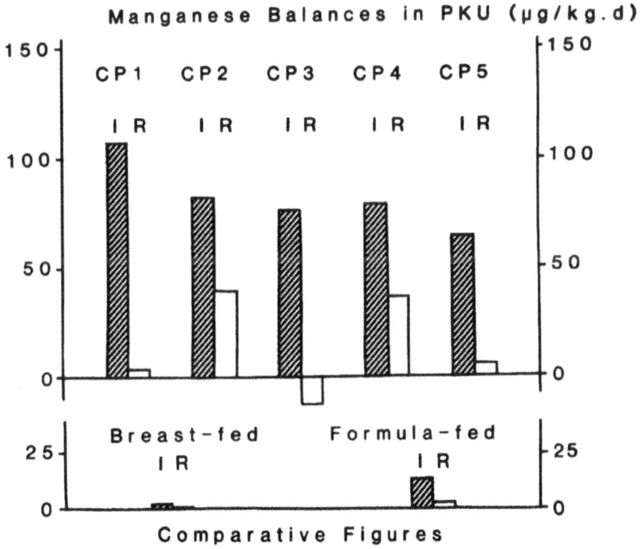

Fig. 3: Manganese intake and retention in an infant with phenylketonuria (CP = collection period, I = intake, R = retention)

Manganese balances in phenylketonuria were studied earlier in 4 infants 1 month old by Alexander et al. (1974). They were fed with a synthetic amino-acid mixture with mineral supplements and an acid hydrolysate of casein respectivly. Intake and retention of the two infants with the hydrolysate nutrition were even higher than in our patient, who had essentially the same intake and retention as the two infants reported by Alexander fed with the synthetic diet with mineral supplementation.

FINAL COMMENTS

As outlined in the introduction, classical balance studies as presented here suffer from several limitations, so the results must not be accepted without criticism. A few facts, however, seem to be certain for our manganese balances.

(1) The intake provided by adapted cow's milk formula is one order of magnitude higher than by breast-milk

(2) The intake provided by the special phenylketonuria diet studied here is in accordance with RDA two orders of magnitude larger than by breast-milk

(3) The amount of manganese retained is higher in formula-fed infants than in breast-fed infants, and the relative retention is related to the intake. Furthermore the absolute retention may not be limited by the degree of the intake.

This results in an average physiological requirement, which seems far lower than supposed in the past. As stated by Keen, Bell and Lõnnerdal earlier this year, a formula-fed infant is less likely to become manganese deficient (which means what?) than manganese intoxicated.

Although the manganese is not a very toxic element and although it has a broad physiological range, the RDA should be re-evaluated once more, since precise data are available on bioavailability of manganese from various types of infant formulae and human milk.

Acknowledgements

This study was supported by Milupa AG, Friedrichsdorf which we gratefully acknowledge.

REFERENCES

Alexander, F.W., Clayton, B.E., Delves, H.T., 1974, Mineral and trace-metal balances in children receiving normal and synthetic diets, Quart.J.Med., New Series XLIII(169): 89

Beisel, W.R., 1979, Metabolic balance studies - their continuing usefulness in nutritional research, Am.J.Clin.Nutr. 32: 271

Dörner, K., Dziadzka, St., Oldigs, H.-D., Schulz-Lell, G., Schaub, J., 1985, Manganese balances in term infants, in:"Composition and Physiological Properties of Human Milk", J.Schaub, ed., Elsevier, Amsterdam, p. 117

Iyengar, G.V., 1982, Elemental composition of human and animal milk, IAEA Report TECDOC-269, Vienna

Iyengar, G.V., Parr, R.M., 1985, Trace element concentrations of human milk from several global regions, in: "Composition and Physiological Properties of Human Milk", J.Schaub, ed., Elsevier, Amsterdam, p. 17

Keen, C.L., Bell, J.G., Lönnerdal, B., 1986, The effect of age on manganese uptake and retention from milk and infant formula in rats, J.Nutr. 116: 395

Keen, C.L., Fransson, G.-B., Lönnerdal, B., 1984, Supplementation of milk with iron bound to lactoferrin using weanling mice. II: Effects on tissue manganese, zinc and copper, J.Pediatr.Gastroenterol.Nutr. 3: 256

Mena, I., 1981, Manganese, in: "Disorders of Mineral Metabolism, Vol.I", F. Brommer, J.W. Coburn, eds., Academic Press, New York, p. 233

Recommended Dietary Allowances, 1980, Committee on Dietary Allowances, Food and Nutrition Board, Commission on Life Sciences, National Research Council, 9th ed., National Academy Press, Washington, DC

Stastny, D., Vogel, R.S., Picciano, M.F., 1984, Manganese intake and serum manganese concentrations of human milk-fed and formula-fed infants, Am.J.Clin.Nutr., 39: 872

Widdowson, E.M., 1969, Trace elements in human development, in: "Mineral Metabolism in Pediatrics", D. Barltrop, W.L. Burland, eds., Blackwell, Oxford, p. 85

FOLATE AND HUMAN MILK

Johan Ek

Pediatric Research Institute
National Hospital of Norway
Oslo, Norway

INTRODUCTION

Folate is necessary for normal cell growth and division. Deficiency of the vitamin in infancy may lead to growth retardation, deranged function of the bone marrow, delayed maturation of the central nervous system, and functional and histological changes in the small intestine; enteritis and other infections have also been described (as reviewed in (1)).

Development assessment of folate status has determined that in the second half of pregnancy the fetus has significantly higher plasma and red blood cell (RBC) folate concentrations than adult control subjects, and six to eight and two times higher concentrations than found in their mothers for plasma and RBC, respectively (2). The plasma folate concentrations in newborn infants and their mothers at parturition are about 40 and 8 nmol/l, respectively, and the RBC folate concentrations in the infants and the mothers are about 650 and 350 nmol/l, respectively. The RBC folate concentrations are correlated to the liver folate concentrations (3), and are thus related to the body stores of the vitamin.

The high plasma and RBC folate concentrations observed in newborn infants are sustained in breastfed term infants (Fig. 1 and 2) (4). The values are significantly higher than those observed in their mothers and adult control subjects during their first 24 months of life.

The observations in newborn infants indicate that there are mechanisms which secure an adequate supply of the vitamin to the fetus during pregnancy, while other mechanisms insure an adequate supply to the infants during the succing period. We have studied the latter hypthesis in a group of women following parturition. Some observations from this study are presented and discussed as follows.

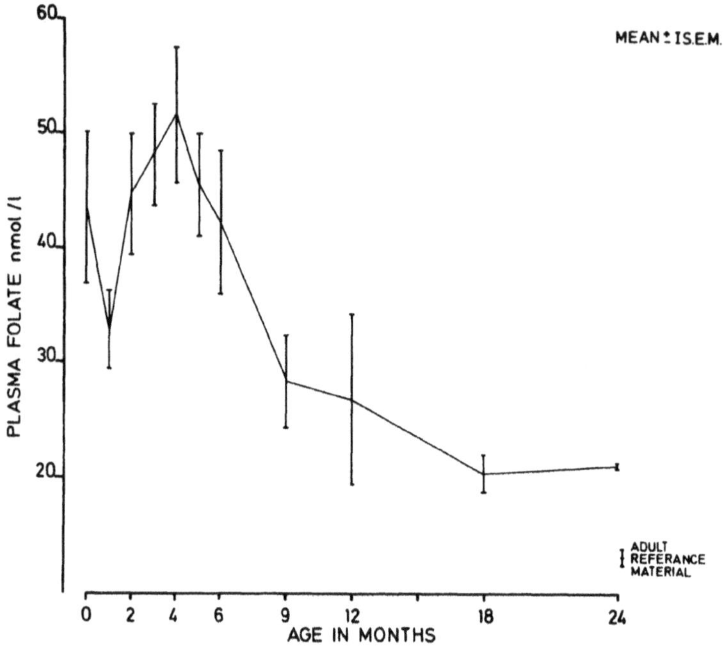

Fig. 1. Plasma folate concentrations in 35 breastfed infants during the first two years of life. Mean values ±1 S.E.M. (Standard Error of the Mean).

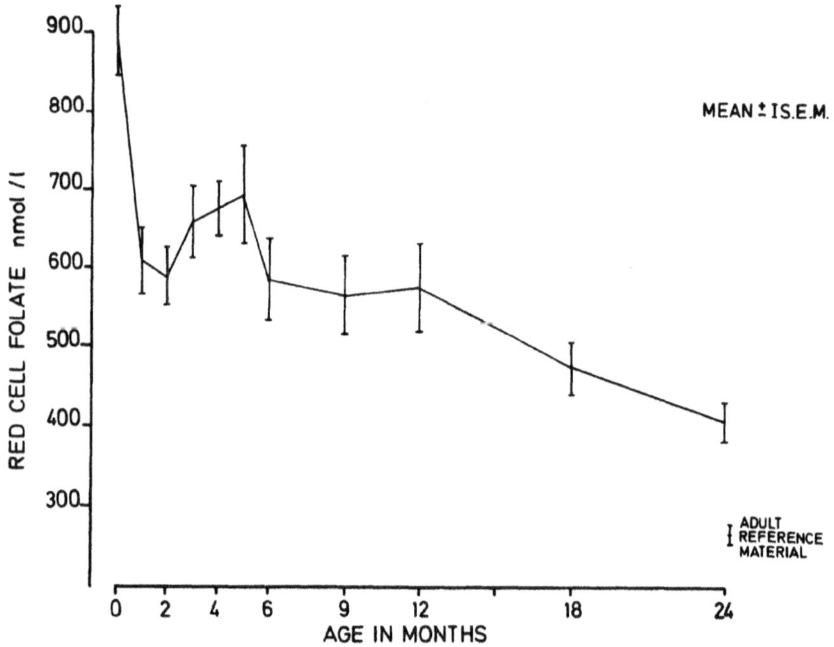

Fig 2. Red blood cell folate concentrations in 35 breastfed infants during the first two years of life. Mean values ±1 S.E.M.

MATERIAL AND METHODS

The women from the same population as in our previous studies were investigated (1,2). They received iron and a multivitamin preparation* from the 3rd month of pregnancy and during the 12 months following parturition. No folic acid supplementation was given. The pregnancies and deliveries were normal. Venous blood sample was taken 3-5 h after the last meal. The first blood samples were obtained within 48 h after parturition; then samples were taken at monthly intervals during the first 6 months of the study, and at 3 month intervals thereafter. Breast milk was obtained at monthly intervals until the infants were weaned. The samples were collected into glass tubes by manual expression during the middle of the meal, and were immediately stored at -20°C. Two milk samples were obtained from 18 women 4-11 months after parturition. The last sample was collected less than 1 month before cessation of lactation. The folate concentrations in plasma, RBC and breast milk were measured microbiologically using Lactobacillus casei (5-7). Comparison of values at various times for the same variable was tested by means of the Wilcoxon signed rank test.

RESULTS

Maternal plasma and RBC folate concentrations in women lactating for 6 months or more (mean, 7.8 months) and the free folate concentrations in breast milk are presented in Fig. 3. Plasma folate concentrations were unchanged during the first 3 months after parturition, thereafter they increased significantly during the next 9 months ($p<0.001$). The RBC folate decreased significantly during the first 2 months after parturition ($p<0.001$), and thereafter increased significantly during the next 10 months ($p<0.001$). The folate concentrations in breast milk increased significantly during the first 3 months following parturition ($p<0.001$); thereafter they remained unchanged. During the first 2 to 3 months of lactation the incorporation of folate into the red cells, and probably also into the liver cells (3), thus seem to be subordinate to the transfer of folate into the breast milk. The observed increase in folate levels in breast milk during the first 3 months of lactation is in accordance with other reports (8-10), although Tamura et al. (11) failed to demonstrate this during early lactation. The reason for this discrepancy is uncertain.

Effects of weaning upon milk folate levels

These observations are presented in Fig. 4. The differences between the second last sample and the last samples were statistically significant ($p<0.001$). The decrease in the milk folate concentrations toward the end of the lactation period was not related to the duration of lactation.

---

*Biovit (R), A/S Farmaceutisk Industri, Oslo, Norway, yielding vitamin A 2500 IU, vitamin D 400 IU, thiamin 2 mg, riboflavin 3 mg, nicotinamide 20 mg, pyridoxine 2 mg, pantothenate 3 mg, and ascorbic acid 50 mg.

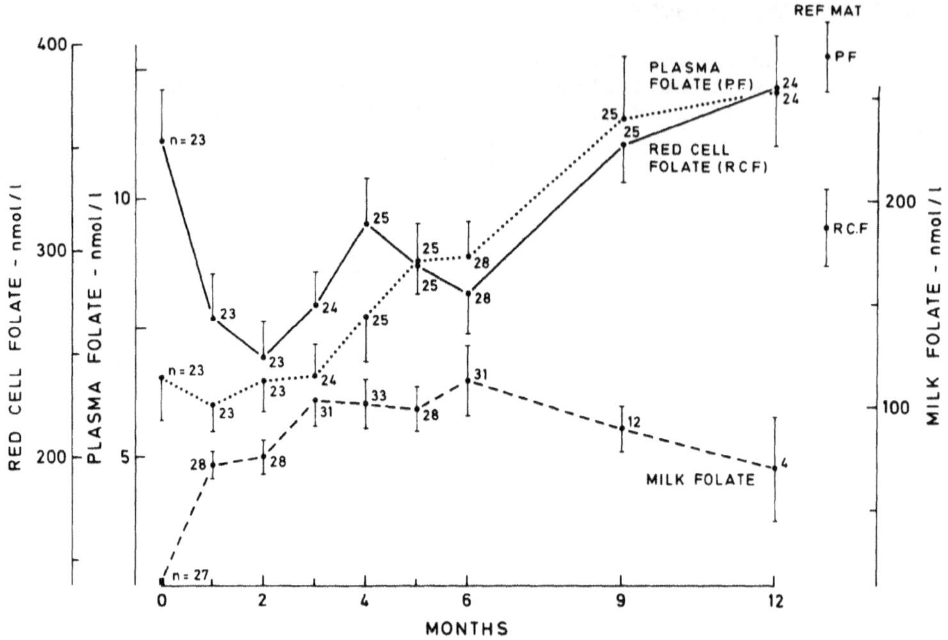

Fig. 3. Plasma and red blood cell folate concentrations in women lactating for more than 6 months, and free folate concentrations in breast milk, expressed as mean values ± 1 S.E.M.

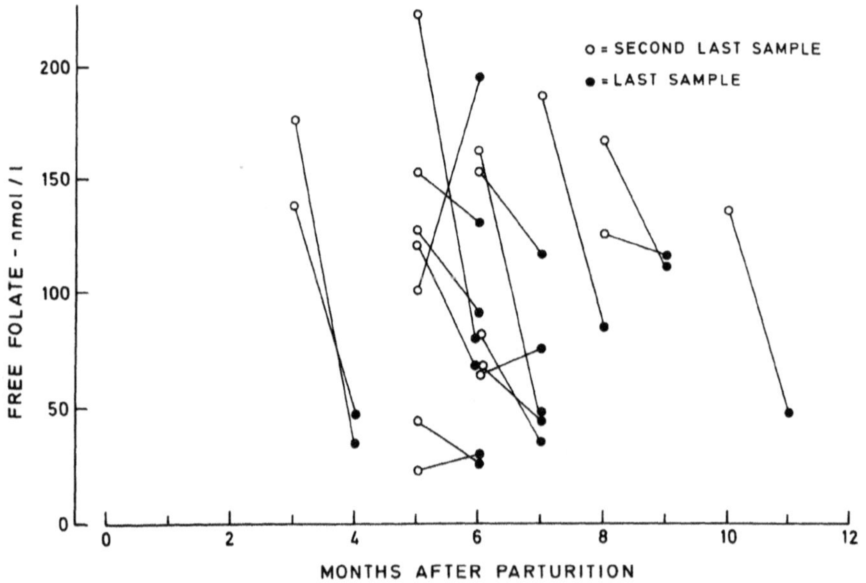

Fig. 4. Free folate concentrations in breast milk towards the end of the lactation period. See text for explanation.

DISCUSSION

The human milk content of different nutrients often forms the basis for calculating the daily requirements of such nutrients. The present study shows that the folate concentrations in human milk increases during the first 3 months of the lactation period, and decrease towards the end of the period. It is therefore important to take into consideration the physiological changes in the breast milk content of the nutrient throughout the lactation period, when the content in human milk forms the basis for calculation of the daily requirement.

Breastfed infants in a well-nourished population may be considered to have an optimal folate status from a nutritional point of view during their first months of life. The amount of folate necessary to maintain plasma and RBC folate levels similar to those found in breastfed term infants (4) may therefore be regarded to be the optimal folate intake in arteficial infant nutrition.

We have studied the nutritional value of the folates in human milk as compared to those in commercially available milk formulas (1). A significant difference was observed, as the folate concentrations in cows' milk based formulas have to be 40 per cent higher than in breast milk to yield plasma and RBC folate concentrations in formula-fed infants similar to those observed in breastfed infants. The reason for this difference is uncertain. However, a human milk factor that facilitates folate uptake by intestinal cells has been found (13), and may contribute to the increased bioavailability of the folates in human milk as compared to the folates in cows' milk based formulas.

During the lactation period no correlation was noted between breast milk folate levels and plasma or RBC folate levels, respectively, in the mothers. In the course of the weaning period, increasing concentrations of plasma and RBC folate coinsided with decreasing milk folate concentrations, thus supporting the hypothesis of a regulatory mechanism maintaining milk folate concentrations at optimum levels (11,12). This is also supported by an earlier report where folic acid supplementation of well-nourished lactating women failed to increase the breast milk folate concentrations (14).

CONCLUSION

The studies provide evidence to the hypothesis that there exist regulatory mechanisms to maintain the folate concentrations in human milk. Breastfed infants have higher plasma and RBC folate concentrations than their mothers and adult reference subjects. These high concentrations are judged to be optimal in infant nutrition. The folate concentrations in cows' milk based formulas have to be 40 per cent higher than in breastmilk to yield plasma and RBC folate concentrations in formula-fed infants similar to those observed in breastfed infants. The higher bioavailability of the folates in human milk is probably related to the presence of a human milk factor that facilitates folate uptake by intestinal cells.

REFERENCES

1. J. Ek and E. Magnus, Plasma and red cell folate values and folate requirements in formula-fed term infants, J Pediatr 100:738 (1982).
2. J. Ek, Plasma and red cell folate values in newborn infants and their mothers in relation to gestational age, J Pediatr 96:288 (1980).
3. A. Wu, I. Chanarin, G. Slavin and A. J. Levi, Folate deficiency in the alcoholic - its relationship to clinical and haematological abnormalities, liver disease and folate stores, Br J Haematol 29:469 (1975).
4. J. Ek and E. Magnus, Plasma and red blood cell folate in breast-fed infants, Acta Paediatr Scand 68:239 (1979).
5. E. Magnus, Folate studies, Scand J Haematol Suppl No 24 1975.
6. E. Magnus, Folate concentrations in top, midle and bottom layer of packed red cells in vitamin B_{12} deficiency. Relation to treatment, Scand J Haematol 24:243 (1980).
7. J. Ek and E. Magnus, Plasma and red cell folacin in cow's milk-fed infants and children during the first 2 years of life: the significance of boiling pasteurized cow's milk, Am J Clin Nutr 33:1220 (1980).
8. B. V. Ramasastri, Folate activity in human milk, Br J Nutr 19:581 (1965).
9. M. J. Cooperman, H. S. Dweck, L. J. Newman, C. Garbarino and R. Lopez, The folate in human milk, Am J Clin Nutr 36:576 (1982).
10. R. Karlin, C. Hours, R. Bertoye, C. Vallier and N. Berry, Etude sur les taux d'acide folique du lai human et du lait bovin, Int J Vit Res 37:334 (1967).
11. T. Tamura, Y. Yoshimura and T. Arakawa, Human milk folate and folate status in lactating mothers and their infants, Am J Clin Nutr 33:193 (1980).
12. J. Metz, R. Zalusky and V. Herbert, Folic acid binding by serum and milk, Am J Clin Nutr 21:289 (1968).
13. N. Colman, N. Hettiarachchy and V. Herbert, Detection of a milk factor that facilitates folate uptake by intestinal cells, Science 211:1427 (1981).
14. M. R. Thomas, S. M. Sneed, C. Wei, P. A. Nail, M. Wilson and E. E. Sprinkle III, The effect of vitamin C, vitamin B_6, vitamin B_{12}, folic acid, riboflavin and thiamin on the breast milk and maternal status of well-nourished women at 6 months postpartum, Am J Clin Nutr 33:2151 (1980).

BODY COMPOSITION AND GROWTH: WORKSHOP SUMMARY

Stephanie A. Atkinson, and Robin K. Whyte

Department of Pediatrics, McMaster University
Medical Centre, Hamilton, Ontario, Canada

The objectives of this workshop were two fold: first, to consider the effects on growth and body composition of feeding human milk or proprietary formulas in both term and premature infants. Secondly, to examine the interrelationships between energy intakes and utilization of the major energy source in milk (i.e., fat) during early developmental stages in infants.

Longitudinal studies (7 to 12 months postnatally) of milk intakes and growth in term breast fed infants were the focus of discussions by Cutberto Garza and Margaret Neville from the U.S.A. In both studies, the carefully quantitated intakes of breast milk averaged 735 to 800 ml/day with about 15% population variance. At a measured energy density of 65 kcal/100 ml, estimated energy intakes of the infants studied averaged well below the current Recommended Dietary Allowances. With respect to growth, both investigators observed that comparison of their data to population growth standards from the National Center for Health Statistics (NCHS) demonstrated a progressive deviation from the 50th percentile of mean weight and length growth especially after the first two to three months of life. In discussing these papers, they and the discussant Kay Dewey from the USA, who has made similar observations with respect to differences in growth patterns of breast and formula fed infants, summarized the questions that must yet be answered: first, since slower growth rates in breast fed infants are not seemingly due to maternally limited milk availability, then what regulates the milk intake of a suckling infant? Secondly, what should be the "gold standard" for estimating energy requirements and normative growth patterns in infants? The future approaches to answer these questions must include careful studies of energy balance in infants which incorporate measurements of basal metabolic rates, energy expenditure for activity and body composition using indirect calorimetry, dual isotope methodology and appropriate analysis of body compartments.

The section on growth and body composition of premature infants was pre-empted by a plenary presentation by Bill Heird of the USA. He demonstrated that at high energy intakes, an increase in energy intake did not provide an increase in weight gain, while a further increase in protein intake did provide an increase in weight gain. These observations could be compared to earlier work in which higher protein intakes without concomittant increases in energy intake had not produced an increase in the rate of weight gain. The message here, reinforced by the observed correlations between the rate of nitrogen storage and the electrolyte retention, was that no one nutrient can be identified as "rate limiting" for growth, but that all nutrients (and therefore an overall increased nutrient density of feedings) must be presented simultaneously for effective growth to take place. This point was reinforced by Guy Putet from France who showed that at moderate levels of metabolizable energy intake, protein supplements have no effect on weight gain, but rather increase the rate of energy expenditure and reduce the rate of energy storage. The net effect is to cause a marked difference in the composition of weight gain, with a relative reduction of lipid storage and a relative increase in protein storage.

Both Robin Whyte from Canada and Putet in presenting their data referred to the similarity in results of energy balance studies conducted with human milk or "standaridzed formulas" in low birthweight infants. Both commented on the relatively slow rates of weight gain experienced by infants in studies of this nature, despite a wide range in metabolizable energy intake. However, when the high energy, high protein and high electrolyte containing formulas were introduced for growing low birthweight infants, both groups reported much higher rates of weight gain. Whyte in describing the composition of this weight gain, attributed most of the increase in weight gain to a more rapid accumulation of electrolytes and water. It is impossible to interpret this change further, but it should be borne in mind that the increase in the rate of water accretion may reflect an increase in weight gain in either the intracellular or extracellular compartment. Further analysis of the composition of this gain would only be obtained by conducting the same kind of electrolyte balances described by Heird.

Combining the work described by Heird, Whyte and Putet, the most consistent unifying conclusion is that at lower energy intakes (i.e when energy is the "limiting nutrient") an increase in energy intake may lead to an increase in weight gain, but this effect is restricted when protein intake becomes limiting. The addition of more protein to a high energy intake results in a further increase in the rate of weight gain, and the rates of weight gain enjoyed by infants on the nutrient dense formulas prepared for low birthweight infants is a reflection of the rapid simultaneous gain of all compartments of the body.

Whyte's presentation included a reference to the role of medium chain triglyceride as a partial replacement for fat in high energy formulas. In a randomized controlled clinical trial replacing half the fat of such a formula produced no increase in energy digestibility, expenditure or storage. This was consistent with findings reported by Bitman from the USA who conducted detailed studies of infants fed with either medium or long chain triglycerides. He presented data indicating that there was gastric hydrolysis of fat and this was greater for the medium than for long chain fats. The data suggested that free fatty acids may be absorbed through the gastric mucosa. However, fat balance studies showed identical absorption rates of MCT or LCT diets. An analysis of fat absorption by postconceptual age suggested that the difference in fat absorption between medium chain and long chain triglycerides tended to exist in favour of medium chain triglycerides in babies of shorter postconceptual age but was in favour of long chain triglyceride in babies of greater postconceptual age. This interesting observation should be evaluated further. It was generally concluded that medium chain triglyceride did not contribute positively to the energy balance of growing low birthweight infants, and there was some discussion with respect to the implications of the organic aciduria which occurred in babies on such diets.

The second objective of the Workshop was met by discussions of the importance of the quality of fat and factors such as lipases and carnitine in human milk to ultimate lipid utilization. Robert Jensen from the USA reviewed the subject of cholesterol in human milk and its potential importance in membrane structure and function. At the absorptive level, the hydrolysis of triglyceride is facilitated by three lipases - pregastric/gastric lipase in the stomach, colipase-dependent lipase from the pancreas and bile salt-stimulated lipase (BSSL), a constituent of human milk. Olle Hernell from Sweden outlined the cascade effect on lipolysis action by the three lipases. While quantitatively pregastric/gastric lipase may not play a major role in the hydrolysis of triglyceride, in vitro studies suggest that it may play an important preparatory (triggering) role for subsequent lipolysis by colipase/lipase and BSSL. The presence of the latter enzyme in raw human milk clearly serves to supplement and complement the action of the other lysases in maximizing complete hydrolysis of milk triglyceride to free fatty acids and thus very efficient fatty acid absorption. Reduced efficiency of fat absorption in infants fed heat treated as opposed to fresh human milk has been attributed to the lipolytic action of BSSL. Stuart Patton from the USA postulated that heat treatment of milk resulted in alterations in milk fat globule membranes that also contributed to a reduced lipid hydrolysis.

At the cellular level of lipid utilization carnitine has been proven an important facilitator for B-oxidation of long and medium chain fatty acids. Peggy Borum from the USA outlined the evidence which supports the concept of carnitine as an essential nutrient for the young neonate. While plasma and red blood cell levels of carnitine are higher in infants fed human milk or carnitine containing formulas as compared to oral intravenous feedings devoid of carnitine, there is no evidence that such an altered carnitine status at the tissue level is physiologically significant.

GROWTH OF THE BREAST-FED INFANT

Cutberto Garza, Janice Stuff, and Nancy Butte

USDA/ARS Children's Nutrition Research Center
Department of Pediatrics, Baylor College of
Medicine and Texas Children's Hospital, Houston, Texas

INTRODUCTION

Nutrient requirements of infants have been based on observations of ad libitum intakes of normal breast-fed populations[1,2] or derived from a factorial approach that sums nutrient needs for growth, maintenance, and activity.[3] Both definitions rely heavily on criteria for normal growth. Estimates derived from ad libitum intakes depend on the identification of populations with normal anthropometric indices. The use of the factorial approach requires a more complex definition of normal growth. Measurements of body composition and anthropometric indices are needed. The combination of body composition and weight gain provides estimates of the net accretion of specific nutrients and the nutrient costs of tissue synthesis and deposition. Current applications of the factorial approach predict that the cost of growth falls from approximately 30 kcal/kg body weight the first month to less than 4 kcal/kg body weight by the end of the first year of life.[4] The decline is not linear, but is more rapid during the first four months of life.

Seldom does either approach for estimating nutrient requirements use growth conceptually as a primary outcome, but instead growth is used as a proxy indicator for functional competence. No one argues credibly for the attainment of specific growth criteria without implying or directly identifying expected functional benefits.[5] Such applications of growth measurements carry an implied warranty of sensitive relationships between growth and function. The sensitivity of growth as a proxy indicator for specific functional competence, however, is not well established.[5-7] The apparent plasticity of growth in response to maternal and paternal genetic endowments, diet, and environment from conception through adolescence complicates the validation of putative relationships.

This discussion will focus on our observations of growth and intake of breast-fed infants 1 to 8 months of age.[8-9] We will emphasize the role of diet as well as the period immediately before and after the introduction of solid foods to an exclusive human milk diet. We also will address briefly the implications of these data in the utilization of energy, composition of tissue gained, and possible effects on behavior.

MATERIALS AND METHODS

Study Design

Three studies were conducted. The first (Study I) was a prospective, longitudinal study of 45 exclusively breast-fed infants during the first 4 months of life.[8] Measurements of milk intake, sampling of milk for compositional studies, and monitoring of infant growth were performed monthly. The second study (Study II) was semi-longitudinal in design.[9] Seventeen infants were studied, nine at months 5 and 6 and eight at months 6 and 7. All were exclusively breast-fed for 5 months. Solid foods were introduced at 6 months. Infants were studied before solid foods were introduced and 30 and 60 days after the addition of solid foods to an exclusive human milk diet. Growth was measured monthly and intakes of milk and other foods were measured for 5 consecutive days.

Intake and growth of infants also have been measured in an on-going study, Study III. Forty-eight of a projected 60 subjects have been recruited. Only women who plan to breastfeed their infants exclusively for 4 to 6 months are eligible and are recruited in the last trimester of pregnancy. Infants are studied longitudinally from birth to at least 3 months after the introduction of solid foods. The timing of the introduction of solid foods is decided by each mother in consultation with the infant's pediatrician. Other than the initial introduction of rice cereal, the sequence in which solid foods are added to the diet is made by the mother and pediatrician. Mothers are asked to offer solid foods only after the infant has nursed to maximize the intake of human milk. Growth measurements are obtained weekly for the first 2 months of life, biweekly during months 3 and 4, and weekly thereafter until 3 months after solid foods have been added. Intakes are measured monthly from 16 weeks of life until 3 months after the introduction of solid foods. All studies have been reviewed by our institutional review boards and informed consent is obtained from each of the participants.

Subjects

Women recruited for these studies represented middle- and upper-income groups who obtain their medical care from private services. Subjects were healthy, on no prescribed or nonprescribed drugs, nonsmokers, and 18 to 35 years of age. Only those with one or two living children were accepted for study. Infants were required to be healthy, term, and appropriate size for gestational age.

Intake Measurements

The amount of milk ingested daily was determined by the test-weighing method during a 24-hour period in Study I.[9-10] Measurements of milk intake were obtained for 5 consecutive days in Studies II and III. Milk samples were collected monthly for compositional studies. An Egnell pump was used to express milk. Alternate breasts were used for feeding and pumping with successive feeds during a 24-hour period.[11] Compositional analyses were performed on a monthly 24-hour milk sample for each subject in Studies I and III. The analyses included the determination of gross energy by bomb calorimetry.[12] In Study II, milk energy was assumed to be 67 kcal/dl. In estimating the intakes of Study III infants, a conservative energy density of 67 kcal/dl was used for milk samples that remain unanalyzed.

Nutrient intakes from solid foods were determined for 5 consecutive days. Mothers were supplied with preweighed jars of food and detailed instructions for their storage after use. Each jar was reweighed after feeding. Nutrient analyses were performed by our laboratory on each lot of the 12 types of solid foods used in Studies II and III.

Table 1. Birth Characteristics of Study Infants

Study	N	Males/Females	Birth Weight	Birth Length
I	44	27/18	3.58 ± 0.45[a]	50.9 ± 2.5
II-A	9	5/4	3.66 ± 0.36	N/A
II-B	8	5/3	3.69 ± 0.52	N/A
III	48	19/29	3.71 ± 0.55	51.62 ± 2.15

[a] Mean ± SD

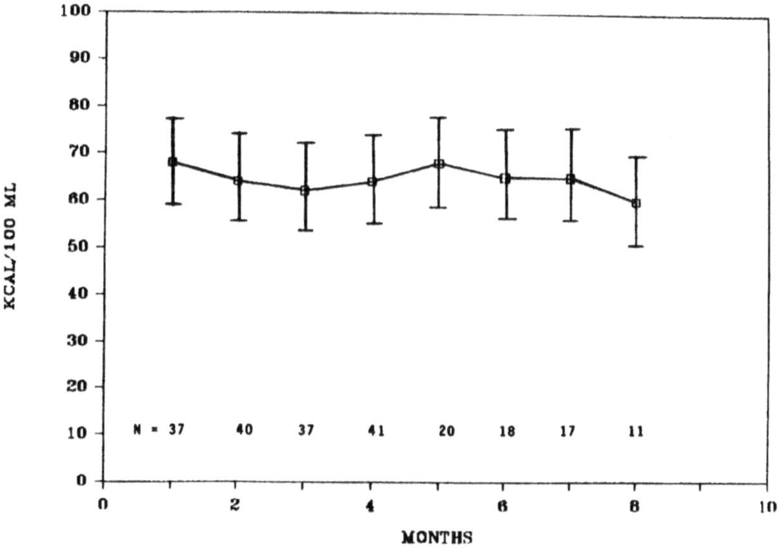

Fig. 1. Concentration of energy (kcal/100 ml) in milk samples collected from month 1 to 8 of lactation. Concentrations of energy from month 5 onward include values for samples from women supplementing their infants' diets with solid foods. Results are expressed as $\bar{X} \pm 1$ SD.

Growth Measurements

Infant weight was recorded from an electronic Sartorius balance before feeding.[7] Infant length was measured on a recumbent infant board.[8] Two research assistants were required to extend the infant satisfactorily.

Data Analysis

Descriptive data for infants enrolled in Studies I and II and for approximately 40 infants enrolled to date in Study III are presented. Z-scores have been calculated for growth indices. Z-scores have been derived from appropriate NCHS standards. These are a measure of relative growth in standard deviation units, i.e., a Z-score of 1 indicates a value of 1 standard deviation above the mean, a value of 0 indicates a value coincident with the population mean, and a value of -1 is one standard deviation below the mean.

RESULTS

Characteristics of study infants at birth are summarized in Table 1. The concentration of energy in 24-hour milk samples analyzed from gross energy from Studies I and III are illustrated in Figure 1. Milk volumes and total energy intakes for Studies I, II, and III from month 1 to approximately month 8 are illustrated in Figures 2 and 3. Z-scores for weight-for-age, length-for-age, and weight-for-length for Studies I and III are illustrated in Figures 4 and 5, respectively. Z-scores for those growth indices for Study III are summarized in Tables 2, 3, and 4 according to age and relative to the time solid foods were introduced into the diet.

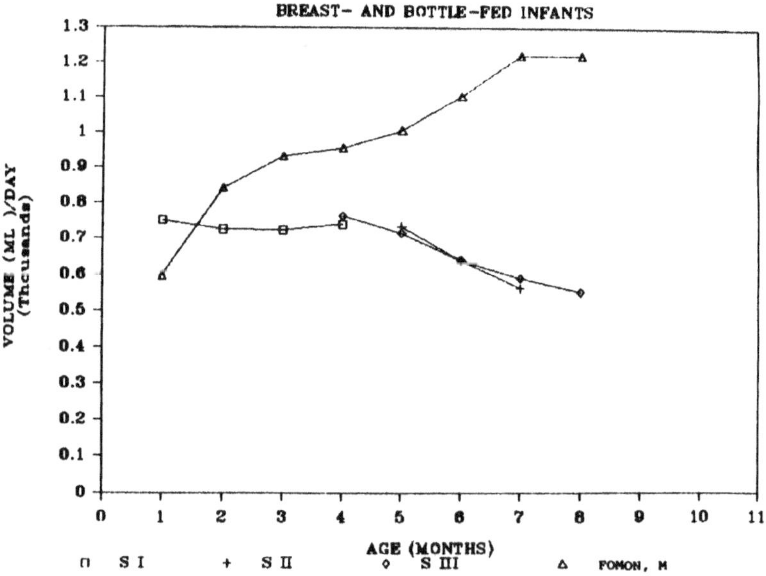

Fig. 2. Milk volumes consumed by breast- and bottle-fed infants. Volumes of breast-fed infants are from Study I (□), Study II (+) and Study III (◇). Volumes of bottle-fed infants (△) are from studies by Fomon et al. (ref. 4).

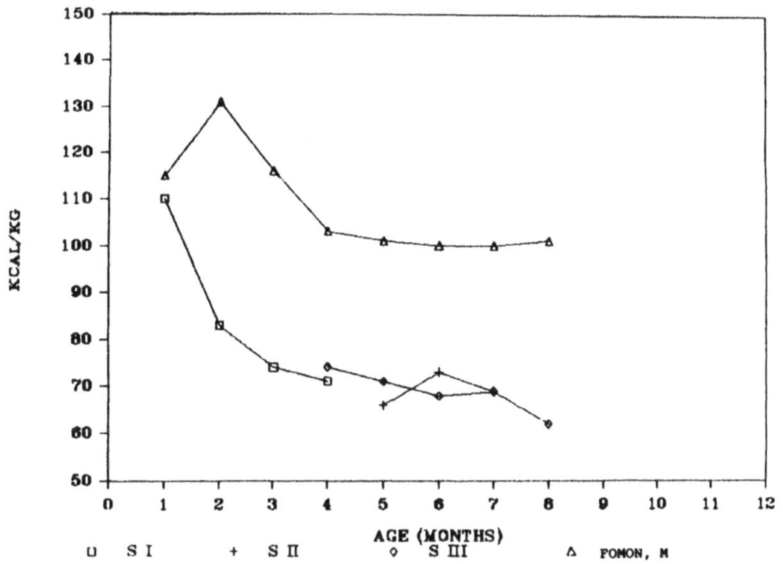

Fig. 3. Energy intakes (kcal/kg) of breast- and bottle-fed infants. Energy intakes of breast-fed infants are from Study I (□), Study II (+), and Study III (◇). Values for bottle-fed infants (△) are from studies of Fomon et al. (ref. 4).

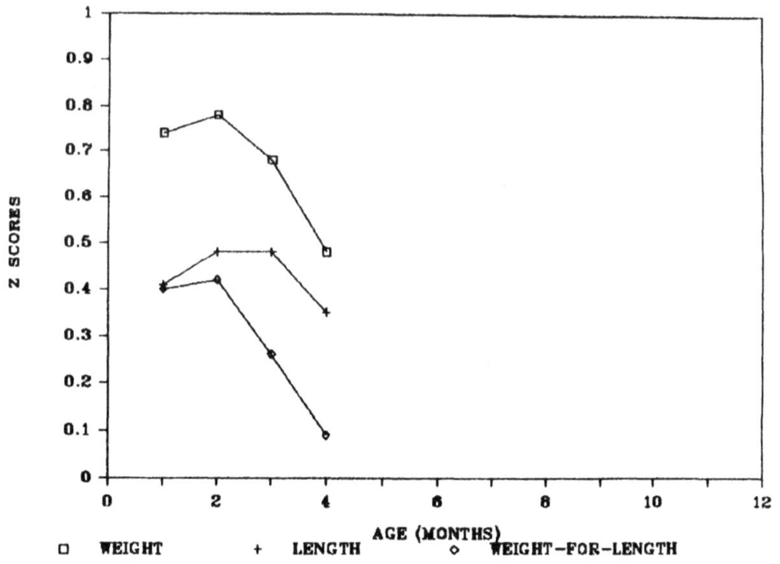

Fig. 4. Growth indices (Z-scores) of infants in Study I, weight-for-age (□), length-for-age (+), and weight-for-length (◇).

Table 2. Weight-for-age (Z-Scores) Tabulated for Age and Timing Relative to Weaning

Age	Before Solid Foods		After Solid Foods		
(wk)	(6 wk)	(2 wk)	(2 wk)	(6 wk)	(10 wk)
16	0.76 ± 0.91(16)[a]	0.72 ± 0.87(17)	-	-	-
20	0.88 ± 0.78(7)	0.69 ± 0.98(14)	0.64 ± 0.88(16)	-	-
24	-	0.64 ± 0.81(7)	0.46 ± 1.05(14)	0.49 ± 0.85(15)	-
28	-	-	0.56 ± 0.69(7)	0.54 ± 1.04(13)	0.34 ± 0.80(15)
32	-	-	-	0.29 ± 0.57(6)	0.45 ± 1.05(12)

[a] Mean ± SD (N of sample size)

Table 3. Length-for-age (Z-Scores) Tabulated for Age and Timing Relative to Weaning

Age	Before Solid Foods		After Solid Foods		
(wk)	(6 wk)	(2 wk)	(2 wk)	(6 wk)	(10 wk)
16	0.41 ± 0.83(16)[a]	0.54 ± 0.88(17)	-	-	-
20	0.34 ± 0.69(7)	0.46 ± 0.58(14)	0.56 ± 0.81(16)	-	-
24	-	0.27 ± 0.61(7)	0.26 ± 0.77(14)	0.60 ± 0.78(15)	-
28	-	-	0.24 ± 0.55(7)	0.16 ± 0.76(13)	0.38 ± 0.78(15)
32	-	-	-	0.05 ± 0.75(6)	0.18 ± 0.75(12)

[a] Mean ± SD (N of sample size)

Table 4. Weight-for-length (Z-Scores) Tabulated for Age and Timing Relative to Weaning

Age	Before Solid Foods		After Solid Foods		
(wk)	(6 wk)	(2 wk)	(2 wk)	(6 wk)	(10 wk)
16	0.41 ± 0.72(16)[a]	0.22 ± 0.87(17)	-	-	-
20	0.64 ± 1.09(7)	0.27 ± 0.75(14)	0.10 ± 1.15(16)	-	-
24	-	0.43 ± 0.98(7)	0.21 ± 0.85(14)	-0.08 ± 1.00(15)	-
28	-	-	0.39 ± 1.00(7)	0.46 ± 0.88(13)	0.01 ± 0.84(15)
32	-	-	-	0.30 ± 0.89(6)	0.38 ± 1.05(12)

[a] Mean ± SD (N of sample size)

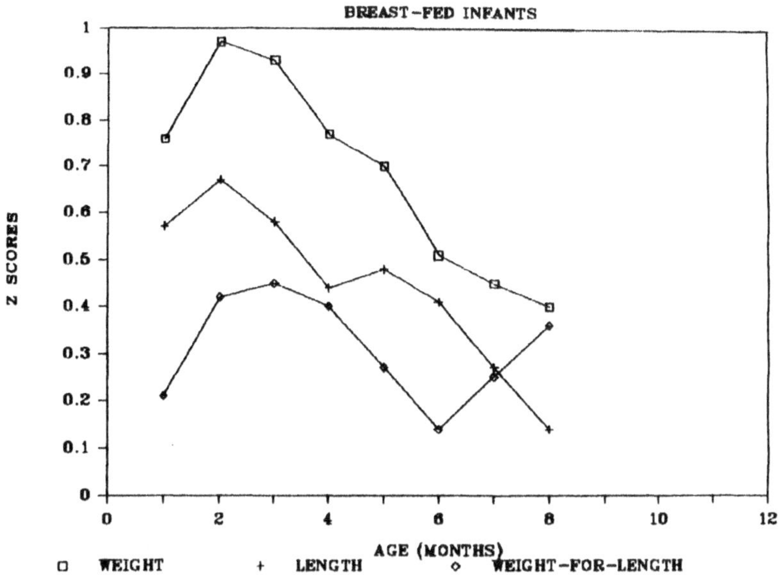

Fig. 5. Growth indices (Z-scores) of infants in Study III; weight-for-age (□), length-for-age (+), weight-for-length (◇).

DISCUSSION

Interpretation of these data presents several issue which relate directly to estimates of nutrient requirements. Examination of Table 2 indicates that energy intakes of breast-fed infants remain at approximately 70 to 75 kcal/kg from month 3 onward. These intakes are substantially below mean energy requirements estimated by the Food and Nutrition Board, National Research Council.[1] Intakes, expressed per kilogram body weight, fell rapidly over the first 3 months of life and remained relatively stable through approximately 8 months, even after ad libitum introduction of solid foods. The gross energy concentration of human milk remained constant (approximately 65 kcal/dl) as milk intake declined following the introduction of solid foods. Three months after solid foods were first added to the diet, the infants' milk intakes fell to approximately 75% of levels observed during exclusive breastfeeding. Energy intakes per body weight basis remained the same, but human milk provided a diminishing proportion of total energy.

The adequacy of an infant's nutrient intake usually is assessed by comparing growth performance with NCHS standards. Comparison of the growth performance of our study infants against this standard leads to a consideration of several issues. If adequate growth is defined on the basis of single cross-sectional observations, the groups' mean weights-for-age, lengths-for-age, and weights-for-length remained above the 50th percentile in all but one case, weight-for-length in Study II at month 7.

This is most likely a reflection of the relatively high socio-economic groups from which the samples were drawn. If longitudinal observations are used to evaluate growth and the expected distribution (as calculated by Z-scores) is considered, breast-fed infants followed for approximately 8 months appeared to grow less rapidly than the NCHS reference population. In Study III the group mean weight-for-age Z-score increased initially from approximately 0.76 to 0.93 over the first 3 months and then fell progressively to 0.40 by the 8th month. The group mean length-for-age Z score declined similarly. This index started at approximately 0.57 the first month and fell to 0.27 on month 7 and 0.14 on month 8. Other groups have made similar observations.[13] Data for head circumference, skinfold thicknesses, individual growth velocities and accelerations have not been related to energy and other nutrient intakes.

Tables 2 to 4 do not indicate any marked change in growth performance after solid foods were introduced into the diet. Although there was no consistent tendency for the weight-for-age Z-scores to rise or fall during the transition from exclusive to partial breastfeeding, length-for-age z scores suggest that delaying the introduction of solid foods may be associated with declining Z-scores. Sample sizes for infants of specific ages at similar stages of feeding were too small, however, to evaluate this tendency confidently.

At this stage of our studies, the timing of the introduction of solid foods does not appear to be associated with a progressively inadequate intake of energy. Though preliminary, our view is based on the stability of energy intakes following the ad libitum availability of energy from solid foods and the progressive declines in the weight-for-age and length-for-age Z-scores. Against this view is the upturn in the weight-for-length Z-scores on the 7th and 8th month. A more definitive evaluation of these trends must await the conclusion of Study III.

If the energy intakes and growth performance of these infants are physiologic optimums, what is the physiologic implication of intakes or growth rates which are substantially higher? A comparison of the gross energy intakes of breast-fed infants in Studies I and II with the mean energy intakes of formula-fed infants published by Fomon et al. demonstrates substantial cumulative differences in intake over the first 8 months of life.[4] By 8 months, breast-fed infants have consumed approximately 30,000 kcal less than bottle-fed infants. A similar expenditure of energy by both groups should result in a mean weight difference between the groups of at least 2.7 kg. This assumes the deposition of cumulative energy difference as fat by bottle-fed infants. Since our study infants remained above the 50th percentile (derived from predominantly bottle-fed infants), it appears that breast-fed infants do not have an energy "deficit" of this magnitude and that bottle-fed infants do not have an energy "surplus" of this amount. Three possibilities, therefore, must be considered: differences in the intake between groups are not as great as our measurements and those of Fomon suggest; energy expenditure is substantially different between feeding groups; or the composition of newly acquired tissue differs between groups.

The validity of intakes presented in this report has been substantiated independently by a fourth experiment presented separately at this conference.[14] Intakes were estimated indirectly by measuring the turnover of labeled water in infants whose mothers had ingested D_2O to label their breast milk. Estimates were 105% of those calculated from concomitant test weighings. Results of a small cross-sectional study of bottle-fed infants conducted by our group agreed well with previous observations made by Fomon.[15] A possibility that we have not evaluated is that the intakes of bottle- and breast-fed infants are not equally consistent. Breast-fed infants may maintain relatively low intakes consistently; bottle-fed infants may experience wider swings in intake presumably during more frequent periods of subclinical and clinical illnesses.[16-17] Decreased intakes and heightened energy needs during these periods would lessen cumulative differences in intakes. We studied breast-fed infants only when healthy; whether Fomon and co-workers also avoided periods of illness in their infant studies is unclear.

If one accepts that the intakes of breast- and bottle-fed infants differ significantly, how may the significance of these intakes be assessed? A first step is to examine energy expenditure. In preliminary studies measuring basal energy expenditure and dietary-induced thermogenesis (DIT), we have been unable to account for observed differences in intake.[18] Our preliminary observations suggest that differences in intake may result in dissimilar expenditures for activity or disparities in body composition. An assumption of similar growth costs between groups suggests that the four-month-old breast-fed infant would have much less energy available for activity than would the bottle-fed infant. Our preliminary observations do not support that possibility.

Application of the dual isotope methodology for assessment of total energy expenditure in children will allow the estimation of energy available for activity and the accretion of tissue.[19] If one has accurate measurements of intake and total energy expenditure, the difference between those quantities will provide an estimate of the sum of energy lost in stools and urine and tissue accretion. Estimates of total energy expenditure, basal energy expenditure, and DIT will provide an indirect estimate of energy expended for activity.

Accounting for possible differences in tissue composition is an important aspect of growth studies. Yet, available techniques for measuring body composition are not readily applicable to infants. We are evaluating estimations of total body water, protein, and fat derived from measurements of total body water and volume.[20] This approach does not require that the water content of lean body mass be known; it assumes that the sum of the volumes of body water, protein, fat, and minerals is equal to total body volume and that the specific gravity of water, protein, fat, and minerals is constant across all ages.[21]

If differences in body composition are detected between feeding groups, one will still be faced with assessing the significance of the observation. Although privileged infants do well on the intakes we have reported, infants in more hostile setting may require more energy to cope with environmental stresses such as increased rates of infection.[5] Growth faltering is reported for breast-fed infants in developing countries at approximately the age that we are detecting changes in weight-for-age and length-for-age Z-scores.[22] Entering the period of weaning or complementation at 4 to 6 months with substantial additional calories in adipose tissue stores may be life-saving in some settings. In contrast, for infants living under more favorable conditions, an "excessive" intake of energy may present a distinct set of risks related to allergic diseases or disorders which normally go unrecognized until adulthood.

We have yet to explore the effects of energy intakes on behavior during infancy. To focus solely on anthropometric measurements would be an error until we better understand the functional outcomes associated with distinct growth patterns. The Minnesota starvation experiments in adults and less controlled studies in children have provided data demonstrating the potential impact of energy on activity and other behaviors.[23,24] The observation that breast- and bottle-fed infants have different intakes and dissimilar growth patterns suggests that differences in activity between groups are likely. The hypothesis that growth performance is affected by nutrient intakes before other functional competences are compromised merits rigorous scrutiny.

ACKNOWLEDGMENTS

This research is supported by a contract (NO 1-HD-2-2814) from the National Institutes of Health and by the USDA/ARS Children's Nutrition Research Center, Department of Pediatrics, Baylor College of Medicine and Texas Children's Hospital, Houston, TX. This project has been funded in part with Federal funds from the U.S. Department of Agriculture, Agriculture Research Service under Cooperative Agreement number 58-7MNI-6-100. The contents of this publication do not necessarily reflect the views or policies of the U.S. Department of Agriculture, nor does mention of trade names, commercial products, or organizations imply endorsement by the U.S. Government.

REFERENCES

1. Committee on Recommended Dietary Allowances, National Academy of Sciences, Washington, D.C. (1980).
2. FAO/WHO Energy and Protein Requirements, WHO Technical Report Series, No. 522. FAO Nutrition Meeting, Report Series, No. 52, Geneva (1973).
3. Protein Requirements, Report of a Joint FAO/WHO Expert Group, WHO Technical Report Series, No. 301, Geneva (1965).
4. S.J. Fomon, "Infant nutrition," 2nd Edition, W.B. Saunders Co., Philadelphia (1974).
5. R. Martorell, Child growth retardation: a discussion of its causes and its relationship to health, in: "Nutritional Adaptation in Man," Sir K. Blaxter and J.C. Waterlow, eds., John Libbey, London, (1985).
6. D. Seckler, "Small but healthy": a basic hypothesis in the theory, measurement and policy of malnutrition, in: "Newer Concepts in Nutrition and Their Implications for Policy," P.V. Sukhatme, ed., Maharastra Association for the Cultivation of Science Research Institute. (1982).
7. G.B. Spurr, M. Barac-Nieto, J.C. Reira, and R. Ramirez, Marginal malnutrition in school-aged Colombian boys: efficiency of treadmill walking in submaximal exercise, Am. J. Clin. Nutr. 39:452 (1984).
8. N.F. Butte, C. Garza, E.O. Smith, and B.L. Nichols, Human milk intake and growth performance of exclusively breast-fed infants, J. Pediatr. 104:187 (1984).
9. J. Stuff, C. Garza, C. Boutte, J.K. Fraley, E.O. Smith, E.R. Klein, and B.L. Nichols, Sources of variation in milk and caloric intakes in breast-fed infants: implications for lactation study design and interpretation, Am. J. Clin. Nutr. 43:361 (1986).
10. M.W. Woolridge, N.F. Butte, K.G. Dewey, A.M. Ferris, C. Garza, and R.P. Keller, Methods for the measurement of milk volume intake of the breast-fed infant, in: "Human Lactation: Milk Components and Methodologies," R.G. Jensen and M.C. Neville, eds., Plenum Press, New York (1985).
11. C. Garza, M.W. Woolridge, N.F. Butte, A. Ferris, and C. Casey, Sampling milk for energy content, in: "Human Lactation: Milk Components and Methodologies," R.G. Jensen and M.C. Neville, eds., Plenum Press, New York (1985).
12. C. Garza, N.F. Butte, and K. Dewey. Determination of the energy content of human milk, in: "Human Lactation: Milk Components and Methodologies," R.G. Jensen and M.C. Neville, eds., Plenum Press, New York (1985).
13. R.G. Whitehead and A.A. Paul, Human lactation, infant feeding and growth: secular trends, in: "Nutritional Needs and Assessment of Normal Growth," M. Gracey and F. Falkner, eds., Raven Press, New York, 1985.
14. N.F. Butte, W.W. Wong, B.W. Patterson, C. Garza, and P.D. Klein, Comparison of the deuterium dilution and test-weighing techniques for the determination of human milk intake, in: "The Effects of Human Milk on the Recipient Infant," A.S. Goldman, ed., Plenum Press, New York (in press 1987).

15. C.M. Montandon, C.A. Wills, C. Garza, E.O. Smith, and B.L. Nichols, Formula intake of one- and four-month old infants, J. Pediatr. Gastroenterol. Nutr. 5:434 (1986).
16. A.S. Cunningham, Morbidity in breast-fed and artificially fed infants. J. Pediatr. 95:685 (1979).
17. R. Daga and H. Pridan, Relationship of breastfeeding versus bottle feeding with emergency room visits and hospitalization for infectious diseases. Eur. J. Pediatr. 139:192 (1982).
18. N.F. Butte, C. Garza, and T.Q. Dang, Basal metabolic rates and diet-induced thermogenesis in breast-fed and formula-fed infants 1 to 4 months of age, Pediatr. Res. 20:236A (1986).
19. P.D. Klein, W.P.T. James, W.W. Wong, C.S. Irving, P. Murgatroyd, M. Cabrera, H. Dellasso, E.R. Klein, and B.L. Nichols, Calorimetric validation of the doubly labeled water method for determination of energy expenditure in man, Hum. Nutr.: Clin. Nutr. 38C:95 (1984).
20. H.P. Sheng, W. Deskins, D. Winter, and C. Garza, Estimation of total body fat and protein by densitometry, Pediatr. Res. 18:212A (1984).
21. W.G. Deskins, D.C. Winter, H.P. Sheng, and C. Garza, Use of a resonating cavity to measure body volume, J. Acoust. Soc. Am. 77:756 (1985).
22. J.C. Waterlow and A.M. Thomson, Observations on the adequacy of breastfeeding, Lancet 2:238 (1979).
23. A. Keys, J. Brozeck, A. Henschel, O. Mickelsen, and H.C. Taylor, The biology of human starvation, Vol II, The University of Minnesota Press, Minneapolis (1950).
24. E. Pollitt, C. Garza, and R.L. Leibel, Nutrition in public policy, in: "Child Development Research and Social Policy," Vol. I., H.W. Stevenson and A.E. Siegel, eds., University of Chicago Press, Chicago (1984).

IS MATERNAL MILK PRODUCTION LIMITING FOR INFANT GROWTH DURING THE FIRST YEAR OF LIFE IN BREAST-FED INFANTS?

Margaret C. Neville, and Jean Oliva-Rasbach

Departments of Physiology and Pediatrics
University of Colorado School of Medicine
Denver, Colorado

Growth rates of breast-fed infants have consistently been found to be slower after the first three months of life than standard growth rates as represented, for example, by NCHS scores. This appears to be true whether one considers infants of well-nourished mothers in developed countries (1-10) or infants in underdeveloped countries (reviewed in 11-13). In this paper we will not deal with the popular argument about whether this slowed growth rate represents growth faltering (6,7) in breast-fed infants. Rather we intend to look more closely at the mother-infant dyad to examine the possible regulatory mechanisms which lead to this slowed growth.

In these times of abundant, indeed over-abundant, food supplies in developed countries the point is sometimes lost that much of human evolution took place under conditions where food supply was marginal, often for extended periods of time. Under such conditions *optimal* growth rates are likely to be those that contribute to survival of both infant and mother and may be less than *maximal* growth rates. For example, Dugdale (14), suggested that slower growth rates would require decreased food intake by the infant; within limits, *i.e.*, so long as food intake was sufficient to maintain the health of the infant, decreased food intake might enhance both maternal and infant survival by decreasing maternal nutritional needs.

In this article we will focus on the mechanisms which produce the most efficient utilization of nutrients to allow survival of the mother-infant dyad. Because this conference focuses on the infant, our comments will be limited primarily to the problem of whether maternal or infant factors limit maternal milk production and the possible mechanisms involved. The interesting and related problem of maternal metabolic adaptation to lactation will be left for a future publication.

GROWTH RATES OF BREAST-FED INFANTS

The question of interest here is not whether breast-fed infants grow more slowly than bottle-fed infants but *whether they grow more slowly than maximum achievable growth rates*. Growth rates of bottle-fed infants vary depending on factors such as formula composition, parent feeding goals, etc. so that comparisons between bottle and breast-fed infants give variable answers depending on the formulas and fashions of the time (11). On the other hand a great deal of evidence (11-13) indicates that breast-fed infants grow more slowly than the NCHS standard curves derived from data collected by the Fels Research Institute between 1929 and 1975 (15). For example, figure 1 depicts the growth rates of 13 Denver infants who were exclusively breast fed at least 5 and up to 9 months. The weight-for-age data are plotted as NCHS percentile scores to underscore the difference between these infants and the reference infants. A curve is shown for each infant to emphasize the range of individual variation. Although infant

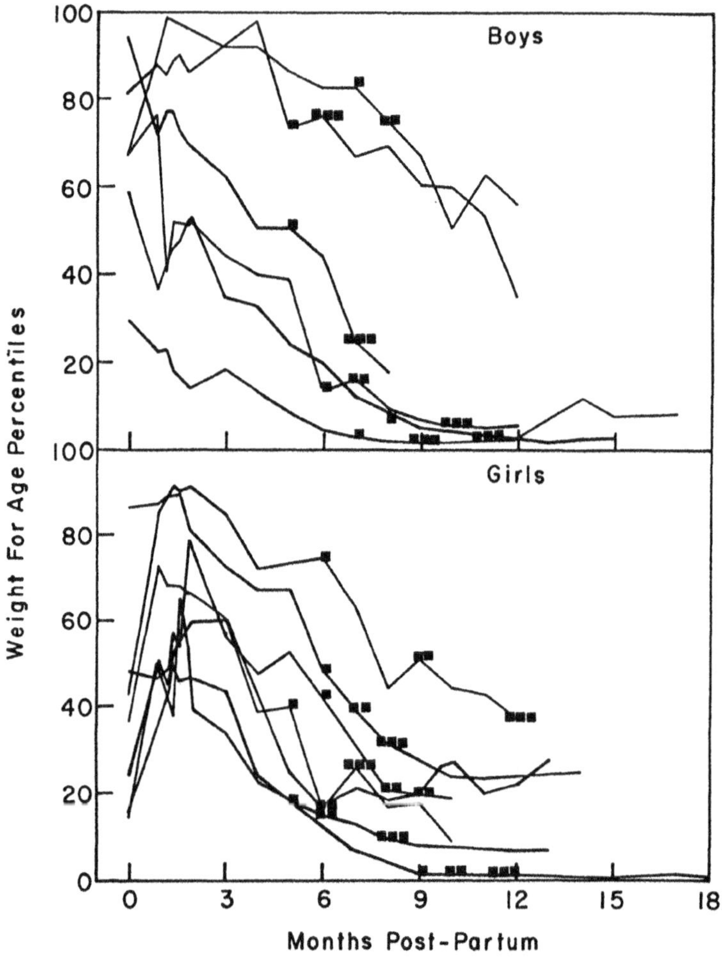

Figure 1. **Growth rates of breast-fed infants in Colorado plotted as NCHS percentile scores.** Infants were weighed naked on an electronic balance at monthly intervals. Monthly three-day diet records were maintained on the infants once supplementary feeds had been introduced. ■, Introduction of solids; ■■, supplementary foods exceeded 100 kcal/day; ■■■, supplementary foods exceeded 200 kcal/day.

length-for-age also declined relative to the NCHS standard, the difference was not as marked (data not shown). Head circumferences closely followed the NCHS standard curves. In the first three months post-partum all the girls and three of the boys gained weight compared to the standards. Three of the boys showed a continuous decline from birth. However, all except one infant were above the 30th percentile at 3 months. Similar results have been found in many studies (11-13). After 3 to 4 months, however, the weight for age percentile declined steadily in all infants, until at 9 months the mean percentile was 24.4 $\pm$ 7. At 12 months it was 20 $\pm$ 6 even though all mothers had started solid foods at least 3 months previously, indicating that catch up growth did not generally occur with the introduction of solid foods. Whitehead and Paul noted a similar phenomenon in their group of Cambridge infants (1). Our data growth differ little from data obtained in other well-documented studies of infants who were exclusively breast-fed for at least the first six months of life (1,2,5,7-10). Thus reports from many communities appear to support the notion that breast-fed infants as a group grow more slowly than the reference, mostly bottle-fed, infants studied up to 1975. *The minimal conclusion must be that after three months of age breast-fed infants do not take in sufficient nutrients to support maximal growth.*[1]

These observations raise a number of important questions: **"Is this limited growth rate of survival value for the species?"** Dugdale (14) has discussed the evidence that the limited growth rate is of survival value to the mother. The question of its survival value to the infant is discussed below. **"Do breast-fed infants show growth faltering after 3 to 4 months of age?" "Should supplementation be started at this age to increase growth rate?"** These questions are of considerable practical importance but cannot be answered without long term prospective data on growth rates in childhood and perhaps even lifetime morbidity and mortality data. **"What are normal growth limits for breast-fed infants?"** Clearly new growth standards more relevant to breast-fed infants are necessary since infant growth is probably the best measure of lactational sufficiency. Beyond this comment, however, this problem is clearly outside the scope of this paper. **"What are the mechanisms which limit breast milk intake?"** This final question will be the focus of the remainder of this paper. In particular we wish to ask whether breast milk intake is limited by the production capacity of the mother. We will begin by examining milk production in the mothers of the exclusively breast-fed infants whose growth data is shown in Figure 1.

BREAST MILK PRODUCTION

Early values for breast milk intake, summarized by Morrison (17) and Prentice *et al.*(13), showed a wide variability from woman to woman. For example, at 5.5 months lactation breast milk production ranged from 200 to 3500 ml/day, the low end representing partial breast-feeding and the high end, wet nurses. More recent values have been more homogeneous ranging from 725 to 850 ml/day during months 2 to 6 when only data from fully breast-fed infants obtained by test-weighing the infant is included (1,3,5-7,13,18-20). Figure 2 shows the average breast milk intake by the infants whose growth curves are shown in Figure 1 up to the time at which supplementary foods exceeded 100 Kcal/day. When the individual trends were analyzed there was

---

[1] It is important to note that, although the percentile score of these breast-fed infants was low, in fact their mean weight at 12 months was 88% of the NCHS standard at this age. Decreases in weight of this severity do not appear to be associated with increased morbidity or mortality (16).

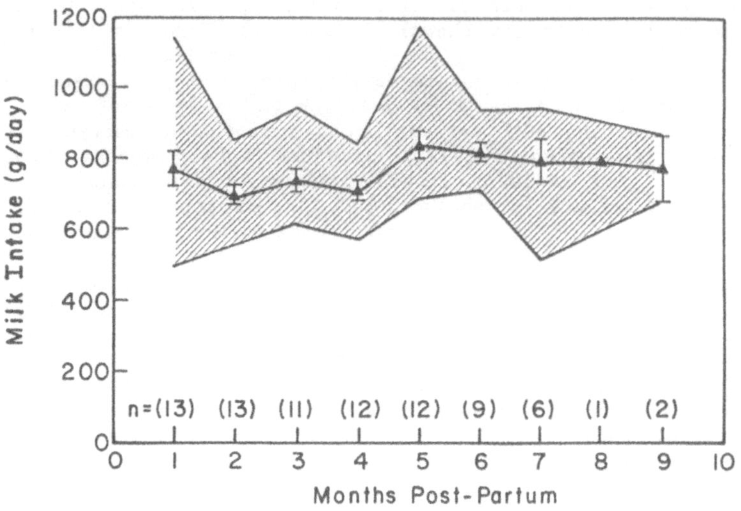

Figure 2. **Milk intake by fully breast-fed infants in Denver, Colorado.** Milk intake is shown for those infants who were documented to be taking less than 100 Kcal/day in supplementary foods. The pattern of supplementation of these infants is indicated by the square symbols in Figure 1. The hatched area indicates the range of values obtained by 24 hour test weighing at monthly intervals.

an average increase in breast milk intake of about 24 ml/month from months 1 to 6 with an average intake around 760 ml/day over this period.

Our observations suggest that nearly maximal milk intake is achieved by 1 month of age, raising the question of whether the limitation on nutrient intake is in the mothers' ability to produce milk or in the infants' appetite. There is rapidly increasing evidence that under many circumstances the infants' appetite is limiting: For example, there is no question that at least some women can produce a great deal more breast milk than did any of our mothers in the course of the study, volumes up to 3.5 liters daily having been recorded in wet nurses (22). Five of the 13 mothers in our study consistently produced more milk than the infant could consume, pumping off at least 100 ml and up to 500 ml/day to relieve pressure in the breast. This phenomenon, also observed in the Gambia (13), was more marked in the early stages of lactation after which milk output presumably adjusted to infant demand. One of the participants in our study is currently breast-feeding a subsequent infant. This infant (weight, 8053 g) is taking in about 860 ml of breast milk daily at 5.5 months postpartum whereas his older brother (weight 6884 g at 5.5 months postpartum) took a mean of 760 ml/day at the same stage of lactation. Many anecdotal observations such as these suggest that at least some mothers are capable of increasing their milk output in response to infant demand.

Further, if milk output were mainly a function of maternal production capacity, it seems likely that maternal rather than infant characteristics would be most closely related to milk intake. However, Dewey, et al (23,24) and Prentice and his coworkers (13) found little or no correlation between maternal factors such as weight-for-height, pregnancy weight gain, nursing frequency, maternal age, and parity and breast milk intake.[2] In contrast infant factors such as weight (hence birth weight and sex) and morbidity have been shown to be strongly correlated with milk intake (reviewed in 13). However such correlations could indicate either that better nourished infants grow faster or that larger infants take more breast milk. Two other approaches suggest that indeed infant demand is the independent variable.

One such approach is to determine whether artificially increasing milk output by using a breast pump to remove additional milk from the breast after feeds increases milk supply. Figure 3 shows the effect of increased emptying of the breasts using an electric breast pump on the volume of milk produced by five lactating women between 4 and 6 months postpartum. During the control period none of these women removed any milk by breast pump; their infants had a mean daily intake of 853 ± 46 (S.E.M) g of milk. By the second and third days of pumping the total production had increased to a mean of 959 ± 34 g/day and by days 6 and 7 the production had increased to 1070 ± 21 g/day. Every women showed an increased production ranging between 15 and 40% during the one week pumping period. There was a small, not statistically significant decline, in breast milk intake by the infants during this period. These data indicate that fully breast-feeding women

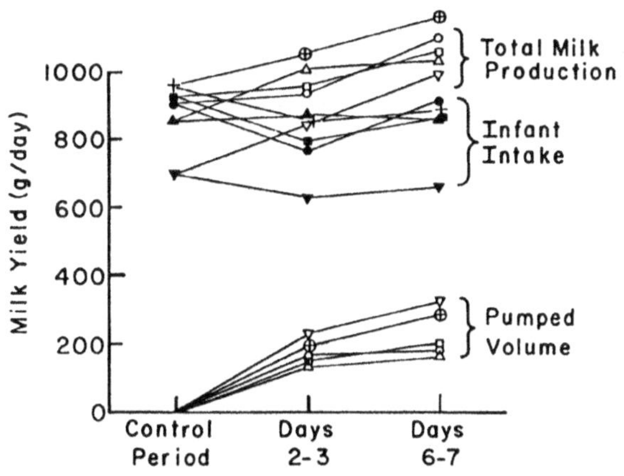

Figure 3. **Effect of milk removal on milk volume production.** Five women who were exclusively breast-feeding their infants 4 to 6 months postpartum were selected for this study. All test-weighed their infants for 48 hours before and after each breast-feed, using an integrating swing type balance designed to minimize movement artifact (Keller and Neville, unpublished). For the next seven days they pumped their breasts for 10 minutes 3 times daily after feeds, then used a drop of nasal oxytocin to produce let-down and pumped a further 5 minutes or until the milk stopped flowing. The amount of milk removed by breast pump was recorded. Forty eight hour test-weighing was carried out on days 2 and 3 and again on days 6 and 7 after commencement of pumping to obtain a complete record of the amount of milk produced.

---

[2]Prentice (21) has shown that parity greater than 9 is associated with reduced milk output and that parity did have an effect on milk composition in mothers in the Gambia. Thus it is improper to conclude that maternal factors have no effect on lactational performance.

are able to increase their breast milk production after their infants have attained 4 months of age, the point at which all infants in our study displayed decreasing weight gain compared to the NCHS standards.

A similar and more extensive study was carried out by Dewey and Lönnerdal (25). Eighteen mothers between 1.5 and 5 months postpartum were asked to express as much extra milk as possible for two weeks. Fourteen of the eighteen mothers increased their milk production during this period by an average of 22%. The infant intake was increased after the expression phase of the study but half these infants returned to baseline intakes within 1 to 2 weeks. The authors conclude that "the wide range in breast milk volume in relatively well-nourished populations appears to be due more to infant 'demand' than to inadequacy of milk production by mothers."

A second approach is to determine breast-milk consumption after supplementary feeds have been introduced. If the infant growth rate is limited by the amount of milk available, then total caloric intake should increase with supplementation. In our longitudinal study it did not; rather breast milk intake declined directly as the amount of supplementary food increased (data not shown). Prentice *et al.* (13) have made a similar observation.

Thus the weight of the evidence to date suggests that the maternal milk supply is not limiting and that some factor regulates infant appetite so that breast-milk intake is sufficient only to produce less than maximal growth. Our studies described were carried out in well-nourished American women. However, effects of supplementation were similar in the Gambia and in Cambridge, England (13). Further, as pointed out by Prentice *et al.* (13) there are data bearing on the question of the relation between maternal food intake and breast milk production which suggest breast milk output in under-nourished populations may be regulated by the same factors as in well-nourished women.

EFFECTS OF FASTING AND UNDERNUTRITION ON MILK PRODUCTION

Mean milk production by exclusively breast-feeding women from the Western world is 725-850 ml/day by the second month of lactation when data obtained by test-weighing the infant are used (1,3,5-7,13,18-20). Values from underdeveloped countries tend to be lower (summarized in 1 and 13). However, in most cases strict control for exclusive breast-feeding has not been maintained; nor has the relation to diet and infant morbidity been stringently examined. An exception is the study carried out by the Dunn Nutrition Laboratory in the Gambia (13). The most important conclusion from this report was that chronic undernutrition *per se* does not affect the volume or macronutrient composition of breast milk.

The effects of acute starvation and water abstention on breast milk composition and volume have also been examined by Prentice and coworkers (26) using the Ramadan fast from sunup (700 hrs to sundown 1900 hours) as a natural experimental situation. They found that milk volume production was little affected by this fast although compositional changes attributed to water deprivation were observed. We have extended this approach to examine milk production and composition after a 20 hour fast in fully lactating woman and found that the mean hourly rate of milk production (Table 1) did not differ from the milk volume production obtained by 48 hour test-weighing a few days earlier. These women were not restricted in water intake and no changes in macronutrient composition were observed.

Table 1

Breast Milk Output by Fasting Women

Subject	Normal Milk Output (g/hr)	Fasting Milk Output (g/hr)
201	36.9	39.3
203	31.6	31.1
204	32.1	36.3
205	31.8	30.6
206	34.6	38.9
Mean +_S.E.	33.4 +_1.0	35.2 +_1.8

Five fully breast-feeding women between 1.5 and 6 months post-partum test-weighed their infants for 48 hrs to obtain a value for milk output. A few days later they were asked to eat a normal dinner and fast from 7 P.M. They were admitted to the Clinical Research Center of University Hospital at 6 A.M. the following morning and their breasts were pumped hourly with an electric breast pump, using nasal oxytocin to obtain maximal emptying of the breasts. They were allowed water and caffeine- and sugar-free beverages *ad libitum* during the course of the day but no food. The experiment was continued until 3 P.M. The fasting milk outputs given are the mean hourly volumes from both breasts from the 2 and 3 P.M. pumpings. We show elsewhere (27) that the pumping procedure gives values for milk output equivalent to those obtained by test-weighing.

Thus data from studies during both acute starvation and chronic undernutrition strongly suggest that *the nutritional status of the mother* (at least under non-extreme conditions) *does not regulate milk output.* Optimal partition of available nutrients between infant and mother therefore appears to depend wholely on regulation of infant appetite to take in the minimal amount of breast milk compatible with good health. These observations are in sharp contrast to data obtained in rodents and goats where 24 hours or less starvation results in sharp decreases in milk production (28,29). Thus it should be emphasized that we are discussing human adaptive mechanisms which may not be valid for other species. These mechanisms appear to operate all the time, not just in periods of famine, since growth of breast-fed infants of both well- and poorly-nourished women proceeds at less than maximal rates.

MECHANISMS FOR LIMITATION OF MILK INTAKE

The regulation of food intake is a complex and poorly understood phenomenon, a review of which would be out of place here. Nonetheless, it may be worth stating the problem and one or two tentative mechanisms to provide a framework for future research. The question can be given some focus if we realize that satiety has at least two components, expressed in the short term by termination of food intake and in the long term by regulation of caloric intake (30). The first of these appears to be a local mechanism requiring interaction between nutrient entry into the duodenum and gastric distension (31) and is mediated by the intestinal peptides cholecystokinin and bombesin (32). The second involves hypothalamic nuclei which probably respond to a variety of hormones and nutrients (33). The effect of the constituents of human milk on these mechanisms is likely to be complex. However, three possibilities present themselves for discussion here.

1. The foods taken in by the "standard" infants were imbalanced, containing an insufficient quantity of a required nutrient. Glucose comes to mind here because the brain, which is relatively large in the human infant, is able to use only this substrate for energy. Further, glucose plays a recognized role in satiety responses; that is, hypoglycemia is a potent stimulus to food intake (33). Because human milk has a higher lactose content than most species (34), ingestion of more calories might be necessary to satisfy the glucose requirement if a high protein, lower lactose containing fluid like cow's milk were ingested. This mechanism might account for the decrease in growth rates of formula-fed infants recently observed as formula composition has begun to approach the composition of human milk (11).

2. Breast-milk could contain a satiety factor which stimulates secretion of cholecystokinin and/or bombesin from the duodenum. Hall, for example suggested that the high fat content of hind-milk may act as a satiety signal and terminate nursing (35). The problem with this hypothesis is that the fat content of hind milk is a maternal, not an infant factor and does not, therefore, provide a mechanism for regulation of intake by infant appetite. Further there is evidence that fat content does not alter milk intake (36). However, other substances or even the pattern of milk intake may play a regulatory role as discussed more fully by Hamosh and Hamosh (37) in this volume.

3. Breast-milk contains or lacks a substance which interacts with the central appetite regulatory mechanisms in the brain (32,33). A possible candidate for this role is zinc. This trace element is present at quite high concentrations (>4 mg/L) in colostrum and early milk (38) but falls through the course of lactation to about 10% of the early level (39). Although it probably remains high enough to supply basic nutritional needs for growth (40), at least in zinc-supplemented mothers, the relatively low intake in later lactation may act as an appetite suppressant. Thus Walravens and Hambidge (41) showed that formula supplemented with zinc to a level of 4 mg/L increased infant growth over unsupplemented formula which contained 1.8 mg/L. The hypothesis is attractive because it would account for the relatively high growth rates during the first three months of lactation when zinc levels in human milk are relatively high. It would be of considerable interest to determine whether breast-fed infants supplemented with zinc from 3 to 6 months of age increased their breast milk intake or rate of growth.

Certainly no clear mechanism for regulation of appetite in breast-fed infants presents itself at the moment. Because of the implications of such a mechanism for regulation of obesity, however, increased research into the problem is of considerable importance.

CONCLUSION

It is clear that fully breast-fed infants grow at less than maximal rates. These growth rates are reflected primarily in a decrease in body weight and to a lesser extent in a decrease in body length. Head circumference is not affected. In general infants appear to maintain good health at these reduced growth rates (although see ref. 6) suggesting that breast feeding provides a regulatory mechanism which maintains infant growth at a minimal rate compatible with survival and normal mental development. As discussed by Dugdale (14) such a growth rate would place a minimal demand on maternal nutrition. If low growth rates are sustained after the first year of life in infants who have been breast-fed longer than 4 months, it is possible that the smaller children and adults which resulted would be better adapted to chronic food shortages. Thus infants who consumed 80% of the amount of food necessary to produce maximal growth might spare their mothers about 100 kcal/day or possibly 30,000 kcal over the course of lactation. However if they grew into smaller adults with only, say, 80% of the nutrient requirements of the maximally grown adult, they could decrease their own yearly caloric requirments by as much as 120,000 kcal or more, a significant saving in the face of marginal food supplies.

We have presented evidence that the growth limitation in breast-fed infants is the result of a limitation of infant demand rather than an inadequacy of maternal milk supply. Although we are already in the realm of hypothesis, a little frank speculation suggests that this adaptive mechanism would operate most efficiently in times of nutritional deprivation when the infant is less likely to be offered supplementary foods. In times of plenty supplementary foods would be available and if given to the infant, one might expect an improvement in appetite and an increase in growth rate. Indeed, this is just what has been observed in the last century in developed countries as food supplies have become constant and abundant. Such considerations lend some urgency to further research into the relation between breast-feeding and ultimate growth and development of the infant.

ACKNOWLEDGEMENT

The research described in this study was supported by contract number HD 2-2801 from NIHCD and by NIH grants HD 19547 and AM 12432. The investigators are profoundly grateful to the subjects of these experiments who were committed and enthusiastic about their roles in this research.

REFERENCES

1. R.G. Whitehead and A.A. Paul. Human Lactation, Infant Feeding and Growth: Secular Trends, in Nutritional Needs and Assessment of Normal Growth. M. Gracey and F. Falkner, Eds. Vevey/Raven Press, New York, (1985).
2. C.H. Ahn and W.C. MacLean. Growth of the exclusively breast-fed infant. Am.J.Clin.Nutr. 33:183-92 (1981).
3. J. Boulton. Nutrition in childhood and its relationships to early somatic growth, body fat, blood pressure and physical fitness. Acta Paediatr. Scand. (suppl.) 284:1-85 (1981).
4. R.L. Jackson, R. Westerfeld, M.A. Flynn, E.R. Kimball, and R.B. Lewis. Growth of "well-born" American infants fed human and cow's milk. Pediatrics 33:642-52 (1964).
5. L. Salmenperä, J. Perheentupa and M.A. Siimes. Exclusively breast-fed healthy infants grow slower than reference infants. Pediatric Res. 19:307-312 (1985).
6. R.K. Chandra. Breast feeding, growth and morbidity. Nutrition Res. 1:25-31 (1981).
7. A. Wallgren. Breast milk consumption of healthy full-term infants. Acta Paediatr. Scand. 32:778-790 (1944-5).
8. G.M. Gwen, P.J. Garry and E.M. Hooper. Feeding and growth of infants. Nutrition Research 4:717-731 (1984).

9. N.E. Hitchcock, M. Gracey and A.I Gilmour. The growth of breast fed and artificially fed infants from birth to twelve months. Acta Paediatr. Scand. 74: 240-245, (1985).
10. G.M. Owen, P.J. Garry and E.M. Hooper. Feeding and growth of infants. Nutrition Res. 4: 727-731 (1984).
11. R.G. Whitehead and A.A. Paul. Growth charts and the assessment of infant feeding practices in the western world and in developing countries. Early Human Devel. 9:187-207 (1984).
12. J.F. Seward and M.K. Serdula. Infant feeding and infant growth. Pediatrics. Supplement 31:728-762 (1984).
13. A.M. Prentice, A.A. Paul, A. Prentice, A.E. Black, T.J. Cole and R.G. Whitehead. Cross cultural differences in lactational performance. in Human Lactation: Maternal and Environmental Factors. M. Hamosh and A. Goldman, Eds. Plenum Press, N.Y. (1986) In press.
14. A.E. Dugdale, Evolution and infant feeding. The Lancet Mar:670-673 (1986).
15. P.V.V. Hamill, NCHS Growth Curves for children, birth to 18 years. U.S. Department of Health, Education and Welfare Publ. No. PHS 78-1650. National Center for Health Statistics, Hyattsville, MD. (1977)
16. A.A.Kielmann and C. McCord. Weight-for-age as an index of risk of death in children. Lancet 1:1247-1250 (1978).
17. S.D. Morrison. Human Milk: Yield, proximate principles and inorganic constituents. Technical Communication # 18. Commonwealth Bureau of Animal Nutrition, Rowett Research Institute, Bucksburn, Aberdeenshire, Scotland (1952).
18. N.F. Butte, C. Garza, E. O'Brian Smith, and B.L. Nichols. Human milk intake and growth in exclusively breast-fed infants. J. Peds. 104:187-195 (1984).
19. Y. Hofvander, U. Hadman, C. Hillervik and S. Sjölin. The amount of milk consumed by 1-3 month old breast- or bottle-fed infants. Acta Paediatr. Scand. 71:953-9589 (1982).
20. K.G. Dewey and B. Lönnerdal. Milk and nutrient intake of breast-=fed infants from 1 to 6 months: relation to growth and fatness. J.Pediatr. Gastroenterol. Nutr. 2:497-506 (1983).
21. A. Prentice. The effect of maternal parity on lactational performance in a rural African community. in Human Lactation: Maternal and Environmental Factors. M. Hamosh and A. Goldman, Eds. Plenum Press, N.Y. (1986) In press.
22. I.G. Macy, Human Milk Studies. X. Daily and monthly variations in milk components as observed in two successive lactation periods. Am J. Dis. Child. 43:1062-1067 (1944).
23. K.G. Dewey, M.A. Strode and B. Lönnerdal. Maternal vs. infant factors related to breast milk volume. Fed. Proc. 44:1675 (1985).
24. K.G. Dewey, D.A. Finley, M.A. Strode and B. Lönnerdal. Relationship of maternal age to breast milk volume and composition. in Human Lactation: II. Maternal and Environmental Factors. M. Hamosh and A. Goldman, Eds. Plenum Press, (1986) pp263-274.
25. K.G. Dewey and B. Lönnerdal. Infant self-regulation of breast milk intake. Acta Paed. Scand. 75: (1986) In Press.
26. A.M. Prentice, W.H. Lamb, A. Prentice and W.A. Coward. The effect of water abstention on milk synthesis in lactating women. Clinical Sci. 66: (1984).
27. M.C. Neville, W.W. Hay, and L Coughlin. Glucose clamp studies in lactating women: effects of glucose and insulin on milk glucose and lactose secretion. Fed. Proc. 45:901 (1986).

28. J.L. Linzell. The effect of frequent milking and of oxytocin on the yield and composition of milk in fed and fasted goats. J. Physiol. 190:333-346 (1967).
29. D.T. Carrick and N.J.Kuhn. Diurnal variation and response to food withdrawal of lactose synthesis in lactating rats. Biochem. J. 174:319-325 (1977).
30. P.R. McHugh and T.H. Moran. The stomach, cholecystokinin and satiety. Federation Proc. 45:1384-1389 (1986).
31. J. Gibbs and G.P. Smith. Satiety: the roles of peptides from the stomach and the intestine. Federation Proc. 45:1391-1395 (1986).
32. S.F. Leibowitz. Brain monoamines and peptides: role in the control of eating behavior. Federation Proc. 45: 1396-1403 (1986).
33. G.A. Bray. Autonomic and endocrine factors in the regulation of energy balance. Federation Proc. 45: 1404-1410 (1986)
34. R. Jenness. The composition of milk, in: Lactation: A Comprehensive Treatise, Volume III (B.L. Larson and V.R. Smith, Eds.) Academic Press, New York pp.3 -107 (1975).
35. B. Hall. Changing composition of human milk and early development of an appetite control. Lancet 1:779-780 (1975).
36. M.W. Woolridge, J.D. Baum and R.F. Drewett. Does a change in the composition of human milk affect sucking patterns and milk intake? Lancet 2:1292-1294 (1980).
37. M. Hamosh and P. Hamosh. Does nutrition in ealy life have long-term metabolic effects? Can animal models be used to predict these effects in the human? in Human Lactation: the Effects of Human milk on the Recipient Infant. A. Goldman and S.A. Atkinson, Eds. Plenum Press, In press.
38. C.E. Casey, K.M. Hambidge, and M.C. Neville. Studies in Human Lactation. Zinc, Copper, Manganese and Chromium in Human milk in the first eight weeks of lactation. Am. J. Clin Nutr. 41: 1193-1200 (1985).
38. K.G. Dewey, D.A. Finley and B. Lönnerdal. Breast milk volume and composition during late lactation (7-20 months). J. Pediatr. Gastroenterol. Nutr., 3:713-720 (1984).
34. N.F.Krebs and K.M. Hambidge. Zinc requirements and zinc intakes of breast-fed infants. Am.J.Clin.Nutr. 43:288-292 (1986).
35. P.A. Walravens Growth of infants fed a zinc supplemented formula. Am. J.Clin.Nutr. 29:1114-1121 (1976).

ENERGY BALANCE AND THE NATURE OF GROWTH IN LOW BIRTHWEIGHT INFANTS

R. K. Whyte, J. C. Sinclair, and H. S. Bayley

Department of Pediatrics, McMaster University, Hamilton
and the Department of Nutrition, University of Guelph
Guelph, Ontario, Canada

Human milk, either from milk banks or as milk expressed by the baby's own mother, is often considered the feed of first preference for low birthweight infants. Infant formulas ("preterm formulas") are now available which have been designed to address the theoretical nutritional needs of growing low birthweight infants, and unlike previous formulas there is a deliberate departure in their design from the composition of human milk. A number of studies have shown that infants fed with banked human milk gain weight more slowly than do the infants fed with a "regular" 2.8 mJ/L (20 kcal/oz) commercial formula (1,2,3,4) and that the advantage in weight gain is even more striking when infants fed with 3.5 mJ/L (24 kcal/oz) formulas are compared to those fed with banked human milk (4,5,6). It has been demonstrated that in many of these studies the methods used for collecting human milk for banking resulted in a banked milk of exceptionally low nutrient content (5,6), and feedings with mother's own expressed breast milk gave rates of weight gain comparable to those of infants fed with regular formula (3). Nevertheless, low birthweight infants fed with high-energy, protein and mineral containing "preterm" formulas appear to experience much greater weight gains than do infants fed with their own mother's expressed breast milk.

It has been recognized for some time that a rapid gain in weight cannot be assumed to be indicative of rapid growth of appropriate composition. In some of the earlier studies of different protein levels in infant feeding (1,4) high rates of weight gain were associated with high protein intakes: later these high rates of weight gain were attributed to excessive gains of salt and water (8,9). Comparisons of gains in body weight are best qualified with measurements of associated changes in body composition.

Body composition is best measured by direct mass measurements of body components thought to be constantly related to the size of various compartments of the composition of the body, an example being the measurement of whole body potassium by external whole body ^{40}K counting (10). None of these direct techniques are accurate enough for measurements of very small infants. Dilutional techniques are becoming more available for use in low birthweight infants now that stable isotope techniques (11,12) are replacing radio-isotope dilution techhiques (13), but volumes of distribution do not always measure the expected compartmental volumes, and true verification data for many of these techniques is lacking. Balance techniques, which are only applicable for the measurement of the composition of growth, have been used extensively in studies of low birthweight infants. The technique involves a number of assumptions described below; like all balance techniques there is a tendency towards overestimation of nutrient retentions (14).

The basis of the technique and its inherent assumptions are these. The rate of energy storage is determined by measuring rates of energy intake, excretion and expenditure, and subtracting the latter two values from the former. Nitrogen balance is established by subtracting excretion from intake. Atwater's values for the heats of combustion of protein, carbohydrate and fat (respectively 24, 16 and 39 kJ/g) are assumed, as is the value of 6.25 g of protein for every gram of stored nitrogen (15). The contribution of carbohydrate storage to the rate of energy storage in the growing fetus is very small (0.02% of the total energy stored by the fetus (16)), and so it can be ignored in the calculations. From these assumptions the following equations are derived:

$$\% \text{ Protein} = \frac{\text{g N ret} \times 6.25 \times 100}{\text{g weight gain}}$$

$$\% \text{ Fat} = \frac{(\text{Energy stored} - (\text{g N ret} \times 0.147))100}{\text{g weight gain} \times 39}$$

$$\% \text{ Water \& minerals} = 100 - \% \text{ Protein} - \% \text{ Fat}$$

Where 'g N ret' is the daily rate of nitrogen retention in mg/d and where daily energy storage and weight gain are expressed as kJ/d and g/d respectively.

Protein growth is at least partly representative of growth of the lean body mass, and is associated with the growth of the body cell mass and with intracellular water. Lipid growth is distributed between central nervous tissue (1%), the structural fats of the body organs (19%), and as storage fat in adipose tissue (80%) (17). Presumably most of the variation in the accumulation of body salt and water is accommodated in the extravascular space.

From estimates of the fetal rate of weight gain and its composition (18) it can be calculated that at 34 weeks gestation the rate of energy accretion is about 8 kJ of energy and about 22 mg of nitrogen per gram of weight gain, which therefore consists of about 17% fat, 14% protein and 69% minerals and water. The term infant in the first six weeks of life has a much higher energy accretion (19), consisting of some 20 kJ of energy and 18 mg of nitrogen per gram of weight gain, which therefore consists of 42% fat, 11% protein and 52% minerals and water.

We conducted a study of the energy balance of growing low birthweight infants of gestational ages ranging from 27 to 34 weeks (20). Nine infants were fed with their own mother's expressed breast milk, and nineteen were fed with a 2.8 mJ/L (20 kcal/oz) formula (figure 1). Both groups received similar gross intakes of nitrogen and energy (mean values 423 mg/(kg.d) and 529 kJ/(kg.d) respectively. We showed that energy digestibility and metabolizability was similar between the two feeds (89% and 86% of gross intake respectively). Although metabolizable energy intake was almost exactly the same for the two groups, the rate of energy expenditure was significantly lower in the human milk fed infants. As a result, the rate of energy storage was slightly greater in the human milk fed infants. As rates of weight gain were similar, the rate of energy

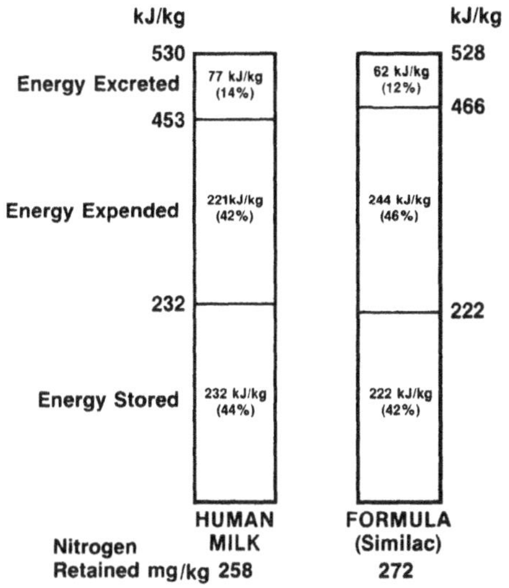

Figure 1. Energy and nitrogen balance of low birthweight infants fed with human milk and with a 2.8 mJ/L (20 kcal/oz) formula (20)

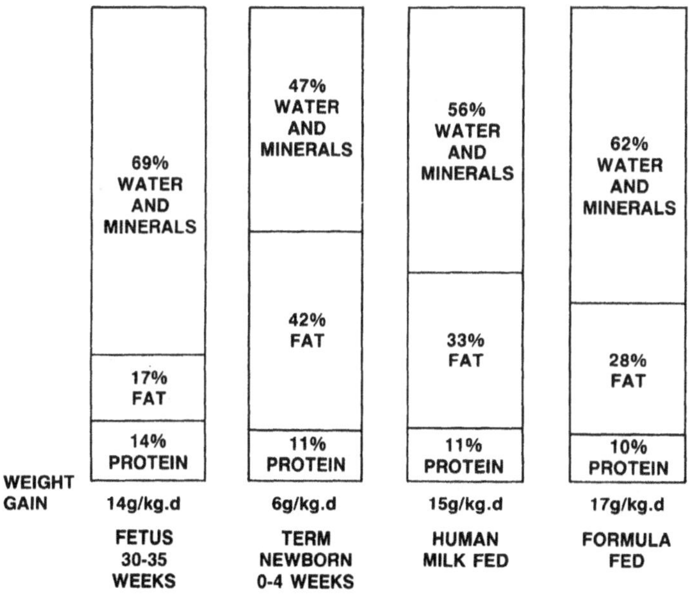

Figure 2. Composition of weight gain of low birthweight infants fed with human milk or with a 2.8 mJ/L formula compared with estimates for the fetus and the term newborn infant (20).

accretion in growing tissue was slightly but significantly higher in the human milk fed infants (15.3 vs 13.2 kJ/g). Nitrogen retention was similar in the two groups and averaged 63% of intake.

We concluded that the composition of growth of human milk fed infants was higher in fat than that of formula fed infants (28% vs 34%), this being a reflection of the slightly higher energy expenditure of the formula fed group. The clinical significance of this small difference in the composition of growth has not been determined: both groups had rates of energy accretion more typical of the growth of term newborn infants than of the fetus.

We speculated that the small difference in the rate of energy expenditure might be a reflection of the fatty acid content of the formula (Similac, Ross Laboratories, Ohio) which was relatively high in laurate (C12:0), the shortest of the long chain fatty acids, when this was compared to human milk (21,22). In animal models, short and medium chain fatty acids are known to be rapidly oxidized rather than stored as energy (23). The introduction of high energy formulas with partial substitution of fat with medium chain triglyceride gave us the opportunity of studying the effects of medium chain triglyceride on the energy balance of growing low birthweight infants (24).

Two formulas were made up by a manufacturer (Ross Laboratories, Columbus, Ohio), modified from their product "Special Care Formula", which, characteristic of other formulas currently designed for use in the growing low birthweight infant, was relatively high in energy, protein, calcium, phosphorus and minerals when compared to human milk. One formula contained 46% and the other contained 4% medium chain triglyceride made up from octanoic and decanoic acids. We conducted a randomized controlled cross-over clinical trial of 15 growing low birthweight infants of mean birthweight 1.45 kg and gestational ages from 26 to 34 weeks. Each infant was fed with Special Care Formula for at least three days, after which they received one of the experimental formulas for five days and the other for a further five days. Energy and excretions were measured over three day balance periods, and energy expenditure was measured by indirect calorimetry twice during each balance period. Growth was measured as gain in weight, length and head circumference.

There were no differences in coefficients of energy digestibility or metabolizability, or in nitrogen retention. There were no differences in rates of energy expenditure or of energy storage attributable to the use of medium chain triglyceride (figure 3). The relationship between metabolizable energy intake and energy storage for the group conforms with that described by our own and with other studies of energy balance (25), in that the increase in the rate of energy storage attributable to the increased rate of metabolizable energy intake is offset by a concomitant increase in energy expenditure. Weight gain was very rapid, typical of that shown by other workers for infants fed with these formulas. The energy accrued in that weight gain was, however, relatively low (11.7 kJ/g). When one considers absolute values for rates of gain in body fat, protein, and water, one finds that minerals and water account for much of the accelerated weight gain attributed to the high energy formula (figure 4).

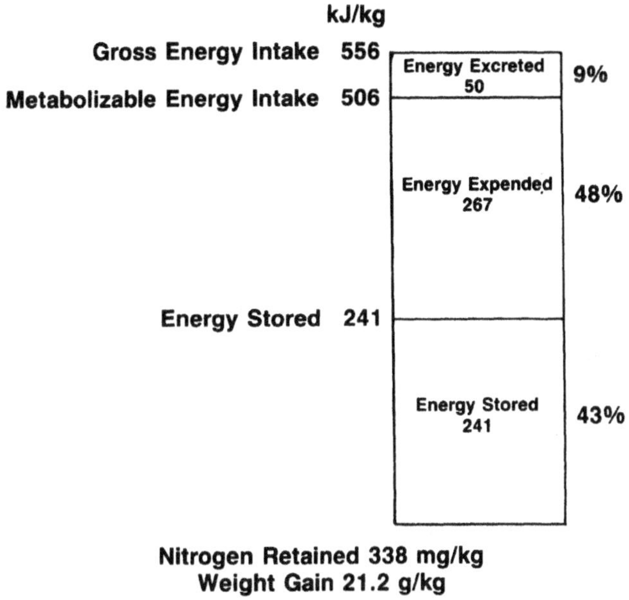

Figure 3. Daily energy balance in low birthweight infants fed a 3.5 mJ/L (24 kcal/oz) formula (24)

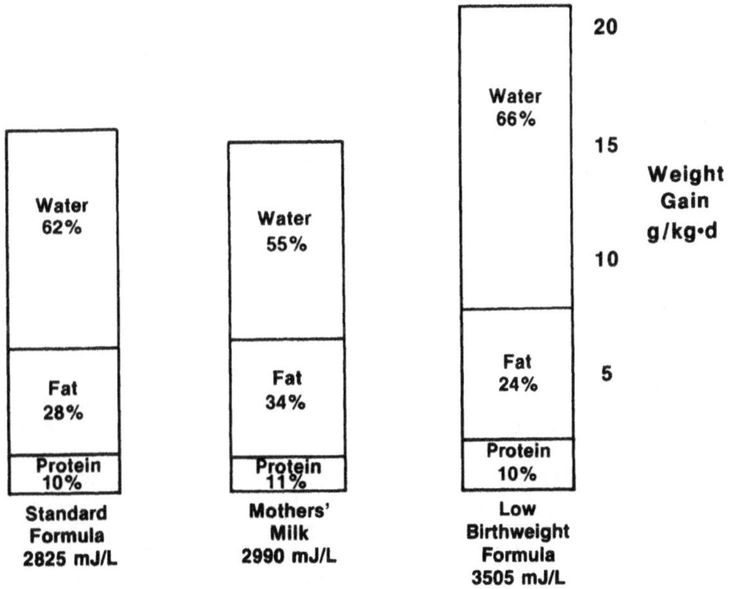

Figure 4. Composition of weight gain of low birthweight infants fed a 3.5 mJ/L (24 kcal/oz) formula (25)

Table 1. COMPOSITION OF WEIGHT GAIN IN ENERGY BALANCE STUDIES

Gross Intake per kg per day						Increment per kg per day				
Energy	Nit*	Na	K	Ca	P	Weight	Fat	Protein	Water	
kJ	mg	mmol				g				
Human milk										Ref #
530	418	2	3	1	1	15.2	5.1	1.6	8.5	(20)
464	484	2	3	1-4[a]	1	15.2	2.3	1.9	10.9	(27)
443	376[b]	3	2	1	2	14.6	4.0	1.6	9.0	(28)
Standard Formulas 2.9 - 3.5 mJ/L										
528	428	2	3	2	2	16.9	4.7	1.7	10.5	(20)
621	510	3	4	5	2	15.4	5.4	1.9	8.1	(26)
Special Preterm Formulas 3.5 mJ/L										
558	482	3	5	6	4	21.5	5.0	2.1	14.4	(24)
536	486	2	3	2	2	22.9	3.4	2.2	17.3	(28)

* Nitrogen.

[a] Only some of the infants were supplemented with calcium lactate.

[b] Human milk in this study was banked, which accounts for the relatively low nitrogen intake.

The results of various energy balance studies have been summarized in the table. It can be seen that most of the variation in weight gain achieved is accountable to non-fat non-protein weight gain. The consistencies with which rates of weight gain are reported for each of the categories of feedings is impressive. Infants fed with a "standard" formula at very high energy intakes (26) (which actually exceed those achieved with "special" formulas) experienced rates of weight gain which were smaller and closer to those of infants fed with expressed mothers' milk (20,27) or regular formula than to those experienced by infants on the preterm formulas (24,28). The reason for the rapid gains in weight so consistently reported with the formulas specially designed for use in preterm infants (4,5,6,24,28) are not solely related to energy intake, and may more reflect mineral and water accretion than gains in body fat or nitrogen.

In addition to showing a lack of expected benefits from the use of medium chain triglyceride on energy balance, we demonstrated an unexpected metabolic effect. Infants fed with medium chain triglyceride were shown to excrete some unusual metabolites of fatty acid metabolism in the urine, which included the ω and the ω-1 oxidized forms of the medium and short chain fatty acids (29). The quantities involved were too small to affect energy balance and the significance of this effect has yet to be determined.

In conclusion, there remains good reason to qualify observations of rates of weight gain with estimates of the composition of that gain, and our present methods, although imperfect, indicate that much of the rapid weight gain seen in infants fed with high energy formulas consists of minerals and water. There seems to be general agreement from all energy balance studies carried out that the composition of weight gain of the low birthweight infant is different from that of the fetal model and approaches that of the term newborn infant. The differences in the composition of weight gain attributed to human milk feeding as opposed to formula feeding are small, and without demonstrated consequence.

The substitution of medium chain for long chain triglyceride in infant feeding has no effect on energy balance or on the composition of weight gain of the growing low birthweight infant. Dietary fat must be thought of as more than just a fuel, and our evolving understanding of the effects of the fatty acid composition of the diet on the function and structure of growing tissue (30) may result in a more sophisticated approach to the formulation of feeds for growing low birthweight infants.

REFERENCES

1. H. H. Gordon, S. Z. Levine, and H. McNamara, Feeding of premature infants: a comparison of human and cow's milk, Am. J. Dis. Child 73:442-452 (1947).
2. D. P. Davies, Adequacy of expressed breast milk for early growth of preterm infants, Arch. Dis. Child 52:296-301 (1977).
3. S. A. Atkinson, H. Bryan, and G. H. Anderson, Human milk feeding in premature infants: protein, fat, and carbohydrate balances in the first two weeks of life, J. Pediatr. 99:617-624 (1981).
4. W. B. Omans, L. A. Barness, C. S. Rose, and P. Gyorgy, Prolonged feeding studies in premature infants, J. Pediatr. 59:951-957 (1961).
5. O. G. Brooke, C. Wood, and J. Barley, Energy balance, nitrogen balance, and growth in preterm infants fed expressed breast milk, a premature infant formula, and two low-solute adapted formulae, Arch. Dis. Child 57:898-904 (1982).
6. J. E. Tyson, R. E. Lasky, C. E. Mize, C. J. Richards, N. Blair-Smith, R. Whyte, and A. Beer, Growth, metabolic response, and development in very-low-birth-weight infants fed banked human milk or enriched formula 1. Neonatal findings, J. Pediatr. 103:95-104 (1983).
7. A. Lucas, S. M. Gore, T. J. Cole, M. F. Bamford, J. F. B. Dossetor, I. Barr, L. Dicarlo, S. Cork, and P. J. Lucas, Multicentre trial on feeding low birthweight infants: effects of diet on early growth, Arch. Dis. Child. 59:722-730 (1984).
8. B. M. Kagan, J. H. Hess, E. Lundeen, K. Shafer, J. B. Parker, C. Stigall, Feeding premature infants:a comparison of various milks, Pediatrics 15:373-382 (1955).
9. N. Raiha, K. Heinonen, D. Rassin, and G. E. Gaull, Milk protein quantity and quality in low-birthweight infants: 1. Metabolic responses and effects on growth, Pediatrics 57:659-674 (1976).
10. L. Novak, Total body potassium in the first year of life determined by whole-body counting of 40K, J. Nucl. Med. 14:550-557 (1973).
11. R. K. Whyte, H. S. Bayley, and H. P. Schwarz, The measurement of whole body water by H_2O^{18} dilution in newborn pigs, Am J. Clin. Nutr. 41:801-809 (1985).
12. F. L. Trowbridge, G. G. Graham, W. Wong, E. D. Mellits, J. D. Rabold, L. S. Lee, M. P. Cabrera, and P. Klein, Body water measurements in premature and older infants using H_2O^{18} isotopic determinations, Pediatr. Res. 18:524-527 (1984).

13. F. D. Moore, K. H. Otesen, J. D. McMuney, V. H. Parker, M. R. Ball, and B. M. Boyden, The body cell mass and its supporting environment, in: "Body Composition in Health and Disease," W. B. Saunders, Philadephia (1963).
14. S. J. Fomon, Comment on metabolic balance studies as a method of measuring body composition, Pediatrics 29:495-498 (1962).
15. A. L. Merrill, and B. K. Watt, Energy value of foods: basis and derivation, "Agricultural Handbook No. 74," United States Department of Agriculture (1973).
16. E. M. Widdowson and J. W. T. Dickerson, Chemical composition of the body, in: "Mineral Metabolism: An Advanced Treatise," C. L. Comar, F. Bronner, eds., Vol. IIA, Academic Press, New York (1964).
17. M. T. Clandinin, J. E. Chappell, T. Heim, P. R. Swyer, and G. W. Chance, Fatty acid utilization in perinatal de novo synthesis, Early Hum. Dev. 5:355-366 (1983).
18. E. E. Ziegler, A. M. O'Donnell, J. E. Nelson, and S. J. Fomon, Body composition of the reference fetus, Growth 40:329 (1976).
19. S. J. Fomon, Body composition of the male reference infant during the first year of life, Pediatrics 40:863-870 (1967).
20. R. K. Whyte, R. Haslam, H. S. Bayley, C. Vlainic, S. Shannon, K. Samulski, D. Campbell, H. S. Bayley, and J. C. Sinclair, Energy balance and nitrogen balance in growing low birthweight fed human milk or formula, Pediatr. Res. 17:891-898 (1983).
21. Ross Laboratories, Similac/Similac with iron, Product monograph, Ross laboratories, Columbus, Ohio (1978).
22. J. E. Chappell, M. T. Clandinin, C. Kearney-Volpe, B. Reichmann, and P. R. Swyer, Fatty acid balance studies in premature infants fed human milk or formula: effect of calcium supplementation, J. Pediatr. 108:439-447 (1986).
23. N. Baba, E. F. Bracco, and S. A. Hashim, Enhanced thermogenesis and diminished deposition of fat in response to overfeeding with diet containing medium chain triglyceride, Am. J Clin. Nutr. 35:678-682 (1982).
24. R. K. Whyte, R. Stanhope, D. Campbell, H. S. Bayley, and J. C. Sinclair, Energy balance of growing low birthweight infants fed with medium or long chain triglycerides, J. Pediatr. 108:964-971 (1986).
25. R. K. Whyte, H. S. Bayley, and J. C. Sinclair, Energy intake and the nature of growth in low birthweight infants, Can. J. Physiol. Pharm. 632:565-570 (1985).
26. B. L. Reichman, P. Chessex, G. Putet, G. J. E. Verrellen, G. Putet, J. M. Smith, T. Heim, and P. R. Swyer, Diet, fat accretion and growth in preterm infants, N. Eng. J. Med. 305:1495-1500 (1982).
27. P. Chessex, G. Reichman, G. Verellen, G. Putet, J. M. Smith, T. Heim, and P. R. Swyer, Quality of growth in premature infants fed their own mothers milk, J. Pediatr. 102:107-112 (1983).
28. G. Putet, J. Senterre, J. Rigo, and B. Salle, Nutrient balance, energy utilization, and composition of weight gain in very-low-birth-weight infants fed pooled human milk or a preterm formula, J. Pediatr. 105:79-85 (1984).
29. R. K. Whyte, D. Whelen, R. Hill, and S. McClorry, Excretion of dicarboxylic and -1 hydroxy fatty acids by low birthweight infants fed with medium-chain triglycerides, Pediatr. Res. 20:122-125 (1986).
30. M. T. Clandinin, C. J. Field, K. Hargreaves, L. Morson, and E. Zsigmond, Role of diet fat in subcellular structure and function, Can. J. Physiol. Pharmacol. 63:546-555 (1985).

THE EFFECT OF PROTEIN INTAKE ON COMPOSITION OF WEIGHT GAIN IN PRETERM INFANTS

G. Putet,[*] J. Rigo,[**] B. Salle,[*] and J. Senterre[**]

(*) Department of Neonatology, Hopital Edouard Herriot 69003 Lyon, France and (**) Department of Pediatrics, University of Liege, Belgium

INTRODUCTION

The comparison between composition of weight gain in the fetus at 32-36 weeks post-conceptional age (1) and weight gain composition of the term infant during the first months of life (2) indicates that the term infant stores much more fat and less water than the fetus. Moreover, published data on weight gain composition of very low birth weight (VLBW) infants fed either human milk (HM) (3-6) or formula (F) (3,4,6) demonstrate that these infants store more fat than the fetus of the same post-conceptional age. It can be assumed that a high fat storage is the normal adaptation to extrauterine life. Alternatively, this high fat storage might also result from the intake of certain nutrients which quantitatively and/or qualitatively may not be adapted to the infant's immature metabolic capacity. Body composition eventually achieved in the premature infant by 41 weeks will be very different from the body composition of a newly-born term infant.

Growth rate of a preterm infant can be increased by increasing his energy and protein intakes (7-11). However, high energy intakes have been associated with fat storage as opposed to increased lean body mass (3,12) and this is not an appropriate goal if one wants to mimic fetal growth. Although some investigators have shown that growth rate can be increased if a higher protein intake is given (13, 14), this has not been a consistent finding in other studies (15,16). The objective of our recent studies was to determine the influence of two levels of protein intake on weight gain, a question which has not been specifically studied in preterm infants. By using the combinated methodologies of nutrient balance and energy expenditure measurement over long periods of time, the estimation of weight gain composition has been facilitated (Fig. 1).

PROTEIN INTAKE AND WEIGHT GAIN COMPOSITION

The influence of protein intake on weight gain composition was studied in two groups of VLBW premature infants fed pooled pasteurised HM, without (HM group; n=8) or with (HM-Pr group; n=8) protein supplementation. The protein supplement was a casein hydrolysate produced by enzymatic hydrolysis of cow's milk casein, thus consisting mainly of peptides. This supplement contents very low amounts of fat (0.01 g/g of powder), only traces of lactose and minerals (per g of powder : 0.08 mg of sodium, 24 mg of calcium and 10 mg of phosphorus). In practice, 1 g of casein hydrolysate was mixed with 100 ml of pooled HM. Infants in both groups were born with similar gestational ages, birth weights, length and head circumference measurements (Table I). Anthropometric data were calculated from measurements taken 2 days before the beginning and two days after the end of the 3 day-nutritional balance. During the study period there was no significant difference in weight, length and head circumference gains between the two groups (Table I).

Table I : Clinical data and protein and energy balance

			HM (n=8)	HM-Pr (n=8)
ANTHROPOMETRIC DATA				
At birth	BW	(g)	1315 ± 122	1391 ± 216
	GA	(wks)	30 ± 1.4	29.9 ± 1.2
At study	W	(g)	1653 ± 267	1799 ± 109
Postnatal age		(d)	31 ± 10	33 ± 11
Weight gain		(g/kg/d)	15.3 ± 2.6	17.1 ± 22
Length gain		(cm/wk)	1.1 ± 0.3	1.2 ± 0.3
H.C. gain		(cm/wk)	1.0 ± 0.1	1.2 ± 0.3
PROTEIN AND ENERGY BALANCE (kg/d)				
Protein absorbed		(g)	2.1 ± 0.4	3.0 ± 0.4*
stored		(g)	1.6 ± 0.3	2.0 ± 0.2*
Energy absorbed		(kcal)	96 ± 8	91 ± 17 *
Energy expended		(kcal)	48 ± 5	58 ± 4 *
Non-protein energy stored		(kcal)	38 ± 8	23 ± 13 *

* student's t test : $p < 0.05$ (results as mean ± SD).

Weight gain composition was studied by the combined technique of indirect calorimetry and nutrient balance (3). The HM-Pr group absorbed 0.8 g/kg/d and stored 0.4 g/kg/d more protein than the HM group (Table I). Energy absorbed (metabolizable) was similar in both groups but energy expenditure was 18 % higher with protein-supplemented HM. As a result, non-protein energy storage was 42 % lower in the HM-Pr group. The comparison of weight gain composition between the two groups can be visualized in Fig. 2. From these data it appears that a small increase in protein intake resulted in a slightly increased weight gain, but a marked difference in weight gain composition between the two groups with much less fat being deposited with HM-Pr. The

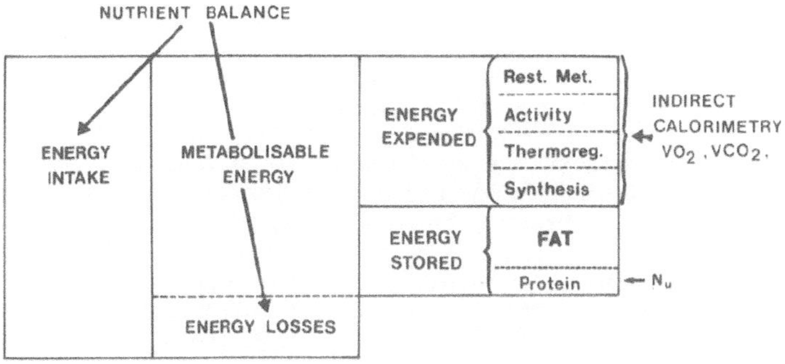

Figure 1 : Estimation of energy balance and composition of weight gain. Energy intake and losses are measured by nutritional balance. Thus, metabolizable (or absorbed) energy is known. By indirect calorimetry measurements over long periods of time, energy expended is assessed ; therefore energy stored can be estimated. Energy storage in a regularly growing premature infant consists mostly of fat and protein. As the amount of protein stored is assessed by the nitrogen balance (Nitrogen retained x 6.25), fat storage can be calculated (non protein energy storage divided by 9.3 kcal = g of fat storage).

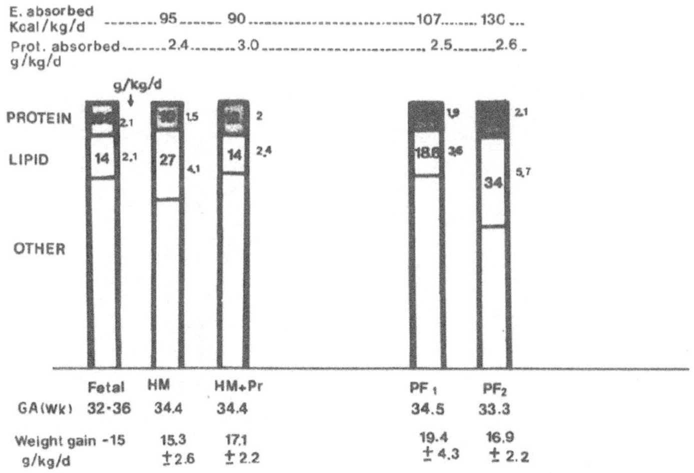

Figure 2 : Composition of weight gain (in percentage of total gain) in VLBW infants fed pooled human milk (HM), protein supplemented-HM (HM-Pr) and two preterm formulas PF_1 (18) and PF_2 (6). The amount of energy and calories absorbed, and weight gain are indicated for each study. For comparison, values for the fetus during the third trimester of gestation (1) are presented.

increased energy expenditure observed in the HM-Pr group (Table I) was due to the higher nitrogen retention and possibly an increased protein synthesis which accounts for a major portion of the metabolic rate.

Published data on nutritional balance and composition of weight gain in premature infants have been summarized in Table II. Most of these studies have been performed on premature infants of similar post conceptional age (32 to 33 weeks) and of similar weight (1200 to 1400g) at the time of the studies. They were fed either pooled human milk (HM) (3), their own's mother milk (4,5), or preterm formulas (F) (4,3,6). The amounts of energy absorbed (metabolizable energy) ranged from 87 to 130 kcal/kg/d and of protein absorbed from 2.1 to 2.8 g/kg/d. Anthropometric data reported in these studies (3-6) did not show significant differences in length or head circumference gains between feeding groups. Weight gains (Table II) do not appear to be associated with the amount of energy absorbed and are similar to those reported in our infants fed HM or HM-Pr (Table I).

Table II : Published data on protein and energy balance and weight gain composition of premature infants fed either human milk (HM) or formula (F)

Study (ref)	A (3)	B (5)	C (4)	D (4)	E (3)	F (6)
Type of milk	HM	HM	HM	F	F	F
AT STUDY (*)						
Weight (g)	1453	1272	1380*	1370*	1577	1271
G.A. (wks)	33.2	33.3	32.7	33.3	33.3	32.3
Weight gain (g/kg/d)	13.6	15.2	15.2	16.9	22.1	16.8
ENERGY (kcal/kg/d)						
absorbed	87	100	108	111	117	130
expended	46	56	53	58	58	63
PROTEIN ABSORBED (g/kg/d)	2.1	2.8	2.2	2.3	2.8	2.6
STORAGE (per kg/d)						
Fat (g)	3.4	3.4	5.0	4.7	5.1	5.7
%†	25	22	33	28	23	34
Prot. (g)	1.6	2.0	1.6	1.7	2.3	2.1
%†	12	13	11	10	10	12

*average weight calculated from reported data
†% of weight gain

Fat storage did not seem to be correlated with the nature of milk. In studies C (human milk) and D (cows milk) (Table II) where comparable amounts of energy and protein are absorbed, fat storage is similar. Indeed fat storage seems to be associated with energy intake when protein intakes are similar. This is well seen when one compares studies B, E, and F (Table II). For similar amounts of protein absorbed (between 2.6 to 2.8 g/kg/d) and retained (between 2.0 to 2.3), an increase in the amount of absorbed energy (100, 117 and 130 kcal/kg/d respectively) leads to increased fat storage (3.4, 5.1 and 5,7 g/kg/d respectively). This is also demonstrated in figure 2 with PF1 (18) and PF2 (6). Increase in energy absorbed raised energy storage without any advantage on weight gain.

From these previous investigations (3-6), it is difficult to evaluate the influence of different levels of protein intake on fat storage since energy intake and energy absorbed were not similar between the studies (Table II). Nevertheless, comparison between studies A and B shows that the increase in protein intake (2.1 vs 2.8 g/kg/d absorbed respectively) did not result in higher fat storage despite the concomitant increase in energy absorbed (87 vs 100 kcal/kg/d) and higher weight gain.

WEIGHT GAIN, PROTEIN INTAKE AND SUBCUTANEOUS FAT DEPOSITION

In a recent prospective study we have compared weight, length, head circumference and subcutaneous fat gains in two similar groups of VLBW infants fed HM or HM-Pr (17). Subcutaneous fat deposition was assessed by skinfold thickness measurements taken at 4 sites as described by Whitelaw (19). Gross energy intake was calculated to be 104 and 106 respectively and protein intake 2.4 and 3.6 g/kg/d. During the study period which lasted 2 weeks, the weight gain was significantly higher with HM-Pr than with HM : 17.7 $\pm$ 3.5 versus 13.5 $\pm$ 4 g/kg/d (this is different from our present data ; it is presumably due to a longer study period (two weeks versus one week) and a slightly higher (but not statistically higher) energy intake with HM-Pr than with HM). There was no difference in length, HC and skinfold thickness gain between the two groups. This indicates that an increase in weight gain was not associated with higher subcutaneous fat deposition. Similar results have been recently published by Kashyap et al (13) ; in their study, an increase in protein intake from 2.2 to 3.6 g/kg/d with no increase in energy intake (115 vs 114 kcal/kg/d) was followed by a significant increase in weight gain (13.9 vs 18.3 g/kg/d) without significant differences in length, head circumference or skinfold thickness gain.

WEIGHT GAIN, PROTEIN AND ENERGY INTAKE

Weight gain was similar in HM-Pr group (Table I) and in studies B, C, D, and E (Table II), whilst energy absorbed increased from 90 to 130 kcal/kg/d. It is evident that storing fat (i.e. 9.3 kcal/g of fat) does not lead to significantly increase weight gain but requires considerable amount of energy. There is probably an optimum energy/protein ratio (which is not yet well defined for a growing premature infant) above which an increase in energy intake results only in increased fat storage.

In conclusion, these data suggest that an increase in energy intake does not have a significant effect on length gain or head circumference gain when protein intake is kept constant ; weight gain can be moderatly increased, but this is mostly the consequence of a higher fat storage. An increase in protein intake can significantly increase the weight gain at the same level of energy intake and, furthermore, is probably followed by a decrease in fat storage, probably since more energy serves as fuel to support tissue synthesis. These studies have demonstrated that a weight gain composition similar to the fetus may be obtained during extra-uterine life. However this does not mean that this is the ideal weight gain composition for a VLBW infant. Clearly, if we want to evaluate correctly the quality of the growth in these infants, the composition of weight gain has to be assessed.

REFERENCES

1 . E. E. Ziegler, A. O'Donnel, S. E. Nelson, S. J. Fomon, Body composition of the reference fetus, Growth 40: 329 (1976).
2 . S. J. Fomon, Body composition of the male reference infant during the first year of life, Pediatrics 40:863-870 (1967).
3 . G. Putet, J. Senterre, J. Rigo, B. Salle, Nutrient balance, energy utilization and composition of weight gain in very-low-birth weight infants fed pooled human milk or a preterm formula, J Pediatr 105:79-85 (1984).
4 . R. K. Whyte, R. Haslam, C. Vlainic, Shannons, K. Samulski, D. Campbell, H. S. Bayley, J. C. Sinclair, Energy balance and nitrogen balance in growing low-birth-weight infants fed human milk or formula, Pediatr Res 17:891 (1983).
5 . P. Chessex, B. Reichman, G. Verellen, G. Putet, J. M. Smith, T. Heim, P. R. Swyer, Quality of growth in premature infants fed their own mother's milk, J Pediatr 102:107 (1983).
6 . B. Reichman, P. Chessex, G. Verellen, G. Putet, J. M. Smith, T. Heim, P. R. Swyer, Dietary composition and macronutrient storage in preterm infants, Pediatrics 72:322-328 (1983).
7 . A. Lucas, S. M. Core, T. J. Cole, M. F. Bamford, J. B. F. Dossetor, I. Barr, L. Dicarlo, S. Cork, P. J. Lucas, Multicentre trial on feeding low birth weight infants : effects of diet on early growth, Arch Dis Child 59:722-730 (1984).
8 . K. A. R Ronnholm, I. Sipila, M. A. Siimes, Human milk protein supplementation for the prevention of hypoproteinemia without metabolic imbalance in breast milk-fed very low-birth weight infants, J Pediatr 101:243-247 (1982).
9 . J. E. Tyson, R. E. Lasky, C. E. Mize, C. J. Richards, N. Blair-Smith, R. Whyte, A. E. Beer, Growth, metabolic response, and development in very-low birth weight infants fed banked human milk or enriched formula. I. Neonatal findings, J Pediatr 103:95-104 (1983).
10 . S. J. Gross, Growth and biochemical response of preterm infants fed human milk or modified infant formula, N Engl J Med 308:237-241 (1983).

11 . O. G. Brooke, C. Wood, J. Barley, Energy balance, nitrogen balance and growth in preterm infants fed expressed breast milk, a premature infant formula and two low-solute adapted formulas, Arch Dis Child 57:898-904 (1982).
12 . G. Moro, I. Minoli, J. Heininger, M. Cohen, G. Gaull, N. C. R. Raiha, Relationship between protein and energy in the feeding of preterm infants during the first month of life, Acta Paediatr Scand 73:49 (1984).
13 . S. Kashyap, M. Forsyth, C. Zucker, R. Ramakrishman, R. B. Dell, W.C. Heird, Effect of varying protein and energy intakes on growth and metabolic response in low birth weight infants, J Pediatr 108:955-963 (1986).
14 . S. G. Babson, J. L. Bramhall, Diet and growth in the premature infant : the effect of dietary intakes of·ash-electrolyte and protein on weight gain and linear growth, J Pediatr 74:890-900 (1969).
15 . N. C. R. Raiha, K. Heinonen, D. K. Rassin, G.E. Gaull, Milk protein quality and quantity in low birth weight infants. I. Metabolic responses and effects on growth, Pediatrics 57:659 (1976).
16 . N. W. Svenningsen, M. Lindroth, B. Lindquist, Growth in relation to protein intake of low birth weight infants, Early Hum Dev 6:47 (1982).
17 . G. Putet, N. Fahmy, J. Senterre, J. Rigo, B. Salle, Le lait de femme dans l'alimentation du prématuré, in : Progrès en Néonatologie 4, Minkowski et Relier ed., Karger publisher, pp. 25-32 (1984).
18 . G. Putet, A. L. Thelin, G. Philipossian, R. Liardon, M. Arnaud, J. Senterre, B. Salle, Medium chain triglycerides as a source of energy in premature infant, in : Lipids in Modern Nutrition. Raven Press (in press).
19 . A. Whitelaw, Subcutaneous fat measurement as an indication of nutrition of the fetus and newborn, in : Nutrition and metabolism of the fetus and infant, H. K. A. Visser ed., Martinus Nijhoff publisher. The Hague, Boston, London, pp. 131-143 (1979).

CHOLESTEROL IN HUMAN MILK

Robert G. Jensen

Department of Nutritional Sciences
University of Connecticut
Storrs, CT 06268

INTRODUCTION

Cholesterol occurs in human milk as a result of the nature of the lactation process (1,2). When the fat globule is extruded from the mammary cell, it is enveloped by membranes from the endoplasmic reticulum of the cell. The membranes contain cholesterol and are the source of most of the cholesterol in milk. About 15% of the cholesterol is esterified asthese compounds are nonpolar, they are found in the hydrophobic core of the fat globule. Cholesterol, relatively hydrophylic because of the free hydroxyl group, is found in the membrane enveloping the globule.

Cholesterol is the precursor of sex hormones, bile acids and vitamin D-3 and is found usually in the membranes of virtually every cell in the body. It is synthesized in the body in addition to the dietary source. High levels of cholesterol in plasma and hence in the diet have been associated with increased risk of atherosclerosis. It has been hypothesized that the "high" quantities of cholesterol in human milk when consumed by the infant might potentiate against the progression of atherosclerosis in adulthood. This is the cholesterol challenge hypothesis and is discussed by the Hamoshs in this volume (3). I will briefly describe this and other nutritional aspects, ie. does the infant need the cholesterol in human milk?

METHODS OF ANALYSIS

Methods which have been used for analyses of cholesterol in human milk are precipitation with digitonin and gravimetry, colorimetry, chromatographic, and enzymatic procedures. However, a useful colorimetric determination, employing o-phthalaldehyde, has been used to determine cholesterol in human milk (4). The milk is saponified directly and the cholesterol extracted with hexane for analysis. This is the best method currently available for routine analysis of cholesterol in human milk since the only instrument needed is a colorimeter. For research purposes, gas (5) and high performance liquid (6) chromatographic procedures are preferred since the traces of sterols other than cholesterol will be detected.

CONTENTS

A range of contents obtained by reliable methods is 10-20 mg of total cholesterol per dl. With a fat content of 4%, these amounts become 250 to 500 mg/100g of fat. Assuming a 24hr milk volume of 800 ml, the infant can consume 80 to 160 mg of cholesterol per day. The lower fat content and volume of colostrum is somewhat compensated by a greater amount of cholesterol per day (7). In contrast, most infant formulas do not contain cholesterol but will have varying amounts of phytosterols.

FACTORS AFFECTING CONTENTS

These are primarily stage-of-lactation and diurnal patterns, both related to changes in fat content and basically, the total surface area of the fat globules. Bitman (7) observed a continual decline in cholesterol from 1.3% of total lipids at day 3 postpartum to 0.4 at day 84 in milks from term and preterm mothers. This was accompanied by an increase in both volume and fat content of the milk. Increases in cholesterol content during the day are due to rises in the fat content (8).

Claims that diet influences the cholesterol content of milk are not supported by any published data (9). Reports that the phytosterol content of milk is raised by consumption of more of these sterols in the diet have not been confirmed (6,10,11). It appears that the identifications were wrong, and, without doubt, sampling was uncontrolled (10). Traces of desmosterol and other sterols have been detected (6,11).

Another potential, and as yet unconfirmed source of variation, is the various lotions that many women apply to their nipples during the time they are breastfeeding. Some of these contain cholesterol, lanosterol and many other lipids and could be the cause of erroneous results. A lump of ointment containing 10 mg of cholesterol entrained into a 50 ml sample of milk, would add 20 mg/dl of the sterol to the milk when analyzed. Investigators should take the obvious measures to exclude those contaminants.

NUTRITIONAL ASPECTS

Cholesterol is a ubiquitous component in almost all cells in the body where it is found mostly in the membranes. The sterol is the precursor of bile acids, sex hormones and vitamin D-3; the latter occuring in skin as a result of exposure to ultraviolet radiation in sunlight. It is also involved in the myelinization of cells in the developing brain and nervous system. In membranes, it provides rigidity in association with phospholipids. The major questions to be considered are can the infant synthesize sufficient cholesterol to provide for its requirements and does dietary cholesterol at this stage in life potentiate the body to process cholesterol more efficiently later? Implicit in the last question is the possible relationship of the exposure of infants to dietary cholesterol and the development of atherosclerosis.

In an attempt to answer these questions, we must know how much cholesterol is consumed by the nursing infant. In colostrum, the total amount is low because of the small volume of milk available although the cholesterol content is relatively high. Even though we do not have data on the cholesterol contents of pooled mature 24 hr milks, we can approximate the quantity of cholesterol in these samples. If we arbitrarily select an amount of 400 mg cholesterol/100g fat and use 800 ml of milk x 4.5% fat we have 36g of fat and 144 mg of cholesterol (1,2,9). This figure is high because the "average" fat content of human milk is 3.0-4.0%.

Both fat content and volume can be much lower. The quantities of cholesterol consumed could then range from 50 to 160 mg/day. The amount in 800 ml of bovine milk will be uniformly about 100 mg because of pooling and most formulas will have very little or none because they are prepared from vegetable oils, in the US usually coconut and soybean. Obviously, many infants have received almost no cholesterol when they were fed only these formulas.

CHOLESTEROL METABOLISM

The sterol from exogenous or endogenous sources is transported in blood mostly as esters in lipoproteins, primarily the low density lipoproteins (LDL) (12,13). The LDL is carried in the blood to specific receptors on cells and incorporated into the cells where after a series of metabolic transformations the endogenous cholesterol is available for the cell's needs. Negative feedback content is exerted on endogenous synthesis of cholesterol via the rate controlling enzyme, 3-hydroxy-3-methyl glutaryl CoA reductase. The number and efficiency of the LDL receptors are mostly genetically controlled, but excess intracellular cholesterol decreases their rate of synthesis (12). Interestingly, a portion of the amino acid sequence of the LDL receptor is identical to that of the epidermal growth factor also found in human milk (12,14).

Much of the LDL cholesterol is transported to the liver where a portion of it is converted to bile acids primarily by the action of 7-alpha-hydroxylase. The bile acids are conjugated in the liver mostly with taurine in breast fed babies and glycine in those fed formula. In many infants, both term and preterm, the pool of bile acids available for efficient absorption of dietary fats is minimal or even below the critical micelle concentration (4mM) required for absorption. The amount of cholesterol needed for synthesis of bile acids is small, since the acids are used repeatedly via the enterohepatic circulation. Watkins et al (15) observed a synthesis rate of 5.1 mg/kg/day of total cholates in infants born preterm who weighed about 2400 gms. Even the small amount of cholesterol in colostrum is more than sufficient for these needs, not considering plasma cholesterol. The authors found that neither cholesterol nor taurine were limiting for bile acids. They also noted that the bile acid pool and intraluminal bile acid concentrations were significantly greater in infants fed human milk at all ages as compared to formula even when the latter was supplemented with cholesterol and taurine. This was attributed in part to the presence in human milk of unique growth and maturation factors.

The composition and profile of the newborn infants' lipoproteins differs from that in the adult (16). By day 30, the pattern begins to resemble the adult distribution, but there are still major differences in the apoproteins which should be investigated. Accompanying these changes was an increase in serum cholesterol of about 40 (newborn) to 130 mg/dl at 112 days (15). Many of the children in the last investigation were tested again at 8 years. Their serum cholesterol contents were about 160 mg/dl with no difference between those who had been breast and bottle fed as infants. The former received up to 100 mg cholesterol and latter, very little during infancy. These and similar data discussed by the Hamosh's in this volume (3) appear to discredit if not disprove the cholesterol challenge hypothesis. Serum cholesterol, however, was about 25 mg/dl higher in the breastfed infants out to day 112 than in those fed formula. It may be that the extra cholesterol in human milk is simply additive since the 100 mg/day consumed by a solely breastfed infant can be accounted for by the 25 mg/dl difference and the increase in blood volume as the baby grows.

The above observation and the suggestion by Brown and Goldstein (11) that the needs of the cells can be met for an adult by 25 to 60 mg of plasma LDL-cholesterol indicate that even the newborn infant synthesizes sufficient cholesterol. LDL-cholesterol in the infant is about 30 mg. Furthermore, an infant at 30 days would consume about 100 mg of milk cholesterol daily which is a small portion of the amount of circulating cholesterol, 136 mg/dl serum. The extra milk cholesterol may be superfluous, unless it is utilized for growth and development of the infant's brain and nervous system.

SUMMARY

The cholesterol content of human milk ranges from about 10 to 20 mg/dl with an approximate average daily consumption by the infant of 100 mg. The content in single samples is usually correlated with the fat content or more likely the surface area of the fat globules since most of the cholesterol is found in the globule membranes. The amount of cholesterol drops as lactation progresses, even though the fat content may rise. About 15% of the total sterol is esterified. Milk lipids also contain some desmosterol and traces of other sterols. The cholesterol content of milk is not altered by diet. Determinations of milk sterols for research purposes are best done by gas-liquid chromatography. Colorimetric analysis with o-phthalaldehyde is suitable for routine testing. Plasma cholesterol in a newborn infant is about 40 mg/dl, rising at day 30 to 136 mg/dl. There is no evidence that consumption of breast milk as an infant provides for more efficient metabolism of the sterol as an adult or that endogenous synthesis is inadequate for the infant's requirements.

ACKNOWLEDGMENTS

Some of the research reported in this article was supported by in part by Federal funds made available through provision of the Hatch Act and by NIH Contract N01-HD-2817. Scientific Contribution No. 1186, Storrs Agricultural Experiment Station, Storrs, CT., 06268.

REFERENCES

1. C.J. Lammi-Keefe, and R.G. Jensen. Lipids in human milk: A review: 2: Composition and fat-soluble vitamins. J. Pediatr. Gastroenterol. Nutr., 3:172 (1984).
2. R.G. Jensen, M.M. Hagerty, and K.E. McMahon. Lipids of human milk and infant formulas. Am. J. Clin. Nutr. 31:990 (1978).
3. M. Hamosh and P. Hamosh. Does nutrition in early life have long term metabolic effects? Can animal models be used to predict these effects in the human?, in Human Lactation: The Effects of Human Milk on the Recipient Infant. A.S. Goldman, S.A. Atkinson, and L.A. Hanson, eds. Plenum Press, New York (1985).
4. G.E. Huston, and S. Patton. Membrane distribution in human milks as revealed by phospholipid and cholesterol analyses, in Human Lactation: Milk Components and Methodologies, R.G. Jensen and M.C. Neville, eds. Plenum Press, New York (1985).
5. R.M. Clark, A.M. Ferris, M. Fey, P.B. Brown, K.E. Hundrieser, and R.G. Jensen. Changes in the lipids of human milk from 2 to 16 weeks postpartum. J. Pediatr. Gastroenterol. Nutr. 1:311 (1982).
6. M. Haug, and G. Harzer. Cholesterol and other sterols in human milk. J. Pediatr. Gastroenterol. Nutr. 3:816 (1984).
7. J. Bitman, D.L. Wood, M. Hamosh, P. Hamosh, and N.R. Mehta. Comparison of the lipid composition of breast milk from mothers of term and preterm infants. Am. J. Clin. Nutr. 38:300 (1983).

8. G. Harzer, and M. Haug. Correlation of human milk vitamin E with different lipids, in: Composition and Physiological Properties of Human Milk, J. Schaub, ed., Elsevier Science Publ., New York (1985).
9. R.G. Jensen. Effect of diet on the lipid composition of human milk, in: Human Lactation 2: Maternal and Environmental Factors, M. Hamosh and A.S. Goldman, eds. Plenum Press, New York (1986).
10. M.J. Mellies, T.T. Ishikawa, P. Gartside, K. Burton, J. MacGee, K. Allen, P.M. Steiner, D. Brady and C.J. Glueck. Effects of varying dietary cholesterol and phytosterol in lactating women and their infants. Am. J. Clin. Nutr. 31:1347 (1978).
11. R.M. Clark, M.B. Fey, R.G. Jensen, and D.W. Hill. Desmosterol in human milk. Lipids. 18:264 (1983).
12. M.S. Brown and J.L. Goldstein. A receptor-mediated pathway for cholesterol homeostasis, Science, 232:34 (1986).
13. Grundy, S.M. Cholesterol and coronary artery disease. J. Am. Med. Assoc. 256:2849 (1986).
14. G. Carpenter. Epidermal growth factor is a major growth promoting agent in human milk. Science, 210:198 (1980).
15. J.B. Watkins, A-L Jarvenpaa, P. Szczpanik-Van Leenwen, P.D. Klein, D.K. Rassin, G. Gaull, and N.C.R. Raiha. Feeding the low birth weight infant: V. Effects of taurine, cholesterol and human milk on bile acid kinetics, Gastroenterology 85:793 (1983).
16. Anonymous. The evolution of serum lipoproteins in infancy. Nutr. Rev. 44:324 (1986).

GASTRIC LIPOLYSIS AND FATTY ACID UTILIZATION IN PRETERM INFANTS

Joel Bitman, Teresa H. Liao, Margit Hamosh, N.R. Mehta, R.J. Buczek, D.L. Wood, L.J. Grylack, and P. Hamosh

U.S. Department of Agriculture, Agricultural Research Service, Beltsville, Maryland 20705
Georgetown University Medical Center, Washington, D.C.

INTRODUCTION

Development of the newborn infant is greatly dependent upon the capability to utilize dietary fat. In human milk and infant formulas, fats provide 50% of the calories. Fat utilization is dependent upon 1) digestion of the dietary lipids by lipases present in the gastro-intestinal tract, and 2) absorption of the hydrolyzed lipid products facilitated by bile salts (1). In the newborn, and especially in the premature infant, because of insufficient maturation, pancreatic lipase activity and bile salt levels are low (2). Consequently intragastric lipolysis by lingual and gastric (3) lipases may play an important role in fat digestion. This is especially critical for the premature or low-birth-weight infant.

Medium chain triglycerides (MCT, containing 8:0 and 10:0 fatty acids) are often used in the nutritional management of premature infants (4-6). Although better fat absorption of MCT has been reported, the benefit of MCT feeding has been questioned recently (7). Most of the previous studies have compared fat absorption and weight gain in separate groups of infants fed either MCT or long-chain triglyceride (LCT) formulas.

The aim of this study was to assess the extent of gastric lipolysis, fat absorption and weight gain in premature infants fed MCT and LCT formulas. In order to minimize inter-infant variability, each infant was his/her own control and was fed each formula (MCT or LCT) for one week.

MATERIALS AND METHODS

SUBJECTS Eight infants were entered into the study when enteral food intake was 80-100 kcal/kg/day. None of the infants had gastrointestinal or other illness at the time of the study. All infants were appropriate for gestational age (AGA), assessed according to Dubowitz et al. (8).

EXPERIMENTAL DESIGN The infants were randomly assigned to start with either MCT or LCT formula. Each formula was fed for one week. The following study protocol was followed on each of the two study weeks: Day 1, body weight; Day 2, fasting gastric aspirate (taken before the 10

TABLE 1. Clinical Data on Infants

Gestational age: wks	28.7 ± 0.6 (26-31)
Birth weight: g	1021 ± 101 (640-1450)
Age at study: wks	6.32 ± 1.22 (2.6-12.4)
Weight at start of study: g	1298 ± 75 (920-1630)

The data are means ± SEM and ranges (in parentheses) of the 8 infants studied

a.m. feeding, usually 3-4 hours after the previous meal); Day 3, gastric aspirates were taken during (about 8 min) and after (about 15 min after starting) the 10 a.m. feeding. During the last 3 days of each regimen fat balance studies were conducted as described previously (9). Stools were collected during a 72 hour period bracketed by charcoal marker. All infants were fed by gastric gavage throughout the study. Parental consent was obtained before enrolling the infants into the study. The experimental protocol was approved by the Committee for Research on Human Subjects of Georgetown University Hospital.

FORMULAS Two infant formulas provided by Mead Johnson & Co. (Evansville, IN) were used. Both formulas had caloric and fat content similar to the infant formula Enfamil available on the market, i.e., 88.3 Kcal/100 ml and 4.4% fat. One formula had its fat component made up entirely of LCT and was Enfamil Premature Formula 3242 containing coconut oil:soybean oil:lecithin, 54:44:2. The other formula had 39% of its lipid supplied by MCT and was Enfamil Premature Formula 3242H containing MCT:corn oil:coconut oil:other fat sources, 39:39:18:4.

TABLE 2. Fatty acid composition of Medium-Chain (MCT) and Long-chain Triglyceride Formulas (weight percent)

Fatty acid	MCT	LCT
8:0	30.3	3.8
10:0	11.8	3.1
12:0	9.3	25.7
14:0	3.7	10.2
16:0	6.5	10.3
18:0	2.0	3.8
18:1	11.6	14.2
18:2	22.0	25.0
20:0	0.5	0.4
18:3	1.2	3.2
Others[a]	1.2	0.8

[a]Others = 6:0, 14:1, 15:0, 16:1, 17:0

MCT = Enfamil premature formula 3242H containing MCT:corn oil: coconut oil:other fat sources, 39:39:18:4
LCT = Enfamil premature formula 3242 containing coconut oil: soybean oil:lecithin, 54:44:2
Each 100 ml contain 88.3 kcal, 2.65g protein, 9.71g carbohydrate and 4.42g fat

ANALYTICAL TECHNIQUES

Lipase assay Lipase activity in gastric aspirates was quantitated by the hydrolysis of tri-^{3}H-olein, as described previously (10). The assay conditions are optimal for lingual and gastric lipase activity.

LIPID ANALYSES

Gastric Aspirates

Lipids from gastric aspirates and infant formulas were homogenized and extracted in 2:1 chloroform/methanol (11). Intragastric hydrolysis was quantified by measuring amounts of partial glycerides and FFA produced (12). Neutral lipid classes were separated by preparative thin-layer chromatography (13). Separated neutral lipid zones were scraped into tubes and methyl esters formed with boron trifluoride complex lipid mix. Fatty acid methyl esters were analyzed by temperature-programmed gas-liquid chromatography (GLC) on 10% SP-2340 on Chromosorb WAW (Supelco Inc., Bellefonte, PA) in a 6 ft x 2 mm i.d. glass column on a model 5840 Hewlett-Packard gas chromatograph (Hewlett-Packard Co., Palo Alto, CA) using conditions described earlier (13). All lipid standards were purchased from Nuchek Prep, Inc., Elysian, MN.

Fecal Fat

Total fecal lipids were extracted and determined gravimetrically according to the methods of Jeejeebhoy et al. (14).

Statistical Analysis

Statistical comparisons were made by analysis of variance. Fisher's Protected least significant difference test (applied only after a significant analysis of variance) was used for mean separation (15).

RESULTS AND DISCUSSION

Clinical data on the infants, gestational age and birth weight, as well as the body weight and the postnatal age at the beginning of the two week study period are given in Table 1.

The composition of the MCT and LCT diets is shown in Table 2. The MCT formula differed greatly from the LCT: 42% of total fat consisted of 8:0 to 10:0 fatty acids vs only 7% in the LCT. Conversely, the LCT formula contained over twice as much saturated long-chain FA, 50% (LCT) vs 22% (MCT). Monounsaturated FA (14:1, 16:1, 18:1) were similar in the two formulas, being 12% (MCT) vs 14% (LCT), as were polyunsaturated FA, 23% (MCT) vs 28% (LCT).

Fat Hydrolysis in the Stomach

Infants were fed MCT or LCT diets in a cross-over design in which each infant served as his own control. The quantity and the distribution of lipids in the gastric contents sampled at 8 min (mid) and 15 min (end) of the meal were very similar on either diet (Table 3 and Figure 1). No differences in patterns of the lipids were observed between the MID and END samples, indicating very rapid lipolysis in the stomach. Triglycerides (TG) were 49%, diglycerides (DG) 12%, monoglycerides (MG) 3% and free fatty acids (FFA) 35% of total lipid.

Detailed examination of the FA composition of the gastric aspirates revealed wide differences, further substantiating the observation that

TABLE 3. The effect of feeding MCT or LCT diets on intragastric lipolysis.

Formula	Percent of total lipid present in: TG	DG	MG	FFA
MCT				
Mid-feed	45.17 ± 6.37	10.85 ± 1.13	3.22 ± 0.59	40.76 ± 6.59
End of feed	48.72 ± 7.26	11.08 ± 1.83	2.79 ± 0.85	37.41 ± 7.31
Formula	86.69	10.39	1.47	1.47
LCT				
Mid-feed	50.42 ± 5.41	12.66 ± 1.48	3.39 ± 0.28	33.52 ± 4.92
End of feed	52.43 ± 4.30	15.19 ± 2.92	3.19 ± 0.42	29.19 ± 4.53
Formula	85.53	8.30	4.39	1.98
Mean, MCT and LCT	49.19	12.45	3.15	35.22
Mean SE	5.84	1.84	0.54	5.84

Values represent mean of 8 infants

TG = triglycerides, DG = diglycerides, MG = monoglycerides, FFA = free fatty acids

the gastric TG had been attacked by lipases (Table 4). Thus, 8:0 and 10:0 were reduced in gastric triglycerides as compared to formula TG (Table 2), indicating preferential hydrolysis. Whereas the triglycerides of the MCT-formula contained 30% 8:0 and 12% 10:0 (Table 2), gastric triglycerides contained only 6% 8:0 and 6% 10:0 (Table 4). Similarly, the triglycerides of the LCT-formula contained 3.8% 8:0 and 3.1% 10:0, whereas gastric triglycerides contained only 0.8% 8:0 and 1.7% 10:0. Relating the proportions of 8:0 and 10:0 in the gastric aspirates to those present in the formulas fed revealed that 82% and 79% of the 8:0 present in MCT- and LCT-diets, respectively, were hydrolyzed. For 10:0, 52% of this fatty acid was hydrolyzed from the triglycerides in the MCT-formula and 45% of that present in LCT. These figures suggested that 8:0 and 10:0 fatty acids were hydrolyzed to the same extent from triglycerides in both MCT- and LCT-formulas.

TABLE 4. Comparison of the fatty acid composition of gastric lipolytic products of infants fed MCT or LCT diets (weight percent)

Fatty acid	TG MCT	TG LCT	DG MCT	DG LCT	MG MCT	MG LCT	FFA MCT	FFA LCT
8:0	5.6	0.8*	14.8	0.3*	13.4	0.1*	11.5	3.4
10:0	5.7	1.7*	5.4	0.5	6.2	0.1*	16.8	2.7*
12:0	13.1	19.7*	8.8	11.6	6.6	7.1	8.4	11.6
14:0	6.6	11.0*	8.1	13.6*	6.7	11.0*	4.0	5.5
16:0	11.5	12.3	12.6	15.0	13.4	18.1*	11.0	9.7
18:0	3.8	4.5	5.0	6.3	6.7	7.8	6.5	4.3
18:1	17.9	16.3	15.6	17.6	16.0	19.1	15.1	21.2*
18:2	32.7	28.8	23.1	27.7	21.9	27.5	20.2	34.0*
20:0	0.5	0.6	0.9	1.1	1.9	2.0	1.1	0.8
18:3	1.2	3.3	2.5	3.6	5.0	3.9	3.6	4.4

*significantly different, $p<0.05$

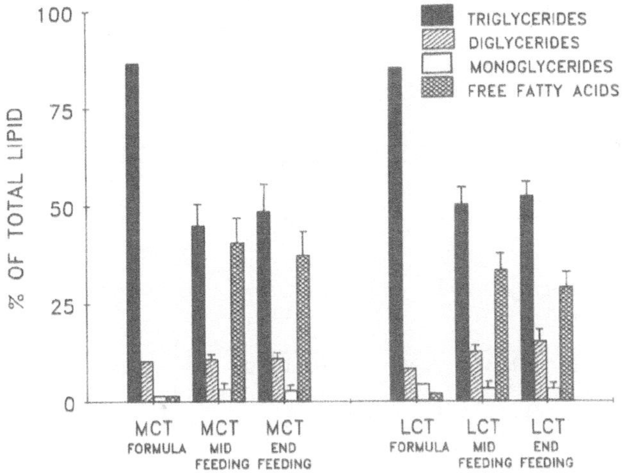

Fig. 1. Hydrolysis of MCT and LCT in the stomach of premature infants. Products of intragastric lipolysis at the middle and end of feeding.

Recalculation of the percentages of FA present in the gastric triglycerides, assuming that only 8:0 and 10:0 were hydrolyzed, yielded values very close to the composition of the MCT- and LCT-formulas that were fed (Table 2). Percentages of 12:0 and longer-chain fatty acids in gastric triglycerides from either MCT or LCT feeding were very similar to those in the formulas fed, indicating minimal hydrolysis of fatty acids from the triglycerides remaining in the stomach. We have interpreted this as very strong evidence that mainly 8:0 and 10:0 were hydrolyzed from the triglycerides in the stomach, supporting the strong preferential role of lipases for the medium-chain fatty acids. Comparison of partial glycerides and free fatty acids (FFA) also demonstrated preferential lipolysis of medium-chain fatty acids (Table 4). The low content of 8:0 in FFA (Table 4) suggested that this acid was directly absorbed in the stomach. These results, indicating preferential and almost exclusive lipolysis of medium-chain fatty acids from the dietary triglycerides, are in close agreement with results we reported earlier for intragastric lipolysis of milk lipids in suckling rats (16).

Lipase Activity in Gastric Aspirates

Lipase activity in gastric aspirates of infants fed MCT or LCT was also quantified (Table 5). The data show that lipase activity in the aqueous phase of the gastric aspirates decreases during feeding, suggesting that the enzyme binds rapidly to the ingested fat. Moreover, the lower lipase levels during MCT feeding as compared to LCT feeding,

TABLE 5. Lipase Activity* in Gastric Aspirates** of Premature Infants during MCT or LCT Feeding

Specimen		Lipase Activity nmol FFA released/ml aspirate/min
Before feeding		180 ± 38
MCT	Mid-feeding***	14.2 ± 9.8
	End-feeding	23.0 ± 11.6
LCT	Mid-feeding	63.3 ± 21.1
	End-feeding	154.3 ± 111.0

* Lipase (lingual and gastric) activity quantitated in vitro by the hydrolysis of tri-^{3}H-olein at pH 5.4.

** Gastric aspirates were taken during the days preceeding the three day fat balance study, while each infant was fed MCT or LCT formula.

*** Mid- and End-feeding specimens were obtained 8 and 15 minutes, respectively, after starting the feed.

further confirm the higher affinity of lingual (10) and gastric (3) lipases for MCT.

Fat Absorption and Weight Gain

The results of the lipid balance study are shown in Table 6 and Fig. 2. The data presented in Fig. 2 contain values for 4 additional infants. Because gastric lipolysis was not studied in these infants, only the data on fat absorption are presented. Neither the type nor the sequence of feeding of the formulas affected the absorption of fat. In general, the infants showed very close values in fat absorption and in weight gain when fed either type of fat. Lipid balance indicated that about 80% of either type of dietary fat was absorbed by the infants. Similarly, weight gain during the study period was very similar (Table 6). There was, however, considerable variation in fat absorption among infants (Fig. 2). Our data suggest that fat absorption is related to the degree of gastrointestinal maturation and is not affected to a great extent by the type of dietary fat. Indeed, in 9 of the 12 infants studied, fat absorption was between 85 and 97%. Only one infant, studied at 31.7 weeks post conception, had much higher fat absorption when fed MCT rather than LCT. Furthermore, above 35 weeks post conception, fat absorption during LCT feeding exceeded or was equal to that observed during MCT feeding.

The remainder of the clinical trial of the present study demonstrates that differences in the type of FA in the formulas fed, and consequent differences in FA of the lipolytic products in the stomach, were without effect on lipid absorption and weight gain in the premature infants. Although in vitro hydrolysis of medium-chain triglycerides by gastric aspirates proceeds at higher rates than long-chain triglycerides favoring increased availability of medium-chain FA, overall fat balance in the in vivo clinical trial was not affected. Indeed, analysis of the FA composition of the lipolytic products in the stomach indicated that, in vivo, the extent of lipolysis of the medium-chain FA from the formulas was the same in infants fed either medium-chain or long-chain triglycerides.

TABLE 6. Fat Absorption and Growth Rate in Preterm Infants fed MCT or LCT Formula

Formula	Infants N	Dietary Fat* g/kg/day	Fat Excret.* g/kg/day	Fat absorpt.* %	Wt gain** g/day
MCT	8	4.29 ± 0.24 (3.36-5.32)	0.80 ± 0.20 (0.10-1.46)	80.88 ± 4.94 (60.12-97.84)	22.39 ± 2.39 (15.71-34.28)
LCT	8	4.21 ± 0.24 (3.26-5.26)	0.75 ± 0.23 (0.13-2.08)	81.31 ± 5.90 (49.39-97.53)	22.96 ± 2.54 (11.42-32.85)

The data are means ± SEM and ranges (in parentheses)

* Fat excretion and absorption during a three day balance study

** Weight gain during one week on either MCT or LCT formula

*** Each infant was fed each formula for one week. Infants were randoml assigned to start with either formula

Intragastric lipolysis probably stimulates fat digestion in the intestine (17). Fatty acids produced in the stomach stimulate the secretion of cholecystokinin, which in turn stimulates biliary and pancreatic secretion. Furthermore, upon reaching the duodenum, free fatty acids accelerate the binding of pancreatic lipase-colipase to fat droplets, thereby enhancing intestinal fat digestion (17,18). In diseases associated with exocrine pancreatic insufficiency (similar to the physiologic pancreatic insufficiency of the newborn), gastric and lingual lipases remain active in the upper small intestine (18,19).

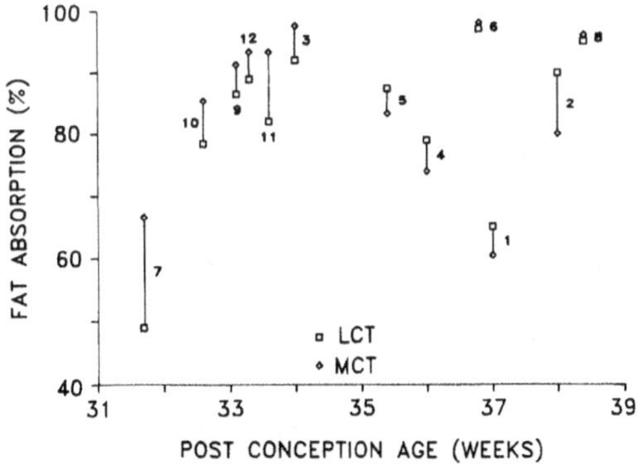

Fig. 2. Comparison of fat absorption in premature infants fed alternately MCT ◇ or LCT □ formula. Numbers alongside symbols designate individual infant numbers.

This study is the first to show that medium-chain fatty acids are absorbed in the stomach of the newborn infant. Thus, while MCT do not seem to improve fat absorption or weight gain (7,20,21), they provide a readily available energy source that can be absorbed much more rapidly than the other energy sources available to the newborn.

SUMMARY

The present studies were designed to determine the extent of gastric lipolysis, fat absorption and infant weight gain in preterm infants fed either medium-chain triglyceride (MCT) or long-chain triglyceride (LCT) formula. The data showed that: 1) There was extensive hydrolysis of formula fat in the stomach during feeding of both MCT or LCT; 2) Lipase activity in the aqueous phase of gastric aspirates was lower during feeding than before the meal, suggesting rapid binding to ingested lipid, especially to MCT; 3) Detailed analysis of fatty acid distribution in glycerides and free fatty acids showed that 8:0 and 10:0 were reduced in gastric triglyceride compared to formula triglyceride, demonstrating preferential lipolysis of medium-chain fatty acids. The low content of 8:0 and 10:0 in gastric free fatty acids indicated that these fatty acids were directly absorbed in the stomach. There was minimal hydrolysis of longer chain (>12 carbon atoms) fatty acids from either MCT or LCT in the stomach; 4) Fat balance studies showed almost identical absorption rates (80.9 ± 4.9 and 81.3 ± 5.9) and weight gain (22.4 ± 2.3 and 22.9 ± 2.5) during feeding of either MCT or LCT. Our study, in which each infant was fed alternately both formulas, clearly shows that although the extent of fat digestion varies among infants, MCT and LCT are absorbed to the same extent by most infants.

ACKNOWLEDGMENT

This study was supported by the National Institutes of Health Grant HD-10823 and by the Mead Johnson Nutrition Division.

REFERENCES

1. Watkins JB. Mechanism of fat absorption and the development of gastrointestinal function. Pediatr Clin N Am 22:721-730 (1975)
2. Hamosh M. A Review, Fat digestion in the newborn: role of lingual lipase and preduodenal digestion. Pediatr Res 13:615-622 (1975)
3. DeNigris SJ, Hamosh M, Kasbekar DK, Fink CK, Lee TC and Hamosh P. Human gastric lipases: secretion from dispersed gastric glands. Biochim Biophys Acta 836:67-72 (1985)
4. Roy CC, Ste-Marie M, Chartrand L, Weber A, Bard H, Doray B. Correction of the malabsorption of the preterm infants with a medium-chain triglyceride formula. J Pediatr 86:446-450 (1975)
5. Tantibhendyangkul P, Hashim SA. Medium-chain triglyceride feeding in premature infants. Effects on fat and nitrogen absorption. Pediatrics 55:359-370 (1975)
6. Shenai PJ, Reynolds JW, Babson SG. Nutritional balance studies in very-low-birth-weight infants: Enhanced nutrient retention rates by an experimental formula. Pediatrics 66:233-238 (1980)
7. Okamoto E, Muttart CR, Zucker CL, Heird WC. Use of medium-chain triglycerides in feeding the low birth weight infant. Am J Dis Child 136:428-431 (1982)
8. Dubowitz LM, Dubowitz V, Goldberg C. Clinical assessment of gestational age in the newborn infant. J Pediatr 77:1-10 (1970)

9. Alemi B, Hamosh M, Scanlon JW, Salzman-Mann C, Hamosh P. Fat digestion in very low birth weight infants: effect of addition of human milk to LBW formula. Pediatrics 68:484-489 (1981)
10. Liao TH, Hamosh P, Hamosh M. Fat digestion by lingual lipase: Mechanism of lipolysis in the stomach and upper small intestine. Pediatr Res 18:402-409 (1984)
11. Folch J, Lees M, Sloane-Stanley GH. A simple method for the isolation and purification of total lipids from animal tissues. J Biol Chem 226:497-509 (1957)
12. Liao TH, Hamosh P, Hamosh M. Gastric lipolysis in the developing rat: Ontogeny of the lipases active in the stomach. Biochim Biophys Acta 754:1-9 (1983)
13. Bitman J, Wood DL, Hamosh M, Hamosh P, Mehta NR. Comparison of the lipid composition of breast milk from mothers of term and preterm infants. Am J Clin Nutr 38:300-12 (1983)
14. Jeejeebhoy KN, Ahmad S, Kozak G. Determination of fecal fats containing both medium and long chain triglycerides and fatty acids. Clin Biochem 3:157-163 (1970)
15. Chew V. Comparisons among treatment means in an analysis of variance. USDA Agricultural Research Publication. ARS/H/6, Beltsville, MD (1977)
16. Bitman J, Wood DL, Liao TH, Fink CS, Hamosh P, Hamosh M. Gastric lipolysis of milk lipids in suckling rats. Biochim Biophys Acta 834:58-64 (1985)
17. Carey MC, Small DM, Bliss CM. Lipid digestion and absorption. Ann Rev Physiol 45:651-677 (1983)
18. Abrams CK, Hamosh M, Hubbard VS, Dutta SK, Hamosh P. Lingual lipase in cystic fibrosis. Quantitation of enzyme activity in the upper small intestine of patients with exocrine pancreatic insufficiency. J Clin Invest 73:374-382 (1984)
19. Abrams CK, Hamosh M, Dutta SK, Hubbard VS, Hamosh P. The role of non-pancreatic lipolytic activity in exocrine pancreatic insufficiency. Gastroenterology 1987; In Press.
20. Huston RK, Reynolds JW, Jensen C, Buist NRM. Nutrient and mineral retention and vitamin D absorption in low birth weight infants: effect of medium-chain triglyceride. Pediatrics 72:44-48 (1983)
21. Whyte RK, Campbell D, Stanhope R, Bayley HS, Sinclair JC. Energy balance in low birth weight infants fed formula of high or low medium-chain triglyceride content. J Pediatr 108:946-71 (1986)

UTILIZATION OF FATTY ACIDS BY THE NEWBORN INFANT

Olle Hernell⁺, Stefan Bernbäck and Lars Bläckberg

Departments of Pediatrics⁺ and Physiological Chemistry
University of Umeå
S-901 87 Umeå, Sweden

Fatty acid is the main energy substrate in human milk. Since 98 % is present as triglyceride, the efficiency of fat utilization is directly dependent on how well the triglyceride is digested and absorbed in the gastrointestinal tract. None the less, our understanding of the principal mechanisms of fat digestion and absorption, and how these are influenced by endogenous and exogenous factors, are still far from complete. There is evidence that this is particularly true for the breastfed infant. In this paper we will focus upon some effects of dietary proteins and phospholipid on digestion of dietary triglyceride.

LIPASES CONTRIBUTING TO NEONATAL FAT DIGESTION

As far as is known, there are three lipases, i.e. triglyceride hydrolyzing enzymes, involved in digestion of human milk triglyceride. Gastric contents from many species, including humans, contain lipase activity.[1,2] The tissue that synthesize and secrete the enzyme protein responsible for this activity differs between species. In man, lipase with acid pH-optimum have been found to be secreted from both gastric[3] and pregastric[4] tissues. The contribution of each tissue is not known. Since there is no evidence that there are molecular differences between the corresponding enzyme proteins, we here refer to this enzyme as pregastric/gastric lipase. Although digestion of dietary triglyceride is now known to be initiated by pregastric/gastric lipase in the stomach, in adults lipid digestion is considered to depend almost exclusively on colipase-dependent lipase secreted from the pancreas.[5] In newborn infants, however, the postprandial intraluminal levels of this pancreatic lipase are only 5 - 10 % of those in adults.[6] These low levels do not necessarily mean that colipase-dependent lipase is unimportant during the neonatal period, but they suggest that newborn infants are quantitatively dependent also on other lipases for efficient utilization of dietary fat. The third lipase that is involved in neonatal fat digestion is a constituent of human but not of cow's milk. This enzyme, bile salt-stimulated lipase (BSSL), is inactive until it is activated by primary bile salts in the duodenum.[7] Today BSSL is considered the essential explanation as to why the coefficient of fat absorption is higher in newborn infants given raw human milk rather than cow's milk based formulas.

Table 1. Human milk triglyceride as substrate for gastrointestinal lipases

Lipase	Substrate	
	Triolein/gum arabic	Milk triglyceride
Colipase-dependent[a] lipase	500	<0.5
Pregastric/gastric[b] lipase	70	80
BSSL[b]	100	<0.5

[a]Human pancreatic juice was used as source of enzyme; values are expressed as μmol/min x ml juice.
[b]Values are expressed as μmol/min x mg enzyme protein. Conditions were chosen to obtain optimal activity for the respective lipases.

HUMAN MILK TRIGLYCERIDE AS SUBSTRATE FOR THE LIPASES

Bearing in mind the high coefficient of fat absorption from human milk, one would expect that milk triglyceride is easily accessible for hydrolysis by the lipases concerned. As shown in Table 1, this is not the case. Only pregastric/gastric lipase hydrolyzes human milk triglyceride, as present in milk, and an artificial emulsion of long-chain triglyceride at equal rates. In contrast, under the same conditions that allow rapid hydrolysis of the artificial emulsion, colipase-dependent lipase and BSSL leave the milk triglyceride virtually intact. This illustrates a unique and, from a physiological point of view, very important function of pregastric/gastric lipase. It also demonstrates that artificial emulsions are different from human milk triglyceride, the natural dietary fat of newborn infants. When extruded from the lactating epithelial cell the triglyceride constitutes the core of a fat globule which becomes enveloped by the apical part of the plasma membrane, the milk fat globule membrane.[8] One way or the other this membrane denies colipase-dependent lipase and BSSL access to the triglyceride core, thus preventing immediate triglyceride digestion by these lipases. Conversely, the membrane is not a barrier for pregastric/gastric lipase. Thus, a direct attack on the globule is possible by this lipase.[9]

DIETARY PHOSPHOLIPID AND PROTEIN AFFECTS LIPOLYSIS

We and other groups have studied the effects of phospholipid and protein, the two principal constituents of the milk fat globule membrane, on hydrolysis by colipase-dependent lipase and to a lesser extent pregastric/gastric lipase.[10-14] Phospholipid and most proteins tested impede the action of colipase-dependent lipase on artificial substrates (Table 2). Gargouri et al. have shown that this effect is not due to a protein-enzyme interaction, but rather to an interaction between the protein and the lipid substrate surface.[13] The degree of inhibition varies between different proteins, β-lactoglobulin being a strong inhibitor while ovalbumin is virtually devoid of inhibitory effect. This difference reflects a difference in surface properties of the respective proteins.

Table 2. Effect of phospholipid and various proteins on activities of pregastric/gastric lipase and colipase-dependent lipase

Addition	Colipase-dependent lipase	Pregastric/gastric lipase
Phospholipid	Inhibits	Stimulates
β-lactoglobulin	Inhibits	Stimulates
Bovine serum albumin	Inhibits	Stimulates
Myoglobin	Inhibits	Inhibits
PIL	Inhibits	Inhibits
Ovalbumin	No effect	Stimulates

The original data are from ref 10-14. Purified colipase-dependent lipase and pregastric/gastric lipase was incubated with different artificial di- or triglyceride substrates, either as emulsion or as monolayer. Since different substrates were used, exact comparison was not possible and therefore no figures given.

There is a strong positive correlation between the initial rate of increase in surface pressure caused by the various proteins at the emulsion particle surface and their respective capacity to inhibit colipase-dependent lipase.[13] Thus, an increase in the surface pressure at the emulsion surface denies the lipase access to its substrate. By analogy these results probably explain why milk fat globule triglyceride is a poor substrate for colipase-dependent lipase, i.e. protein and phospholipid of the milk fat globule membrane denies lipase access to the triglyceride core.[15] Only preliminary comparative studies have been done with BSSL. Thus, the reason why this enzyme cannot hydrolyze native milk triglyceride is not yet known. The obligatory requirement of detergent, i.e. bile salts, for activity makes it difficult to fully apply the results obtained with colipase-dependent lipase to the milk lipase. However, induced binding does not necessarily result in hydrolysis even in the presence of bile salts.

As shown in Table 2 the activity of pregastric/gastric lipase behaves very differently following the addition of phospholipid and protein. Rather than being inhibitory, phospholipid and dietary protein enhances hydrolysis by this lipase. In fact, purified human pregastric/gastric lipase has no activity by itself against an emulsion of tributyrin. This is due to irreversible inactivation of the enzyme protein at the substrate surface. However, when albumin and β-lactoglobulin, the same proteins that inhibit colipase-dependent lipase, are added to the incubation pregastric/gastric lipase remains active and hydrolysis proceeds. The explanation of these events is that these proteins decrease the surface pressure at the emulsion surface to a level that is close to optimal for hydrolysis by pregastric/gastric lipase. This, however, is not true for all proteins. Myoglobin and PIL (protein inhibiting lipase), a protein purified from soy meal, reduce the surface pressure below that required for activity of this lipase.[14] In fact, this soy protein is a potent inhibitor also of colipase-dependent lipase.[13]

In light of this strong inhibitory effect of PIL some recent data from Gargouri et al. are of interest.[16] Rats were fed three different diets. One was a stock diet based on casein, the two others were based on soybean flour. The flour was either used raw or autoclaved at 135° C for 10 minutes before used in the diet. After 21 days on the respective diet the rats were killed, the pancreatic glands removed and total concentration of enzymes in the gland were estimated. The rats fed soybean flour had much lower concentrations of colipase-dependent lipase and colipase, but not of the proteolytic enzymes. The rats fed raw soy flour also had a pancreatic hypertrophy and less gain in body-weight while on the diet. This supports the view that dietary proteins may have diverse effects on fat utilization of which we have only little information as yet.

GASTRIC LIPOLYSIS TRIGGERS INTRADUODENAL LIPID DIGESTION

With this perspective it may seem paradoxical that fat digestion in the breastfed newborn is very efficient. Although milk triglyceride digestion is initiated in the stomach by pregastric/gastric lipase, the quantitative importance should not be overestimated in the healthy newborn infant. In the stomach this lipase is very susceptible to inhibition by long-chain fatty acids, the main product of lipolysis. Hence the enzyme becomes inactive after a few per cent of the triglycerides have been hydrolyzed. If the enzyme reaches the duodenum it rapidly becomes inactive by action of pancreatic proteases and bile salts.[17] This, however, does not imply that pregastric/gastric lipase is of low physiological significance. Results obtained with crude enzyme preparations have suggested an important qualitative function for this lipase.[10,18,19] By use of purified bovine enzyme we have confirmed and extended these findings. If only a few per cent of milk fat globule triglyceride is first hydrolyzed by pregastric/gastric lipase, the resulting globule becomes a substrate now immediately accessible for hydrolysis by colipase-dependent lipase (Fig. 1). The same principal result is obtained when an artificial emulsion of long-chain triglyceride covered with protein is used as substrate (data not shown). The mechanism behind this triggering effect of pregastric/gastric lipase depends on the release of a low concentration of free fatty acid since the enzyme can be replaced by addition of pure free fatty acid (data not shown). Long-chain fatty acid enables colipase-dependent lipase to bind to the substrate surface, probably by altering the properties of the particle surface.[10,12]

Similarily, activity of BSSL can also be triggered by such a limited prehydrolysis by pregastric/gastric lipase. The effect is, however, less pronounced than that obtained with colipase-dependent lipase.[20] How the triggering of BSSL activity is obtained is unclear because it can not be replaced by free fatty acid alone. Preliminary experiments indicate that BSSL, unlike colipase-dependent lipase, can bind to native milk fat globules provided bile salts are present. This binding is for unknown reasons not necessarily catalytically productive. Suffice it here to say that the combination of pregastric/gastric lipase and BSSL may represent a route of complete lipid digestion unique to breastfed infants.

THE SEQUENTIAL STEPS OF MILK TRIGLYCERIDE DIGESTION

In the physiological situation all three enzymes are operating together. Fig. 1 illustrates an <u>in vitro</u> experiment under conditions aiming at mimicking this situation. Pregastric/gastric lipase triggers the activity of colipase-dependent lipase as illustrated in the experiment using pasteurized milk as substrate.

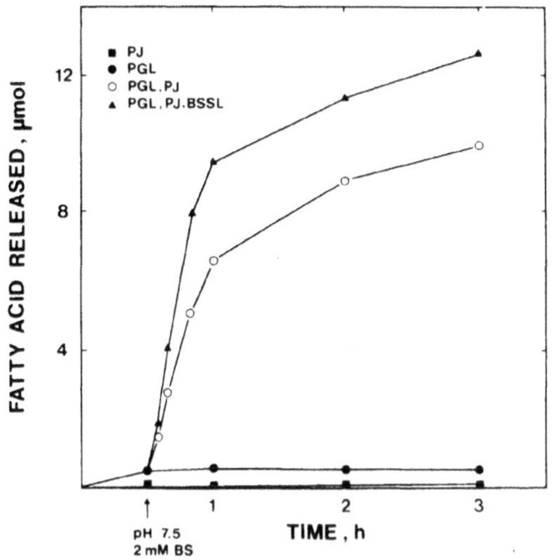

Fig. 1 Sequential steps of milk fat digestion. The assay mixture contained 20 % human milk, 0.2 % bovine serum albumin, 0.15 M NaCl and 5 mM $CaCl_2$. The pH was 6.0 for the first 30 min and was then raised to 7.5. Purified[22] bovine pregastric/gastric lipase (2 µg/ml; PGL) and human pancreatic juice (20 µl/ml; PJ) was added at 0 min and 30 min respectively as indicated. The milk used was either pasteurized (inactive BSSL) or fresh (active BSSL). Incubations were carried out at 37° C.

Pasteurization inactivates BSSL completely. When using raw human milk, the addition of BSSL results in an initial increase in the rate of lipolysis as well as in higher final degree of lipolysis. Thus, it is evident that the most efficient lipolysis is obtained when all three lipases are present. Other experiments with artificial emulsions have shown that BSSL's function also should be considered in qualitative terms. Colipase-dependent lipase generates two free fatty acids and one monoglyceride for each triglyceride. The monoglycerides are an excellent substrate for BSSL which hydrolyzes each monoglyceride to one free glycerol and another free fatty acid.[21] Thus, BSSL promotes complete triglyceride digestion partly by supplementing the endogenous lipases and partly by complementing them, i.e. by altering the composition of final products. The release of glycerol, which is unavailable for reesterification to di- and triglycerides, will shift the equilibrium between hydrolysis and reesterification catalyzed by colipase-dependent lipase towards hydrolysis. The formation of glycerol and fatty acid, rather than monoglyceride and fatty acid, may also have an impact on product absorption. There are several indirect pieces of evidence supporting a view that under conditions of low intraluminal bile salt concentration, as in the neonatal period, fatty acids are more readily absorbed than monoglycerides.[23,24] It is possible that the reason for this is that fat absorption occurs from unilamellar vesicles rather than from mixed micelles.[25,26] This interesting possibility needs to be explored.

ACKNOWLEDGEMENTS

This study was supported by the Swedish Medical Research Council (19X-05708), the Medical Research Council of Swedish Life Insurance Companies, the Swedish Nutrition Foundation, the Swedish Society of Medicine and the Medical Faculty, University of Umeå

REFERENCES

1. B. Fredrikzon, O. Hernell, and L. Bläckberg, Lingual lipase. Its role in lipid digestion in infants with low birthweight and/or pancreatic insufficiency, Acta Pediatr. Scand. 296:75 (1982).
2. M. Hamosh, Lingual lipase, in: "Lipases", eds. B. Borgström and H. L. Brockman, Elsevier, Amsterdam (1984).
3. Y. Gargouri, Les lipases du tractus gastrointestinal, Doctoral thesis, University of Aix-Marseille (1985).
4. M. Hamosh, and W. A. Burns, Lipolytic activity of human lingual glands (Ebner), Lab. Invest. 37:603 (1977).
5. J. S. Patton, Gastrointestinal lipid digestion, in: "Physiology of the gastrointestinal tract", ed. L. R. Johnsson, Raven Press, New York (1981).
6. B. Fredrikzon, and T. Olivecrona, Decrease of lipase and esterase activities in intestinal contents of newborn infants during test meals, Pediatr. Res. 12:631 (1978).
7. O. Hernell, L. Bläckberg, B. Fredrikzon, and T. Olivecrona, Bile salt-stimulated lipase in human milk and lipid digestion during the neonatal period, in: "Gastroenterology and nutrition in infancy", ed. E. Lebenthal, Raven Press, New York (1981).
8. S. Patton, and T. W. Keenan, The milk fat globule membrane, Biochim. Biophys. Acta 415:273 (1975).
9. J. S. Patton, M. W. Rigler, T. H. Liao, P. Hamosh, and M. Hamosh, Hydrolysis of triacylglycerol emulsions by lingual lipase. A microscopic study, Biochim. Biophys. Acta 712:400 (1982).
10. L. Bläckberg, O. Hernell, G. Bengtsson, and T. Olivecrona, Colipase enhances hydrolysis of dietary triglycerides in the absence of bile salts, J. Clin. Invest. 64:1303 (1979).
11. B. Borgström, and C. Erlanson, Interactions of serum albumin and other proteins with porcine pancreatic lipase, Gastroenterology 75:382 (1978).
12. B. Borgström, The importance of phospholipids, pancreatic phospholipase A_2 and fatty acids for the digestion of dietary fat, Gastroenterology 78:954 (1980).
13. Y. Gargouri, G. Pieroni, C. Riviere, A. Sugihara, L. Sarda, and R. Verger, Inhibition of lipases by proteins: a kinetic study with dicaprin monolayers, J. Biol. Chem. 260:2268 (1985).
14. Y. Gargouri, G. Pieroni, P.A. Lowe, L. Sarda, and R. Verger, Human gastric lipase. The effect of amphiphiles, Eur. J. Biochem. 156:305 (1986).
15. L. Bläckberg, O. Hernell, and T. Olivecrona, Hydrolysis of human milk fat globules by pancreatic lipase: role of colipase, phospholipase A_2 and bile salts, J. Clin. Invest. 67:1748 (1981).
16. Y. Gargouri, K. Khalifa, J. Belleville, and L. Sarda, Effect of long-term feeding soybean flour on pancreatic lipase and colipase levels in rat, J. Nutr. In press.

17. S. Bernbäck, O. Hernell, and L. Bläckberg, Bovine pregastric lipase: a model for the human enzyme. Properties relevant to its site of action, In preparation.
18. T. Olivecrona, O. Hernell, T. Egelrud, Å. Billström, H. F. Helander, G. Samuelsson, and B. Fredrikzon, Studies on the gastric lipolysis of milk lipids in suckling rats and human infants, in: "Dietary lipids and postnatal development", eds. G. Jacini and A. Pecile, Raven Press, New York (1973).
19. M. Hamosh, Breast milk fat: origin and digestion, in: "Human milk: its social and biological value", eds. S. Freier and A. I. Edelman, Excerpta Medica ICS No 518 (1980).
20. O. Hernell, L. Bläckberg, and S. Bernbäck, Milk lipases and in vivo lipolysis, in: "Protein and non-protein nitrogen in human milk", eds. B. Lönnerdahl and S. Atkinson, CRC Press, New York, In press.
21. O. Hernell, and L. Bläckberg, Digestion of human milk lipids: physiological significance of sn-2-monoacylglycerol hydrolysis, Pediatr. Res. 16:882 (1982).
22. S. Bernbäck, O. Hernell, and L. Bläckberg, Purification and molecular characterization of bovine pregastric lipase, Eur. J. Biochem. 148:233 (1985).
23. R. G. H. Morgan, and B. Borgström, The mechanism of fat absorption in the bile fistula rat, Q. J. Exp. Physiol. 54:228 (1969).
24. R. O. Scow, P. Desnuelle, and R. Verger, Lipolysis and lipid movement in a membrane model, J. Biol. Chem. 254:6456 (1979).
25. M. C. Carey, D. M. Small, and C. M. Bliss, Lipid digestion and absorption, Ann. Rev. Physiol. 45:651 (1983).
26. O. Hernell, L. Bläckberg, and S. Bernbäck, Digestion and absorption of human milk lipids, in: "Perinatal nutrition", Bristol-Meyers Nutrition Series No 6, Academic Press, New York, In press.

CARNITINE IN RELATION TO FEEDING INFANTS

Peggy R. Borum, Janet K. Baltzell and Alesia Patera

Food Science and Human Nutrition Department
University of Florida
Gainesville, FL 32611

INTRODUCTION

The importance of carnitine in facilitating long chain fatty acid oxidation and the importance of long chain fatty acid oxidation in energy metabolism of the infant are both well documented (1). Carnitine is therefore an important nutrient for the infant. However, the physiological processes in infant metabolism facilitated by carnitine may also include oxidation of medium chain fatty acids, catabolism of branched chain amino acids, thermogenesis, ketogenesis, utilization of ketone bodies, gluconeogenesis, prevention of hyperammonemia, prevention of accumulation of toxic concentrations of acyl CoA, and regeneration of free coenzyme A (2,3). The mechanism of carnitine action in many of these physiological processes remains to be elucidated.

IS CARNITINE AN ESSENTIAL NUTRIENT?

The question of whether or not carnitine is an essential nutrient for the infant has been discussed by a number of investigators during the past decade (4,5,6). The neonate may have both reduced tissue stores of carnitine and a reduced carnitine biosynthetic capability compared to the adult. Human milk (but not all forms of nutritional support) contain significant quantities of carnitine in a highly bioavailable form.

In the absence of any accompanying pathophysiology, the simple documentation of low blood or tissue carnitine concentrations in an infant compared to the concentrations found in an infant reference population should elicit the response "So what!". Establishment of the essentiality of carnitine for the infant requires the identification of some pathophysiology induced by the lack of exogenous carnitine in an otherwise healthy infant that is resolved by carnitine supplementation. A number of infants do have documented low tissue carnitine concentrations in the presence of documented pathophysiology such as nonketotic hypoglycemia. These infants have responded to carnitine supplementation with improvement of both the tissue carnitine concentrations and the clinical symptoms. Many of these infants may be suffering from inborn errors of metabolism (1).

In addition there is a much larger group of infants who have low carnitine concentrations in blood and/or tissues but do not have any pathophysiology which can be readily associated with carnitine's function in facilitating the oxidation of long fatty acids. However, these infants have clinical symptoms which may be associated with the other functions of carnitine mentioned above that are only now being recognized.

A better understanding of the various physiological functions of carnitine is needed in order to adequately address, for the infant, the question "Is carnitine an essential nutrient?".

CARNITINE IN THE INFANT'S DIET

Human milk contains carnitine at concentrations of 50-100 nmoles per milliliter (7). The milk of all mammals studied contains carnitine which appears to be used by the infant to increase tissue carnitine concentrations. Administration of a radioactive precursor of carnitine to a lactating rat dam is followed by the appearance of radioactive carnitine in her milk and later by radioactive carnitine in the tissues of her pups (8). Infant formula prepared from bovine milk contains the carnitine which is carried through the formula processing procedure. Since carnitine is water soluble, it is found in the whey fractions during formula manufacture. Thus, for most formula a larger percentage of whey is associated with a higher concentration of carnitine (9). However, treatment of the whey with a procedure such as dialysis to remove small molecular weight compounds will result in severe reductions in the carnitine (molecular weight of 162 daltons) concentrations of the formula. Two to four fold variations in the formula carnitine concentrations have been observed when manufacturing procedures or suppliers of raw whey product are changed. Most milk based formula contain carnitine concentrations that are equal to or greater than the concentration found in human milk.

Soybeans do not contain carnitine. Therefore, infant formula prepared from soy protein contains no carnitine (9) unless exogenous carnitine is added during manufacture. Isomil (manufactured by Ross Laboratories), and Prosobee (manufactured by Mead Johnson) are infant formulas commercially available in the United States which have been supplemented with exogenous carnitine to give a final carnitine concentration that is similar to the concentration in human milk.

Parenteral nutrition solutions contain no carnitine (1). Thus, an infant maintained on total parenteral nutrition receives no dietary carnitine.

ASSESSMENT OF THE INFANT'S CARNITINE STATUS

Assessment of the carnitine status of an infant should include carnitine determinations of plasma, red blood cell, urine, liver, skeletal muscle, and heart samples. The invasive nature of tissue biopsy prohibits its use in most infants. Blood and urine are often the only samples available and often they are in limited quantities.

Table 1 lists the normalized values for plasma carnitine concentration, red blood cell carnitine concentration, and the percentage of whole blood carnitine found in the red blood cells of infants at different ages and fed different diets. All of the blood carnitine determinations were performed in our laboratory using the same methodology. The infants are from three different cities in the United States and were studied over a period of five years. The values have been normalized based on cord blood carnitine values of healthy full term infants set equal to a value of 1.0. The plasma and red blood cells of cord blood from preterm infants have higher carnitine concentrations than the cord blood from fullterm infants. However, the percentage of cord blood carnitine found in the red blood cells was the same for preterm and fullterm infants (10). It should be noted that of all the populations investigated, the newborn (cord blood) tended to have the highest percentage of whole blood carnitine in the red blood cells.

Fullterm infants fed either human milk or infant formula containing carnitine for the first one to two weeks of life (see Table 1) have plasma carnitine concentrations and red blood cell carnitine concentrations that are 1.5 to 2 times higher than the concentrations in cord blood of full term infants (11).

Although the preterm infants had higher carnitine concentrations in cord blood than those of fullterm infants, after one to two weeks on carnitine free total parenteral nutrition both the plasma and red blood cell carnitine concentrations of the preterm infants are approximately half of the concentration found in fullterm cord blood (see Table 1). If the preterm infants continue to receive carnitine free total parenteral nutrition until three to four weeks of age, the plasma carnitine concentration and red blood cell carnitine concentration remain very low (12). If one to two week old preterm infants receive carnitine supplemented total parenteral nutrition for a two week period, the plasma carnitine concentration increases to a level as high as or higher than that of fullterm infants fed human milk. However the red blood cell carnitine concentrations remain low in the preterm infants even though they received carnitine supplemented total parenteral nutrition for two weeks (12).

Full term infants fed human milk or carnitine containing infant formula for three months maintain plasma and red blood cell carnitine concentrations that are two to three times the concentrations found in cord blood (see Table 1). Alternatively, if the fullterm infant is fed a carnitine free infant formula for the first three months of life, the plasma carnitine concentration is similar to that of full term cord blood. The red blood cell carnitine concentrations appear to be higher than that of fullterm cord blood but not as high as that of three month old infants who receive dietary carnitine (11).

Infants receiving no dietary carnitine for as long as nine months (long term total parenteral nutrition patients) have very low plasma and low red blood cell carnitine concentrations (13).

Blood carnitine parameters of a young adult reference population are also listed in Table 1 to allow comparison of infant and adult values. There were no differences in any blood carnitine parameters between male infants and female infants. The 17-20 year old male population had a higher plasma carnitine concentration than the 17-20 year old female population but there was no significant difference between their red blood cell carnitine concentrations (14).

SPECULATION ON THE COMPARTMENTALIZATION AND METABOLISM OF EXOGENOUS CARNITINE BY THE INFANT

Investigations in dogs and rats have indicated that there are at least two, and perhaps more, metabolic compartments of carnitine in the body (15,16,17,18). There is a critical need for a better understanding of how the body compartmentalizes dietary carnitine administered by either the oral or intravenous route. We need to know how many carnitine metabolic compartments exist in the body and the interrelationships of these compartments in order to adequately assess the effect of human milk carnitine on the infant. The data in Table 1 were obtained by sampling the plasma carnitine compartment and the red blood cell carnitine compartment. Data from our laboratory indicate that these two compartments are separate and distinct from one another and that in adult surgical patients both compartments are poor indicators of the carnitine concentration in tissue compartments such as skeletal muscle and liver (19). Even with the cautious interpretation required for the data in Table 1, it is clear that fullterm infants fed human milk or carnitine containing formula have significantly higher carnitine concentrations in the plasma compartment and the red blood cell compartment than is found in the same compartments in the cord blood of fullterm infants. If fullterm infants are fed a carnitine free formula, the same increase in carnitine concentration is not observed. Preterm infants maintained on carnitine free total parenteral nutrition have very low carnitine concentrations in both the plasma and red blood cell compartments. Only the plasma carnitine compartment responds to supplementation of the parenteral nutrition solutions. The mechanism for the difference in effect of exogenous carnitine on these two compartments remains unexplained. Unfortunately we cannot extend the blood carnitine data of Table 1 to evaluate the possible effects of carnitine in human milk on the infant's tissue carnitine compartments. Tissue samples must be used for this evaluation.

SUMMARY AND SUGGESTIONS FOR FUTURE RESEARCH

The mechanism of compartmentation of dietary carnitine administered by either the oral or intravenous route remains to be elucidated. Data presented here demonstrate that the compartmentation by the neonate of oral carnitine appears to differ from that of intravenous carnitine and that compartmentation of carnitine by the very preterm neonate appears to differ from the compartmentation of carnitine by the fullterm neonate. Future investigations should include evaluation of tissues in addition to evaluation of blood samples. The techniques for obtaining tissue biopsies are too invasive to be performed on human infants. Therefore an animal model must be used to investigate the effect of milk carnitine on the carnitine status of infants. The animal model must have a similar anatomy and metabolism and must be at a similar stage of development to that of the human infant. The neonatal piglet has been shown by several laboratories to have suitable anatomy and metabolism to be used as a model for nutritional questions pertaining to the infant (20,21). Our laboratory has also demonstrated that the pattern of tissue carnitine accretion during gestation for the piglet is similar to what is known about the human infant. We have developed procedures to care for colostrum-deprived neonatal piglets that are similar to the techniques used in a Neonatal Intensive Care Unit (22,23). The neonatal piglet model is being used in our laboratory to extend our knowledge of the effect of carnitine in relation to feeding infants.

TABLE 1
EFFECT OF DIET AND AGE ON BLOOD CARNITINE PARAMETERS

AGE	DIET	NORMALIZED BLOOD CARNITINE PARAMETERS (Plasma)	(Red Blood Cells)	(Percent In Red Blood Cells)
Newborn (Cord Blood)				
Fullterm	None	1.00 ± 0.30 (n=72)	1.00 ± 0.60 (n=72)	1.00 ± 0.17 (n=72)
Preterm	None	1.29 ± 0.58 (n=53)	1.71 ± 1.04 (n=53)	1.00 ± 0.17 (n=53)
1-2 Weeks				
Fullterm	HM	1.75 ± 0.56 (n=13)	1.60 ± 0.73 (n=13)	0.81 ± 0.21 (n=11)
Fullterm	CS-Form	2.03 ± 0.68 (n=19)	1.81 ± 0.96 (n=20)	0.85 ± 0.24 (n=10)
Preterm	CF-TPN	0.60 ± 0.25 (n=8)	0.47 ± 0.33 (n=8)	0.67 ± 0.22 (n=8)
3-4 Weeks				
Preterm	CF-TPN	0.35 ± 0.18 (n=8)	0.32 ± 0.17 (n=8)	0.75 ± 0.31 (n=8)
Preterm	CS-TPN	3.92 ± 1.57 (n=8)	0.49 ± 0.31 (n=8)	0.24 ± 0.11 (n=8)
3 Months				
Fullterm	HM	3.00 ± 0.59 (n=13)	2.56 ± 1.04 (n=13)	0.61 ± 0.13 (n=13)
Fullterm	CS-Form	2.73 ± 0.51 (n=12)	2.24 ± 0.85 (n=12)	0.61 ± 0.14 (n=11)
Fullterm	CF-Form	1.03 ± 0.60 (n=10)	1.69 ± 1.09 (n=10)	0.86 ± 0.14 (n=10)
1-9 Months	CF-TPN	0.40 ± 0.29 (n=7)	0.74 ± 0.72 (n=7)	0.83 ± 0.24 (n=7)
17-20 Years	U.S. Diet	2.30 ± 0.46 (n=39 Males) 1.90 ± 0.46 (n=54 Females)	1.11 ± 0.44 (n=29)	0.51 ± 0.10 (n=8)

HM = Human Milk; CS-Form = Carnitine Supplemented Formula; CF-TPN = Carnitine Free Total Parenteral Nutrition; CS-TPN = Carnitine Supplemented Total Parenteral Nutrition; CF-Form = Carnitine Free Formula

REFERENCES

1. P. R. Borum. Carnitine. in: "Annual Review of Nutrition," William J. Darby, Harry P. Broquist and Robert E. Olson eds., Annuals Reviews, Palo Alto, 3:233-259 (1983).
2. P. R. Borum. Disturbances in Carnitine Metabolism. Biochemical Society Transactions 14:681-683 (1986).
3. P. R. Borum. Carnitine Function. in: "Clinical Aspects of Human Carnitine Deficiency," Peggy R. Borum ed., Pergamon Press, New York, pp. 25-36 (1986).
4. P. R. Borum. Possible Carnitine Requirement of the Newborn and the Effect of Genetic Disease on the Carnitine Requirement. Nutrition Reviews 39:385-390 (1981).
5. C. J. Rebouche. Is Carnitine an Essential Nutrient for Humans? J. Nutr. 116:704-706 (1986).
6. P. R. Borum and S. G. Bennett. Carnitine as An Essential Nutrient. J. Amer. College Nutr. 5:177-182 (1986).
7. P. R. Borum, J. J. Chapman, K. Portier, M. J. Macey and J. Keegan. Carnitine Concentration of Breast Milk During the First 16 Weeks of Lactation. Fed. Proc. 45:364 (1986).
8. C. Robles-Valdes, J. D. McGarry and D. W. Foster. Maternal-Fetal Carnitine Relationships and Neonatal Ketosis in the Rat. J. Biol. Chem. 251:6007-6012 (1976).
9. P. R. Borum, C. M. York and H. P. Broquist. Carnitine Content of Liquid Formulas and Special Diets. Am. J. Clin. Nutr. 32:2272-2276 (1979).
10. J. P. Shenai, P. R. Borum, P. Mohan and S. C. DonLevy. Carnitine Status at Birth of Newborn Infants of Varying Gestation. Pediatr. Res. 17:579-582 (1983).
11. A. Patera, M. Neylan and P. R. Borum. Reference Values of Blood Carnitine Parameters for Neonates. Fed. Proc. 45:615 (1986).
12. J. P. Riddell, M. Behnke, J. Neu, T. G. Baumgartner, M. T. Gersovitz and P. R. Borum. L-Carnitine Supplementation in Very Low Birth Weight Premature Infants Receiving Total Parenteral Nutrition. in: "Clinical Aspects of Human Carnitine Deficiency," P. R. Borum, ed., Pergamon Press, NY, pp. 161 (1986).
13. P. R. Borum and R. A. Helms. Effect of Oral Carnitine Administration on Blood Carnitine Parameters of Infants Maintained on TPN. Amer. J. of Clin. Nutr. 43:680 (1986).
14. P. R. Borum and S. G. Bennett. Assessment of Carnitine Status Using Blood Parameters. Fed. Proc. 44:2063 (1985).
15. K. T. N. Yue and I. B. Irving. Fate of Tritium-labeled Carnitine Administered to Dogs and Rats. Am. J. of Physiology 202:122-128 (1962).
16. C. J. Rebouche and A. G. Engel. Kinetic Compartmental Analysis of Carnitine Metabolism in the Dog. Archives of Biochem. and Biophysics 220:60-70 (1983).
17. D. E. Brooks and J. A. McIntosh. Turnover of Carnitine by Rat Tissues. Biochem. J. 148:439-445 (1975).
18. G. Cederblad and S. Lindstedt. Metabolism of Labeled Carnitine in the Rat. Archives of Biochem. and Biophysics 175:173-180 (1976).
19. P. R. Borum, T. O. Rumley and E. Taggart. Caution Required in Clinical Use of Plasma Carnitine Concentration for Assessment of Carnitine Status. Eighth Congress of the European Society of Parenteral and Enteral Nutrition Book of Abstracts, pp. 76 (1986).
20. J. E. Cooper. The Use of the Pig as an Animal Model to Study Problems Associated With Low Birthweight. Lab. Animals 9:329-336 (1975).
21. W. J. Dodds. The Pig Model for Biomedical Research. Fed. Proc. 41:247-256 (1982).

22. J. K. Baltzell, F. W. Bazer, S. G. Miguel and P. R. Borum. The Neonatal Piglet as an Animal Model for Carnitine Supplementation of Human Neonates. Fed. Proc. 45:615 (1986).
23. P. R. Borum, J. K. Baltzell and A. Patera. Care of Colostrum Deprived Neonatal Piglets for Nutritional Investigations. J. Nutr. 116:000 (1986).

HORMONES IN MILK: THEIR PRESENCE AND POSSIBLE PHYSIOLOGICAL SIGNIFICANCE

O. Koldovský, A. Bedrick, P. Pollack, R.K. Rao, and W. Thornburg

Department of Pediatrics
University of Arizona College of Medicine
Tucson, Arizona 85724

INTRODUCTION

The presence of hormones in breast milk and their physiological significance was suggested by Schein (1895) and Mosse and Cathala (1898) during the last century and described five decades ago by Heim (1931a; 1931b) and Yaida (1929). In the last decade we witnessed an increased interest in this area. In our presentation we will review two aspects of the presence of hormones in breast milk, along with a brief discussion of clinical studies considering their functional significance for the suckling infant, as well as experiments exploring absorption of some milk hormones in laboratory animals.

PRESENCE OF HORMONES IN BREAST MILK

Table 1 lists hormones detected in breast milk. Practically all of them were found to be present in milk of various experimental and farm animals (for references on animal data see Koldovský and Thornburg, 1987b). Some of the hormones are present in milk in concentrations similar or lower than in blood (plasma, serum). It is noteworthy that several of them exhibit much higher concentrations in milk than in plasma (see Table 2).

Clinical interest was concentrated mainly on the presence of thyroid hormones in milk, and intriguing studies were recently published dealing with the metabolic role of acid soluble nucleotides present in milk.

Thyroid Hormones

At the end of the last century, several reports had already described the ameliorative effect of breast feeding on the immediate or delayed clinical sequalae of congenital hypothyroidism, but some publications did not confirm this (for review of older literature see Robertson, 1945; and Koldovský and Thornburg, 1987b). Less than ten years ago two studies supported the positive effect of breast milk on infants with congenital hypothyroidism (Tenore et al., 1977; Bode et al., 1978). These studies were unconfirmed in later studies (Letarte et al., 1980; Mizuta et al., 1983; Banagale and Erenberg, 1984) where the investigators observed neither clinical nor biochemical differences in hypothyroid infants which were breast or bottle fed. Only small and unsystematic effects of milk intake on thyroid parameters were found in a prospective study of 88 healthy infants which were followed from two weeks of age during the first year of life (Štrbák et al., 1983). In contrast to this observation, lower levels of serum thyroid hormones were reported in smaller groups of two- to three-week-old normal bottle fed infants in comparison with those which were breast fed (Hahn et al., 1983; Banagale and Erenberg, 1984; Oberkotter et al., 1985). It is of interest that according to two reports, breast feeding does not interfere with neonatal screening for hypothyroxinemia (Abassi and Steinour, 1980; Banagale and Erenberg, 1984).

The values of concentration of thyroid hormones as reported from various laboratories vary considerably: thyroxine, from 13 μg/dl (Strbák et al., 1976); 0.7 μg/dl (Sack et al., 1978); and 0.07 μg/dl (Jansson et al., 1983); and to undetectable (Mizuta et al., 1983; Sato and Suzuki, 1979; Mallol et al., 1982; and Varma et al., 1978). Triiodothyronine values were reported from 10 ng/dl (Sato and Suzuki, 1979) to 400 ng/dl (Sack et al., 1979). Various methodological problems such as extraction of milk or removal of interfering substances can be responsible for these differences (Mizuta et al., 1983; Vigouroux et al., 1980; Oberkotter and Tenore, 1983; Slebodzinski and Gawecka, 1983; Moller et al., 1983; Vigouroux and Rostaqui, 1980; Sato and Suzuki, 1979; Mallol et al., 1982; Jansson et al., 1983). Therefore, due to discrepancies in determined levels of thyroid hormones in milk, it is not surprising that their opinion regarding significance varies for the breast fed infant. Some investigators concluded the evidence as positive (Sack et al., 1978; Bode et al., 1978; Oberkotter and Tenore, 1983; Štrbák et al., 1976; Varma et al., 1978; Sack et al., 1979; Sack et al., 1977; Salakhova et al., 1980) whereas others reported negative conclusions (Mizuta et al., 1983; Moller et al., 1983; Sato and Suzuki, 1979; Mallol et al., 1982; Jansson et al., 1983). Of interest are studies from two laboratories reporting relatively high values of thyroxine in breast milk that did not detect thyroxine in various infant nutrition formulas (Štrbák et al., 1976; Salakhova et al., 1980).

Nucleotides

Presence of various acid soluble nucleotides has been reported in breast milk (see Table 1) and also in bovine milk (Kobata and Suzuoki-Zird, 1962; Gil and Sanchez-Medina, 1981a,b; Denamur et al., 1959; Johke and Goto, 1962; Johke, 1963; Richardson et al., 1980; Gil and Sanchez-Medina, 1981a,b; 1982a). Industrial processing has been found to decrease their concentration (Gil and Sanchez-Medina, 1982b). Because of the qualitative and quantitative differences in nucleotide patterns between bovine and breast milk, studies were performed to compare the effect of feeding healthy newborns with human milk infant milk formula and nucleotide supplemented milk formula (NSMF) during the first month of life. Plasma polyunsaturated fatty acids with more than 18 carbons of the ω 6 family were significantly increased in NSMF fed groups as opposed to groups fed unsupplemented formula. The values of the NSMF group did not differ from those fed breast milk (Gil et al., 1986), and in addition, this dietary modification influenced plasma lipoprotein pattern in these infants (Sanchez-Pozo et al., 1986).

Table 1. Hormones and Hormone Related Substances Present in Breast Milk

Hormone	Reference
Adrenal steroids	Alexandrová and Macho, 1983
Bombesin	Ekman et al.,1985 Jahnke and Lazarus, 1984
Calcitonin	Bucht et al., 1983; Werner et al., 1982
EGF and other growth factors	Kidwell et al, 1986; Thornburg and Koldovský, 1986 (also see Table 2)
Erythropoietin	Bielecki et al., 1972
GnrH	Amarant et al., 1982; Gupta, 1983; Sarda and Nair, 1981
GRF	Werner et al., 1986
Insulin	Ballard et al., 1981, 1982 Cevreska et al., 1975; Ekman et al., 1985; Kulski and Hartmann et al., 1983; Read et al., 1984, 1985
Neurotensin	Ekman et al., 1985; Werner et al., 1982
Nucleotides (includes cAMP, cGMP)	Deutsch and Nilsson, 1960; Gil and Sanchez-Medina, 1981a,b, 1982a,b; Janas and Picciano, 1982; Kobata and Suzuoki-Zird, 1962; Skala et al., 1981
Oxytocin	Leake et al., 1981
Ovarian steroids	Alexandrová and Macho, 1983 McGarrigle and Lachelin, 1983 Wolford and Argoudelis, 1979 Yaida, 1929
Prolactin	Adamopoulos and Kapolla, 1983 Gala et al., 1975; Gala and van de Walle, 1977; Healy et al., 1980; Kleinberg et al., 1977
Prostaglandins	Chappell et al., 1983; Craig-Schmidt et al., 1984; Evans and Johnson, 1977; Johnson and Evans, 1978; Lucas and Mitchell, 1980 Reid et al.,1980
Somatostatin	Werner et al., 1985
Triiodothyronine, thyroxine	Jansson et al., 1983; Karimova et al., 1983; Mallol et al., 1982 Mizuta et al., 1983; Moller et al., 1983; Sack et al., 1977 Salakhova et al., 1980; Sato and Suzuki, 1979; Štrbák et al.,1974, 1976; Tenore et al., 1977; Varma et al., 1978
TRH	Amarant et al., 1982; Hazum et al., 1977; Sack et al., 1978
TSH	Tenore, et al., 1981

Table 2. List of Hormones and Hormone Related Substances Present in Breast Milk in Concentration Higher than in Blood (plasma/or serum)

Hormone	Reference
Calcitonin	Werner et al., 1982
cAMP, cGMP	Skåla et al., 1981
EGF	Beardmore et al., 1983 Carpenter, 1980; Hirata et al. 1980; Moran et al.,1983 Starkey and Orth, 1977 Thornburg et al., 1984a
GRF	Werner et al., 1986
GnrH	Gupta, 1983; Nair et al., 1983; Sarda and Nair, 1981
Insulin	Ballard et al., 1981
Neurotensin	Ekman et al., 1985
Oxytocin	Leake et al., 1981
Prostaglandins	Craig-Schmidt et al., 1984 Lucas and Mitchell, 1980 Materia et al., 1984; Reid et al., 1980
Relaxin	Lippert et al., 1981
Somatostatin	Werner et al., 1985

EXPERIMENTAL STUDIES ON ABSORPTION OF HORMONES FROM THE GASTROINTESTINAL TRACT

In the preceding section, we have reviewed the presence of hormones in breast milk and discussed clinical studies concerning the possible role of thyroid hormones and nucleotides in the breast milk for the human neonate. In this section, studies will be reviewed concerning the question of absorption of other hormones in laboratory rodents. As can be seen from Table 3, in some studies the authentic hormone was detected either in circulation (EGF, NGF, prolactin, prostaglandins) or in peripheral organs (EGF, NGF, prostaglandins); other studies reported a functional effect on the suckling (ACTH, erythropoietin, insulin, TRH, TSH). In the following, we will discuss the results concerning three of these hormones in detail, namely, prostaglandins, insulin and EGF.

Prostaglandins

Breast (Tables 1, 2), bovine (Manns, 1975; Hansel et al., 1976), and rat milk (Bedrick and Holtzapple, 1986) contain prostaglandins, but according to one study they were not detected in infant nutrition formulas (Reid et al., 1980). This, as well as their known cytoprotective effect (Robert, 1979), led us to explore their handling by the gastrointestinal tract of the suckling rat.

In our first study (Revsin et al., 1982), we administered orogastrically radiolabelled (^{3}H) PGE_2 to suckling rats. Authentic prostaglandin was found 30 and 60 min later in intestinal lumen and wall, and also in liver and kidney. In following experiments (Bedrick and Koldovský, 1986a,b), suckling and weanling rats were given another radiolabelled prostaglandin, $PGF_{2\alpha}$, in a dose corresponding to the average daily consumption in suckling rats (Revsin et al., 1982). Animals were killed two hours after prostaglandin administration, and lipid extracts of various tissues were subjected to silicic acid column chromatography followed by thin layer chromatography silica G thin layer chromatography. Characterization of radioactivity present in stomachs from animals of both age groups revealed low prostaglandin $F_{2\alpha}$ metabolism. The amount of unmetabolized prostaglandin present in intestinal segments (i.e., lumen and wall combined) in suckling rats ranged from 11 to 14% of the total counts separated on the silica thin layer chromatography. In weanling animals the percentage of authentic prostaglandin counts in the distal ileum was even higher (26%).

Analysis of liver tissue demonstrated delivery of authentic prostaglandin. In sucklings it represented 11% of the total tissue counts analyzed on silica thin layer chromatography, and in weanlings, 7% ($p < 0.01$). Out of the original dose given to each animal, there was a six-fold difference in the presence of unmetabolized $PGF_{2\alpha}$ in sucklings compared to weanlings (1.2% versus 0.2%, $p < 0.01$). Characterization of metabolites showed that weanling rats' liver had more $PGF_{2\alpha}$ degradation products of lesser polarity than sucklings. It is noteworthy that a similar pattern was seen in animals killed either earlier or later (i.e., after one or three hours).

These studies thus show that sucklings and weanlings are able to absorb orally delivered prostaglandins. The cytoprotective role of milk prostaglandins for the suckling is an intriguing question for further studies, especially due to the high risk of necrotizing enterocolitis in the human premature neonate.

Table 3. Absorption of Gastrically Administered Hormones in Suckling Rats and Mice

Hormone	Effect	Reference
ACTH	corticosterone*	Tenore et al., 1980; Vaucher et al., 1983
EGF	detected in organs	Thornburg et al., 1984a,b
Erythropoietin	erythropoiesis	Bielecki et al., 1973 Carmichael et al., 1978
Insulin	glycemia*	Mosinger et al., 1959
NGF	NGF*	Aloe et al., 1982
Prolactin	prolactin*	Mulloy et al., 1979 Whitworth and Grosvenor, 1978
Prostaglandins	detected in organs	Bedrick and Koldovský, 1986a,b Revsin et al., 1982
TRH	to mothers; TSH in pups*	Štrbák and Macho, 1977
TSH	T_3, T_4*	Tenore et al., 1980; Vaucher et al., 1983

* = changes in serum (blood)

Insulin

This hormone is present in human milk and colostrum (Cevreska et al., 1975; Kulski and Hartmann, 1983; Read et al., 1984), and it was also detected in bovine milk (Falconer et al., 1984). Interestingly, orogastric administation to suckling rats evoked a hypoglycemic effect; no effect was observed in weaned rats (Mosinger et al., 1959). Administration of insulin directly into the lumen of the small intestine was effective in both age groups (Hirsovå and Koldovský, 1969); therefore, we concluded that the absence of effect after orogastric administration in weanling rats is caused by peptic activity in the stomach. This activity is very low in sucklings and high in weanlings (Mosinger et al., 1959). Orally administered insulin was effective also in piglets (Asplund et al., 1962) and calves (Pierce et al., 1964).

Epidermal Growth Factor

The presence of epidermal growth factor (EGF) in milk of various species has been demonstrated in a number of laboratories (for review see Koldovský and Thornburg, 1987a,b; Kidwell, 1986: this conference proceedings). Of note, this polypeptide is pepsin, trypsin and chymotrypsin resistant. Its presence in milk, combined with a very low concentration in the salivary gland (one of the sites of its production) in suckling mice-compared to adults by several orders of magnitude (Byyny et al., 1972), led us to explore whether EGF is absorbed in rats after oral administration. These studies were further prompted by the fact that several infant formulas do not contain EGF (Carpenter, 1980; Tapper et al., 1979).

In the first experiments (Thornburg et al., 1984b) suckling rats were fed mouse ^{125}I-EGF. Using Sephadex G-25 column chromatography, immunocolumn chromatography and receptor assay, we have demonstrated that luminal content of stomach and small intestine contained "intact" EGF and very little degradation products within 30 or 60 min after EGF administration. In the stomach and intestinal wall, EGF was processed in a time and dose dependent fashion. Analysis of plasma, lung and liver confirmed that "intact" EGF is absorbed and delivered to peripheral organs. In another group of experiments (injecting ^{125}I-EGF into isolated stomach, jejunum or ileum) we have shown that the major site of EGF absorption is the small intestine, both in suckling and weanling rats (Thornburg et al., 1986). Further studies were performed to compare suckling and weanling rats (Thornburg et al., 1984a). Rats were killed 30 min after orogastric administration of ^{125}I-EGF. Results (summarized in Table 4) confirmed previous experiments with sucklings. In addition, they clearly demonstrated the absorption (although to some extent diminished) of "intact" EGF in weanling rats. Surprising, too, was the higher delivery of EGF to the skin of weanlings as compared to sucklings. It is noteworthy that immunoreactive labelled EGF was detected also in the brain.

Various effects of EGF (given in pharmacological doses to adult or suckling rodents) on the gastrointestinal tract have been described (Thornburg and Koldovský, 1986; Menård, 1986). The question arose whether EGF effect can be demonstrated in suckling animals when given orally and in doses corresponding to those found in rats' milk (Thornburg et al., 1984b). To prevent any "background" effect of the EGF present in milk, 10-14-day-old rats were hand gavage fed an artificial milk diet with or without EGF every 3 hours (Pollack et al., 1986). Caloric intake was the same (about 30 Cal/100 g body weight); the growth of animals in both groups was practically identical. Several parameters determined in the small intestine (wet weight, protein content, sucrase and lactase activity) were not influenced; but in the colon, especially the distal half, the EGF treated group exhibited considerably increased DNA content and reduced protein content. These data support a physiological role of breast milk EGF (and possibly other growth factors as well) in the gastrointestinal development.

Table 4. Distribution of Orogastrically Administered ^{125}I-EGF to Suckling and Weanling Rats

ORGAN	OC/TC(a)		% as A(b)		A as %TC(c)		% as IR(d)		% as A431(e)	
	S	W	S	W	S	W	S	W	S	W
Stomach										
Wall	7.2	5.6	62	44*	4.8	2.6	59	39*	63	26*
Content	40	32	66	45*	26	15	66	48*	64	55
Small Intestine										
Wall	5.2	1.6*	64	23*	3.5	0.4*	49	15*	26	3*
Content	7.2	7.2	71	39*	51	28	56	23*	44	20*
Liver	0.3	0.7	21	13*	0.05	0.1*	17	9	21	0*
Lung	0.2	0.2	38	23	0.09	0.05	47	8*	26	4*
Skin	0.8	17.2*	23	20	0.14	4.1*	18	11	0	0
Plasma	0.1	0.2	23	8	0.01	0.01	0	0	0	0

Animals were killed 30 min after orogastric administration of labelled EGF (80-85 ng/animal). (a) = % of total counts (acid extractable) given per organ. (b) = Sephadex G-25 peak A fraction as % of total counts per organ. (c) = Sephadex G-25 peak A fraction as % of total counts (EGF) given to rats. (d) = % of immunoreactive counts per organ. (e) = % of counts binding to A431 cell receptors per organ. S = 13-15-day-old sucklings. W = 29-31-day-old weanlings. Only means are given; * denotes significant statistical differences ($p < 0.02$, unpaired t-test) between corresponding values of sucklings and weanlings; N = 4/group (Thornburg, et al., 1984b).

CONCLUSION

It is quite clear that many hormone and hormone related substances are present in human milk and in milk of various species. To approach the question of the physiological significance of their presence in milk, we have summarized some of the data obtained in our laboratory on gastrointestinal processing of prostaglandins, insulin, and epidermal growth factor in the developing rat. Further experiments are needed to analyze these processes in more detail with the inclusion of comparative studies of other species to determine if these findings are applicable to humans. The question posed several years ago (Koldovský, 1980), namely, whether we should consider hormones in milk "equally" important components of infant nutrition as calories and minerals remains open for further investigation.

ACKNOWLEDGEMENTS

This work was supported by National Institutes of Health Grant AM27624 and Nestlé Grant Programme Award. The authors wish to thank Miss Melita Stine for her expertise in computerization of references included in this review.

REFERENCES

Abbassi, V., and Steinour, T. A., 1980, Successful diagnosis of congenital hypothyroidism in four breast-fed neonates, J. Pediatr., 97: 259-261.

Adamopoulos, D. A., and Kapolla, N., 1983, Prolactin concentration in milk and blood of patients with galactorrhoea, Acta Endocrinol., 261: 5-7.

Alexandrová, M., and Macho, L., 1983, Glucocorticoids in human, cow and rat milk, Endocrinol. Exper., 17: 183-189.

Aloe, L., Calissano, P., and Levi-Montalcini, R., 1982, Effects of oral administration of nerve growth factor and of its antiserum on sympathetic ganglia of neonatal mice, Dev. Brain Res., 4: 31-34.

Amarant, T., Fridkin, M., and Koch, Y., 1982, Luteinizing hormone-releasing hormone and thyrotropin-releasing hormone in human and bovine milk, Eur J. Biochem., 127: 647-650.

Asplund, J. M., Grummer, R. H., and Phillips, P. H., 1962, Absorption of colostral gamma-globulins and insulin by the newborn pig, J. Ani. Sci., 21: 412-413.

Ballard, E. J., Nield, M. K., Francis, G. L., Dahlenburg, G. W., and Wallace J. C., 1982, The relationship between the insulin content and inhibitory effects of bovine colostrum on protein breakdown in cultured cells, J. Cell Physiol., 110: 249-254.

Ballard, F. J., Nield, M. K., Francis, G. L., and Knowles, S. E., 1981, Regulation of intracellular protein degradation by insulin and growth factors, Acta Biol. Med. Ger., 40: 1293-1300.

Banagale, R. C., and Erenberg, A. P., 1984, Serum T_4 level in term newborns: comparison between breast-fed and formula-fed infants, Nutr. Res., 4: 353-355.

Beardmore, J. M., Lewis-Jones, D. I., and Richards, R. C., 1983, Urogastrone and lactose concentrations in precolostrum, colostrum, and milk, Pediatr. Res., 17: 825-828.

Bedrick, A. D., and Holtzapple, P. G., 1986, Indomethacin fails to induce ulceration in the gastrointestinal tract of newborn and suckling rats, Pediatr. Res., in press.

Bedrick, A. D., and Koldovský, O., 1986a, Hepatic metabolism of orally administered prostaglandin $F_{2\alpha}$ in suckling and weanling rats, Biol. Neonate, 48: 351-356.

Bedrick, A. D., and Koldovský, O., 1986b, Gastrointestinal processing of gastrically administered prostaglandin $F_{2\alpha}$ in suckling and weanling rats, J. Pediatr. Gastro. Nutr., in press.

Bielecki, M., Lazewska, M., Wojtowicz, A., and Gruszecki, W., 1973, The effect of orally administered erythropoietin on erythropoiesis in experimental animals, Acta Physiol. Pol., 24: 351-356.

Bielecki, M., Przala, F., and Lazewska, M., 1972, Level of erythropoietin in the woman milk, Acta Physiol. Pol., 23: 435-439.

Bode, H. H., Vanjonack, W. J., and Crawford, J. D., 1978, Mitigation of cretinism by breast-feeding, Pediatrics, 62: 13-16.

Bucht, E., Arver, S., Sjoberg, H. E., and Low, H., 1983, Heterogeneity of immunoreactive calcitonin in human milk, Acta Endocrinol., 103: 572-576.

Byyny, R. L., Orth, D. N., and Cohen, S., 1972, Radioimmunoassay of epidermal growth factor, Endocrinology, 90: 1261-1266.

Carmichael, R. D., Gordon, A. S., and Lobue, J., 1978, The effects of maternal phlebotomy and orally-administered erythropoietin (EP) on erythropoiesis in the suckling rats, Biol. Neonate, 33: 119-131.

Carpenter, G., 1980, Epidermal growth factor is a major growth-promoting agent in human milk, Science, 210: 198-199.

Cevreska, S., Kovacev, V. P., Stankovski, M., and Kamamaras E., 1975, The presence of immunologically reactive insulin in milk of women during the first week of lactation and its relation to changes in plasma insulin concentration, God. ZB. Med. Fak. Skopje, 21: 35-41.

Chappell, J. E., Clandinin, M. T., Barbe, G. J., and Armstrong, D. T., 1983, Prostanoid content of human milk: relationships to milk fatty acid content, Endocrinol. Exper., 17: 351-358.

Craig-Schmidt, M. C., Weete, J. D., Faircloth, S. A., Wickwire, M. A., and Livant, E. J., 1984, The effect of hydrogenated fat in the diet of nursing mothers on lipid composition and prostaglandin content of human milk, Am. J. Clin. Nutr., 39: 778-786.

Denamur, R., Fauconneau, G., and Guntz, G., 1959, Les nucleotides acido solubles des laits de brebis, vache, chevre, et truie, Rev. Esp. Fisiol., 15: 301-310.

Deutsch, A., and Nilsson, R., 1960, Uber die saurelöslichen Nucleotide der Frauenmilch, Hoppe-Seylers Z. Physiol. Chem., 321: 246-251.

Ekman, R., Ivarsson, S., and Jansson, L., 1985, Bombesin, neurotensin and progamma-melanotropin immunoreactants in human milk, Reg. Pep., 10: 99-105.

Evans, G. W., and Johnson, P. E., 1977, Prostaglandin E_2: the zinc-binding ligand in human breast milk, Clin. Res., 25: 536 (abstract).

Falconer, J., Sheldrake, R. F., and Robinson, J. S., 1984, Insulin in ovine milk, Proc. Aust. Soc. Reprod. Biol., 16: 103.

Gala, R. R., Singhakowinta, A., and Brennan, M. J., 1975, Studies on prolactin in human serum, urine and milk, Horm. Res., 6: 310-320.

Gala, R. R., and van de Walle, C., 1977, Prolactin heterogeneity in the serum and milk during lactation, Life Sci., 21: 99-104.

Gil, A., and Sanchez-Medina, F., 1981, The determination of acid-soluble nucleotides in milk by improved enzymic methods: a comparison with the ion-exchange column chromatography procedure, J. Sci. Food Agric., 32: 1123-1131.

Gil, A., Sanchez-Medina, F., 1981, Acid-soluble nucleotides of cow's, goat's and sheep's milks at different stages of lactation, J. Dairy Res., 48: 35-44.

Gil, A., and Sanchez-Medina, F., 1982a, Acid-soluble nucleotides of human milk at different stages of lactation, J. Dairy Sci., 49: 301-307.

Gil, A., Sanchez-Medina, F., 1982b, Effects of thermal industrial processing on acid-soluble nucleotides of milk, J. Dairy Res., 49: 295-300.

Gil, A., Pita, M., Martinez, A., Molina, J. A., and Sanchez-Medina, F., 1986, Effect of dietary nucleotides on the plasma fatty acids in at-term neonates, Hum. Nutr.: Clin. Nutr., 40C: 185-195.

Gupta, D., 1983, Hormones and human milk, Endocrinol. Exper., 17: 359-370.

Hahn, H. B., Spiekerman, M., Otto, W. R., Otto, W. R., and Hossalla, D. E., 1983, Thyroid function tests in neonates fed human milk, Am. J. Dis. Child., 137: 220-222.

Hansel, W., Hixon, J., Shemesh, M., and Tobey, D., 1976, Concentrations and activities of prostaglandins of the F series in bovine tissue, blood, and milk, J. Dairy Sci., 59: 1353-1365.

Hazum, E., Friedkin, M., Baram, T., and Koch, Y., 1977, Gonadotropin-releasing hormone in milk, Science, 198: 300-301.

Healy, D. L., Rattigan, S., Hartmann, P. E., Herington, A. C., and Burger, H. G., 1980, Prolactin in human milk: correlation with lactose, total protein, and β-lactabumin levels, Am. J. Physiol., 238: E83-E86.
Heim, K., 1931a, Hormonale Wirkungen der Frauenmilch, Klin. Wochenschr., 10: 357.
Heim, K., 1931b, Brustdrüse und Hypophysenvorderlappen, Klin. Wochenschr., 10: 1598.
Hirata, Y., Moore, G. W., Bertagna, C., and Orth, D. N., 1980, Plasma concentrations of immunoreactive human epidermal growth factor (Urogastrone) in man, J. Clin. Endocrinol. Metab., 50: 440-444.
Hiršová, D., and Koldovský, O., 1969, On the question of the absorption of insulin from the gastrointestinal tract during postnatal development, Physiol. Bohemoslov., 18: 281-284.
Jahnke, G. D., and Lazarus, L. H., 1984, A bombesin immunoreactive peptide in milk, Proc. Natl. Acad. Sci. USA, 81: 578-582.
Janas, L. M., and Picciano, M. F., 1982, The nucleotide profile of human milk, Pediatr. Res., 16: 659-662.
Jansson, L., Ivarsson, S., Larsson, I., and Ekman, R., 1983, Tri-iodothyronine and thyroxine in human milk, Acta Paediatr. Scand., 72: 703-705.
Johke, T., 1963, Acid-soluble nucleotides of colostrum, milk, and mammary gland, J. Biochem., 54: 388-397.
Johke, T., and Goto, T., 1962, Acid-soluble nucleotides in cow's and goat's milk, J. Dairy Sci., 45: 735-741.
Johnson, P. E., and Evans, G. W., 1978, Identification of a prostaglandin E_2-zinc complex in human breast milk, and porcine and rat duodenum, Fed. Proc., 37: 889 (abstract).
Karimova, S. F., Turakulov, Y. K., Salakhova, N. S., and Gulamova, F. Y., 1983, Content in thyroid hormones in human and animal milk and in cow milk based infant formulas, Endocrinol. Exper., 17: 237-242.
Kidwell, W. R., Salomon, D. D., and Mohanam, S., 1986, Production of growth factors by normal human mammary cells in culture, in: Effects of Human Milk on the Recipient Infant; International Meeting on Human Lactation held in Konstanz, Germany, to be published by Plenum Press: New York.
Kleinberg, D. L., Noel, G. L., and Frantz, A., 1977, Galactorrhea; a study of 235 cases, including 48 with pituitary tumors, N. Engl. J. Med., 296: 589-600.
Kobata, A., and Suzuoki-Zird, K. M., 1962, The acid-soluble nucleotides of milk. 1. Quantitative and qualitative differences of nucleotides constituents in human and cow's milk, J. Biochem., 51: 277-287.
Koldovský, O., 1980, Hormones in milk, in: Proceedings of the 79th Ross Conference on Pediatric Research: Feeding the Neonate $<$ 1500 grams-Nutrition and Beyond. Columbus, Ohio: Ross Laboratories, pp. 62-65.
Koldovský, O., and Thornburg, W., 1987a, Peptide hormones and hormone-like substances in milk, in: Proteins and Non-Protein Nitrogen in Human Milk, eds, Atkinson, S., and Lonnerdal, B., Boca Raton, Florida, CRC Press, Inc., in press.
Koldovský, O., and Thornburg, W., 1987b, Hormones in milk, J. Pediatr. Gastro. Nutr., in press.
Kulski, J. K., and Hartmann, P. E., 1983, Milk insulin, GH and TSH: relationship to changes in milk lactose, glucose and protein during lactogenesis in women, Endocrinol. Exper., 17: 317-326.
Leake, R. D., Weitzman, R. E., and Fisher, D. A., 1981, Oxytocin concentrations during the neonatal period, Biol. Neonate, 39: 127-131.

Letarte, J., Guyda, H., Dussault, J. H., and Glorieux, J., 1980, Lack of protective effect of breast-feeding in congenital hypothyroidism: report of 12 cases, Pediatrics, 65: 703-705.

Lippert, T. H., God, B., and Voelter, W., 1981, Immunoreactive relaxin-like substance in milk, IRCS Med. Sci., 9: 295.

Lucas, A., and Mitchell, M. D., 1980, Prostaglandins in human milk, Arch. Dis. Child., 55: 950-952.

Mallol, J., Obregon, M. J., and de Escobar, G. M., 1982, Analytical artifacts in radioimmunoassay of L-thyroxin in human milk, Clin. Chem., 28: 1277-1282.

Manns, J. G., 1975, The excretion of prostaglandin $F_{2\beta}$ in milk of cows, Prostaglandins, 9: 463-474.

Materia, A., Jaffe, B. M., Money, S. R., Rossi, P., De Marco, M., and Basso, N., 1984, Prostaglandins in commercial milk preparations. Their effect in the prevention of stress-induced gastric ulcer, Arch. Surg., 119: 290-292.

McGarrigle, H. H. G., and Lachelin, G. C. L., 1983, Oestrone, oestradiol and oestriol glucosiduronates and sulphates in human puerperal plasma and milk, J. Steroid Biochem., 18: 607-611.

Menård, D., 1986, Epidermal and neural growth factors, in: Biology of Human Milk, Proceedings of 14th Nestlé Nutrition Workshop, Athens, Greece, in press.

Mizuta, H., Amino, N., Ichihara, K., Harada, T., Nose, O., Tanizawa, O., and Miyai, K., 1983, Thyroid hormones in human milk and their influence on thyroid function of breast-fed babies, Pediatr. Res., 17: 468-471.

Moller, B., Bjorkhem, I., Falk, O., Lantto, O., and Larsson, A., 1983, Identification of thyroxine in human breast milk by gas chromatographymass spectrometry, J. Clin. Endocrinol. Metabol., 56: 30-34.

Moran, J. R., Courtney, M. E., Orth, D. N., Vaughan, R., Coy, S., Mount, C. D., Sherrell, B. J., and Greene, H. L., 1983, Epidermal growth factor in human milk: daily production and diurnal variation during early lactation in mothers delivering at term and at premature gestation, J. Pediatr., 103: 402-405.

Mosinger, B., Placer, A., and Koldovský, O., 1959, Passage of insulin through the wall of the gastro-intestinal tract of the infant rats, Nature, 184: 1245-1246.

Mosse, A., and Cathala, N., 1898, Guerison du goitré congenital d'un nourrisson per l'alimentation thyroidienne de la nourrice, Bull. Acad. Med., 39: 420-423.

Mulloy, A. L., Keen, S. J., and Malven, P. V., 1979, Absorption of orally administered bovine prolactin by neonatal rats, Biol. Neonate, 36: 148-153.

Nair, R. M. G., Sarda, A. K., Barnes, M. A., and Phansey, S., 1983, Elevated LHRH levels in human milk, Endocrinol. Exper., 17: 335-342.

Oberkotter, L. V., Pereira, G. R., Paul, M. H., Ling, H., Sasanow, S., Farber, M., 1985, Effect of breast-feeding vs formula feeding on circulating thyroxine levels in premature infants, J. Pediatr., 106: 822-825, 1985.

Oberkotter, L. V., and Tenore, A., 1983, Separation and radioimmunoassay of T_3 and T_4 in human breast milk, Horm. Res., 17: 11-18.

Pierce, A. E., Risdall, P. C., and Shaw, B., 1964, Absorption of orally administered insulin by the newly born calf, J. Physiol. 171: 203-215.

Pollack, P. F., Goda, T., Thornburg, W., Edmond, J., Korc, M., and Koldovský, O., 1986, Effect of enterally-fed epidermal growth factor (EGF) on neonatal rat colon, Gastroenterology, 90: 1588 (abstract).
Read, L. C., Upton, F. M., Francis, G. L., Wallace, J. C., Dahlenberg, G. W., and Ballard, F.J., 1984, Changes in the growth-promoting activity of human milk during lactation, Pediatr. Res., 18: 133-139.
Read, L. C., Francis, G. L., Wallace, J.C., and Ballard, F.J., 1985, Growth factor concentrations and growth-promoting activity in human milk following premature birth, J. Dev. Physiol., 7: 135-145.
Reid, B., Smith, H., and Friedman, Z., 1980, Prostaglandins in human milk, Pediatrics, 66: 870-872.
Revsin, B., Lemen, R., and Koldovský, O., 1982, Fate of prostaglandin E_2 in suckling rats after intragastric administration, Biochim. Biophys. Acta, 711: 101-106.
Richardson, T., McGann, T. C. A., and Kearney, R. D., 1980, Levels and location of adenosine 5'-triphosphate in bovine milk, J. Dairy Res. 47: 91-96.
Robert, A., 1979, Cytoprotection by prostaglandins, Gastroenterology, 77: 761-767.
Robertson, J. D., 1945, The preparation and biological effects of iodinated proteins. 5. The effect on basal metabolism of milk from cows fed with iodinated protein, J. Endocrinol. 4: 300-304.
Sack, J., Amado, O., and Lunenfeld, B., 1977, Thyroxine concentration in human milk, J. Clin. Endocrinol. Metab., 45: 171-173.
Sack, J., Frucht, H., Amado, O., Brish, M., and Lunenfeld, B., 1979, Breast milk thyroxine and not cow's milk may mitigate and delay the clinical picture of neonatal hypothyroidism, Acta Paediatr. Scand., 277 (suppl): 54-56.
Sack, J., Frucht, H., Amado, O., and Lunenfeld, B., 1978, Thyroxine and thyrotropin-releasing hormone in human milk, Isr. J. Med. Sci.,14: 408-409.
Salakhova, N.S., Karimova, S. F., and Kallikorm, A. P., 1980, Regulation of lactation and the content of thyroid hormones in the milk of puerperants and in food products for supplementary feeding of children, Problem Endocrinol. (in Russian), 26: 21-25.
Sanchez-Pozo, A., Pita, J. L., Martinez, A., Molina, J. A., Sanchez-Medina, P., and Gil, A., 1986, Effects of dietary nucleotides upon lipoprotein pattern of newborn infants, Nutr. Res., in press.
Sarda, A. K., and Nair, R. M. G., 1981, Elevated levels of LRH in human milk, J. Clin. Endocrinol. Metab., 52: 826-828.
Sato, T., and Suzuki, Y., 1979, Presence of triiodothyronine, no detectable thyroxine and reverse triiodothyronine in human milk, Endocrinol. Jpn., 26: 507-513.
Schein, M., 1895, Das Schilddrusensekret in Milch, Wiener Med. Wochenschr., 5: 513-515.
Skåla, J. P., Koldovský, O., and Hahn, P., 1981, Cyclic nucleotides in breast milk, Am. J. Clin. Nutr., 34: 343-350.
Slebodzinski, A. B., and Gawecka, A., 1983, Passage of thyroid hormone into milk in rabbits, Endocrinol. Exper., 17: 243-254.
Starkey, R. H., and Orth, D. N., 1977, Radioimmunoassay of human epidermal growth factor (Urogastrone), J. Clin. Endocrinol. Metab., 45: 1144-1153.
Štrbåk, B., and Macho, L., 1977, Increase of serum thyrotropin (TSH) of suckling rats after thyroliberin (TRH) injection to lactating mothers, Biol. Neonate, 32: 331-335.

Štrbák, B., Macho, L., Knopp, J., and Struhárová, L., 1974, Thyroxine content in mother's milk and regulation of thyroid function of suckling rats, Endocrinol. Exper., 8: 59-69.
Štrbák, V., Macho, L., Kovác, R., Skultétyová, M., and Michalicková, J, 1976, Thyroxine (by competitive protein binding analysis) in human and cow milk and in infant formulas, Endocrinol. Exper., 10: 167-174.
Štrbák, V., Macho, L., Skultétyová, M., Michalicková, J., and Pohlová, G., 1983, Thyroid hormones in milk;: physiological approach - a review, Endocrinol. Exper., 17: 219-235.
Tapper, D., Klagsbrun, M., and Neumann, J., 1979, The identification and clinical implication of human breast milk mitogen, J. Pediatr. Surg., 14: 803-808.
Tenore, A., Oberkotter, L. V., Koldovský, O., Parks, J. S., and Vanderberg, C. M., 1981, Thyrotropin in human breast milk, Horm. Res., 14: 193-200.
Tenore, A., Parks, J. S., and Bongiovanni, A. M., 1977, Relationship of breast feeding to congenital hypothyroidism, in: "Proceedings of Symposium of Recent Progress in Pediatric Endocrinology", pp. 213-222.
Tenore, A., Parks, J., Gasparo, M., and Koldovský, O., 1980, Thyroidal response to peroral TSH in suckling and weaned rats, Am. J. Physiol., 238: E428-E430.
Thornburg, W., and Koldovský, O., 1986, Growth factors in milk: their effects on the developing gastrointestinal tract, in: Prenatal and Perinatal Biology and Medicine, Vol. II, eds., Kretchmer, N., Quilligan, J., Johnson, J., New York: Gordon and Breach Science Publishers, in press.
Thornburg, W., Magun, B., Matrisian, L., and Koldovský, O., 1984a, Effect of maturation on gastrointestinal absorption of epidermal growth factor in rats, Pediatr. Res., 18: 215A.
Thornburg, W., Matrisian, L., Magun, B., and Koldovský, O., 1984b, Gastrointestinal absorption of epidermal growth factor in suckling rats, Am. J. Physiol., 246: G80-G85.
Thornburg, W., Rao, R. K., Grimes, G., and Koldovský, O., 1986, Absorption of epidermal growth factor from isolated stomach and intestinal segments of rats, Pediatr. Res., 20: 250A.
Varma, S. K., Collins, M., Row, A., Haller, W. S., and Varma, K., 1978, Thyroxine, tri-iodothyronine, and reverse tri-iodothyronine concentrations in human milk, J. Pediatr., 93: 803-806.
Vaucher, Y., Tenore, A., Grimes, J., Krulich, L., and Koldovský, O., 1983, Absorption of TSH and ACTH in biologically active form from the gastrointestinal tract of suckling rats, Endocrinol. Exper., 17: 327-333.
Vigouroux, E., and Rostaqui, N., 1980, Particular aspects of thyroid function development in the postnatal rat with special reference to interrelationships between mother and young, Reprod. Nutr. Dev., 20: 209-215.
Vigouroux, E., Rostaqui, N., and Fenerole, J. M., 1980, Estimation of hormonal and non-hormonal iodine uptake from maternal milk in suckling rats, Acta Endocrinol., 93: 332-38.
Werner, H., Amarant, T., Fridkin, M., and Koch, Y., 1986, Growth hormone releasing factor-like immunoreactivity in human milk, Biochem. Biophys. Res. Comm., 135: 1084-1089.
Werner, H., Amarant, T., Millar, R. P., Fridkin, M., and Koch, Y., 1985, Immunoreactive and biologically active somatostatin in human and sheep milk, Eur. J. Biochem., 148: 353-357.

Werner, S., Widstrom, A-M., Wahlbert, V., Eneroth, P., and Winberg, J., 1982, Immunoreactive calcitonin in maternal milk and serum in relation to prolactin and neurotensin, Early Hum. Dev., 6: 77-82.
Whitworth, N. S., and Grosvenor, C. E., 1978, Transfer of milk prolactin to the plasma of neonatal rats by intestinal absorption, J. Endocrinol., 79: 191-199.
Wolford, S. T., and Argoudelis, C. J., 1979, Measurement of estrogen in cow's milk, human milk, and dietary products, J. Dairy Sci., 62: 1458-1463.
Yaida, N., 1929, Ovarial hormone in blood of pregnant women, of pregnant animals; ovarial hormone in urine of pregnant women; ovarial hormone in milk of pregnant animals, Trans. Jap. Path. Soc., 19: 93-101.

SUMMARY OF WORKSHOP: GROWTH FACTORS, HORMONES AND INDUCERS

Otakar Koldovský

Department of Pediatrics
University of Arizona College of Medicine
Tucson, Arizona 85724

Growth Factors, Hormones, and Inducers

The role of two important groups of substances (growth factors and casomorphins) with hormone or hormone-like effects present in breast milk was discussed in this session.

Kidwell and coworkers reviewed the production of various growth factors using normal mammary cells in culture; Dai et al demonstrated that growth factors isolated from milk given subcutaneously to mice increased gastric and duodenal mucus formation; and Read and coworkers reported the absorption of epidermal growth factor from the small intestine in sheep occurred via blood circulation.

Teschemacher and coworkers summarized their older and recent studies on interesting groups of substances with opioid effects, casomorphins, digestive products of casein. Read, in her discussion, reported data on intestinal absorption of casomorphins using the same sheep model as that used in epidermal growth factor studies.

INTESTINAL ABSORPTION OF EPIDERMAL GROWTH FACTOR IN NEWBORN LAMBS

Leanna C. Read , Susan M. Gale* and Carlos George-Nascimento**

Dept of Animal Sciences, University of Adelaide, Waite Agricultural Research Institute, Glen Osmond, S.A., 5064, Australia. *Dept of Biochemistry, University of Adelaide, Adelaide, S.A., 5000 Australia **Chiron Corporation, Emeryville, CA 94608

INTRODUCTION

Milk contains high concentrations of numerous growth factors including epidermal growth factor (EGF)[1] and insulin-like growth factor-1 (IGF-1)[2,3,4], observations that have led to the suggestion that milk growth factors may play an important role in infant growth and development.[1,4] This would be possible only if growth factors survive digestion in the neonatal gastrointestinal tract. Moreover, direct actions on tissues other than the gut mucosa would further require that substantial amounts of milk growth factors are absorbed intact into the neonatal circulation.

On the basis of the above criteria, EGF may represent a very important growth factor in human milk. Firstly, the EGF concentration in human milk (30-60 µg EGF/l) is 10-fold that required for maximal mitogenic responses in cultured cells.[1] Secondly, EGF is an acid-stable protein that can probably survive gastrointestinal digestion, at least in the newborn rat[4,5] and finally, EGF is mitogenic for gut[6] and lung,[7] tissues that are frequently immature following premature birth.

In the present study, we have determined in newborn lambs, both the proportion of EGF that reaches the small intestine intact and the <u>in vivo</u> rates of intestinal absorption of EGF into blood and lymph. The rates of absorption of EGF (molecular weight 6000 D) have been compared with those of a much larger protein, human immunoglobulin (IgG, 150000 D).

METHODS

Intestinal Processing of EGF

Synthesis of recombinant human EGF, production of a polyclonal antiserum against human EGF and the method of protein iodination are described elsewhere[8,9]. Chronic gastric and re-entrant intestinal catheters were inserted into anaesthetized lambs of 5-10 days of age, using surgical techniques similar to those described by Brown et al.[10] The re-entrant catheter was placed near the midpoint of the small intestine. Lambs were returned to their mothers after they regained consciousness and generally they resumed suckling within several hours.

To determine the intestinal processing of EGF, ^{125}I-labelled EGF (2-4 MBq, 1-2 µg EGF) was infused as a bolus (EGF infusate) into the gastric catheter, the re-entrant catheter was disconnected and the entire intestinal contents were collected over the next 3 h into tubes kept on ice. No sedation of lambs was required during the 3 h experimental period. All samples were centrifuged at 4°C for 30 min at 40000 x g and the fat-free infranatant recovered. For determination of the percentage of immunoreactive EGF, aliquots (100 ul) of infranatant or EGF infusate were incubated at 4°C with polyclonal antiserum to EGF (1/5000 dilution). Buffers and precipitation of bound radioactivity were as described previously[9]. The ratio of the % of radioactivity bound to antiserum in samples and infusate was used to determine the proportion of immunoreactive ^{125}I-labelled EGF in intestinal and gastric samples.

Intestinal Absorption of Proteins

The in vivo auto-perfused intestine, described by Windmueller and Spaeth in adult rats[11], was adapted to newborn lambs. Briefly, lambs were anaesthetized with halothane and a segment of small intestine, drained by a single vein, was exteriorized into a 38°C bath. The segment ends and apposing veins were ligated, following placement of a luminal catheter. Catheters were also placed in the main intestinal lymph duct for quantitative collection of lymph from the segment, in the carotid artery for blood pressure monitoring, and in the jugular vein for transfusion of heparinized sheep blood. Finally, the single vein draining the segment was cannulated to allow collection of the entire venous output and measurement of the blood flow rate. Following an equilibration period of approximately 15 min, 1 ml of a Earle's balanced salt solution containing EGF (50 µg, 8.3 nmol) and IgG (5 mg, 33 nmol) was infused into the lumen of the closed segment. Blood and lymph samples were then collected for the next 30-60 min into tubes kept on ice. Samples were centrifuged at 900 x g (4°C) for 30 min, after which the plasma and lymph supernatants were stored frozen for later EGF and IgG radioimmunoassays[9].

RESULTS AND DISCUSSION

Intestinal Processing of EGF

Determination of the immunoreactivity of ^{125}I in gastric samples taken at the end of the experimental period indicated that EGF was not degraded significantly in the stomach of 7-13 day lambs (results not shown). Moreover, Fig. 1 shows that over the entire 3 h period, 70-90% of the radioactivity reached the lamb small intestine as immunologically-intact EGF. Several interpretations of these results should be considered. Firstly, they could indicate that very little degradation occurred in the lamb gastrointestinal tract. Alternatively, EGF may have been degraded to fragments which retained immunological reactivity, but not necessarily biological activity. This explanation appears unlikely because preliminary evidence suggests that ^{125}I-EGF in intestinal samples also retained the ability to bind to cell-surface receptors (results not shown). Lastly, if substantial amounts of radioactivity were absorbed from the gut at sites proximal to the intestinal catheter, the data may not reflect the true percentage of EGF reaching the small intestine intact. However, over 50% of infused radioactivity was recovered in gastric and intestinal samples, indicating that a minimum of 30-50% of intragastric EGF would still have reached the small intestine intact. Although further studies will be needed to verify the structural integrity of ingested EGF, it appears likely that a large proportion of orally-derived EGF survives digestion in the neonatal gastrointestinal tract, a conclusion supported by results obtained by Koldovsky's group in studies with suckling rats[4,5].

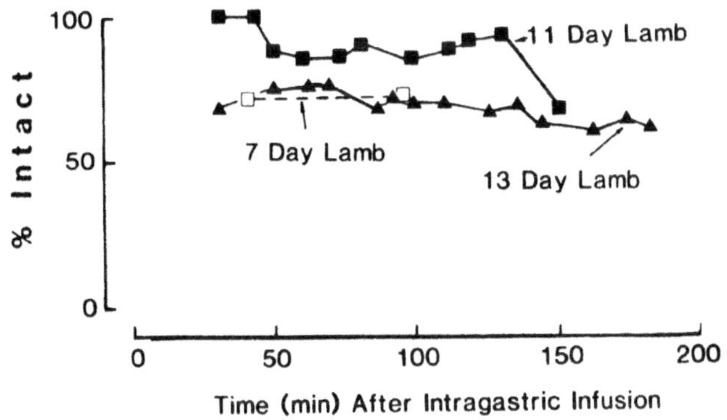

Fig. 1. % ^{125}I-EGF reaching the small intestine intact in lambs. ^{125}I-EGF was infused into the stomach at 0 min. Data are shown for one lamb at each age.

If the results of the present study can be extrapolated to the human infant, the EGF concentration present in the lumen of the small intestine would be in excess of 10 µg/l. Considering that similar concentrations of EGF have been shown to exert mitogenic effects on neonatal intestinal mucosal cells in culture[6], it appears likely that milk-derived EGF could play an important role in the growth and maturation of the intestine through a direct topical action on mucosal cells. Furthermore, it is conceivable that luminal EGF could also induce mitogenic responses indirectly by stimulating the release of gastrointestinal peptides such as gastrin and enteroglucagon, which in turn enhance the growth or maturation of various tissues in addition to the gut mucosa.

Intestinal Absorption of EGF

A representative intestinal perfusion is illustrated in Fig. 2. During the experimental period, blood pressure and heart rate remained steady (results not shown), while rectal and intestinal segment temperature were maintained at 40°C and 38°C, respectively. In most experiments, blood flow rates and the rates of absorption of EGF and IgG into blood remained steady throughout the experiment (Fig. 2). The mean rates of absorption of EGF and IgG into blood and lymph are given in Table 1 for lambs of ages varying between 1 and 84 days of age.

As expected from previous studies on the intestinal absorption of immunoglobulins in newborn lambs,[12] substantial amounts of intact IgG were taken across the auto-perfused intestine in the day 1 lamb, mostly into lymph. Intestinal closure was apparent by day 3. In contrast, EGF was absorbed into blood, not lymph, and although the rate of uptake varied considerably between lambs, there was no dependence on age (Table 1).

The mean EGF concentration in efferent venous blood averaged 2 ng/ml (results not shown). Considering that this concentration would produce marked mitogenic responses in cultured cells, it appears that physiologically-active amounts of EGF may be absorbed across the neonatal small intestine. However, further studies are needed to verify that immunological reactivity truly reflects biologically-active EGF and to determine whether similar rates of absorption would occur following luminal infusion of lower, more physiological amounts of EGF.

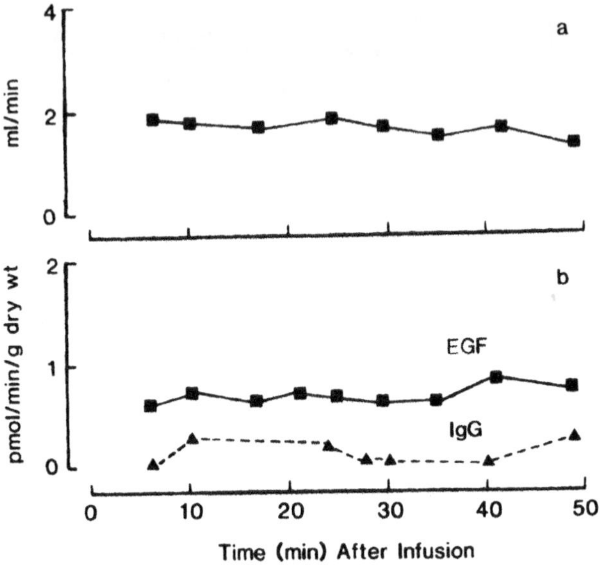

Fig. 2 Auto-perfused lamb intestinal segment. Representative experiment (3 day-old lamb) showing: (a) blood flow rate; (b) rates of absorption of IgG and EGF into blood.

Table 1. Rates of absorption of IgG and EGF into blood and lymph from the auto-perfused lamb intestine.

Age of Lamb (days)	IgG Uptake (pmol/min/g)		EGF Uptake (pmol/min/g)	
	Lymph	Blood	Lymph	Blood
1	4.90	0.60	0	0.47
3	0.01	0.01	0	0.70
4	0	0	0	0.87
6	0.01	0	0	0.10
17	0	0	0	1.69
35	0	0.01	0	0.72
84	0	0	0	1.01

Rates (pmol EGF or IgG absorbed /min/g dry wt intestine) are the means over the entire experimental period, and were calculated by (concentration x flow rate of blood or lymph)/dry wt of intestine. Data are given for one lamb at each age.

CONCLUSIONS

We have obtained evidence suggesting that 60-90% of intragastrically-administered EGF reaches the small intestine intact in newborn lambs. Furthermore, significant amounts of immunologically-intact EGF are absorbed from the small intestine into blood, but not lymph, by mechanisms that appear to differ from those responsible for uptake of IgG. Milk may therefore provide an important source of EGF to the suckling infant.

ACKNOWLEDGEMENTS

The authors would like to thank Jamie McNeil and Callum Gillespie for their skilled and enthusiastic technical assistance. S.M. Gale is the recipient of a NH & MRC (Australia) Biomedical Postgraduate Award. The work was supported by a grant from the Channel 10 Children's Hospital Medical Research Foundation of S.A.

REFERENCES

1. L. C. Read, F. M. Upton, G. L. Francis, J. C. Wallace, G. W. Dahlenburg, and F.J. Ballard, Changes in the growth-promoting activity of human milk during lactation, Pediatr. Res. 18: 133 (1984).
2. G. L. Francis, L. C. Read, F. J. Ballard, C. J. Bagley, F. M. Upton, P. M. Gravestock, and J. C. Wallace, Purification and partial sequence analysis of insulin-like growth factor 1 from bovine colostrum, Biochem. J. 233: 207 (1985).
3. R. C. Baxter, Z. Zaltsman, and J. R. Turtle, Immunoreactive somatomedin-C/insulin-like growth factor 1 and its binding protein in human milk, J. Clin. Endoc. Metab. 58: 955 (1984).
4. O. Koldovsky, A. Bedrick, P. Pollack, R. K. Rao, and W. Thornburg, Hormones in milk: their presence and possible physiological significance, in: Effects of Human Milk on the Recipient Infant; International Meeting on Human Lactation held in Konstanz, Germany, to be published by Plenum Press, New York.
5. W. Thornburg, L. Matrisian, B. Magun, and O. Koldovsky, Gastro-intestinal absorption of epidermal growth factor in suckling rats, Amer. J. Physiol. 246: G80 (1984).
6. J.-F. Beaulieu, and R. Calvert, The effect of epidermal growth factor (EGF) on the differentiation of the rough endoplasmic reticulum in fetal mouse small intestine in organ culture, J. Histochem. Cytochem. 29: 765 (1981).
7. H. W. Sundell, M. E. Gray, F. S. Serenius, M. B. Escobedo, and M. T. Stahlman, Effects of epidermal growth factor on lung maturation in fetal lambs, Amer. J. Pathol. 100: 707 (1980).
8. M. S. Urdea, J. P. Merryweather, G. T. Mullenbach, D. Coit, U. Heberlein, P. Valenzuela, and P.J. Barr, Chemical synthesis of a gene for human epidermal growth factor urogastrone and its expression in yeast, Proc. Natl. Acad. Sci., USA 80: 7461 (1983).
9. L. C. Read, L. Summer, S. M. Gale, C. George-Nascimento, F. J. Ballard, and J. C. Wallace, Immunological, receptor-binding and biological properties of synthetic-gene reconmbinant human epidermal growth factor: comparison with the natural growth factor from human urine and milk, J. Endoc. 109: 245 (1986).
10. G. F. Brown, D. G. Armstrong, and J. C. MacRae, The establishment in one operation of a cannula into the rumen and re-entrant cannulae into the duodenum and ileum of the sheep, Brit. Vet. J. 124: 78 (1968).

11. H. G. Windmueller, and A. E. Spaeth, Vascular perfusion of rat small intestine: metabolic studies with isolated and in situ preparations, Fed. Proc. 36: 177 (1977).
12. F. W. R. Brambell, The passive immunity of the young animal, Biol. Rev. 33: 120 (1958).

EFFECTS OF HUMAN MILK GROWTH FACTOR ON GASTRIC AND DUODENAL MUCUS FORMATION

S. Dai, M. Klagsbrun*, C. W. Ogle, and Y. Shing*

Department of Pharmacology, Faculty of Medicine
University of Hong Kong, Hong Kong
*Departments of Biological Chemistry and Surgery
Harvard Medical School and The Children's Hospital
Boston, MA 02115

INTRODUCTION

A polypeptide growth factor has been isolated from human milk, and designated human milk growth factor III (1-3). It is chemically related to epidermal growth factor-urogastrone, having a molecular weight of about 6,000 and an isoelectric point of about 4.5 (1-3). EGF-urogastrone is a potent mitogen for a variety of cells (4) and is known to stimulate the growth and maturation of the intestinal mucosa (5, 6). The detection and isolation of epidermal growth factor (EGF) in human milk have also been reported by several other laboratories (7,8,9,10). Besides being a mitogen, EGF-urogastrone is an inhibitor of gastric acid secretion (11) and has been used to promote healing of ulcers (12). Thus, it is conceivable that human milk growth factor III (HMGF III) may also have urogastrone-like effects on the gastrointestinal tract. In a previous study, we demonstrated that HMGF III was capable of dose-dependently protecting mice against the formation of duodenal ulcers induced by cysteamine (13). These results are in agreement with those reported by Kirkegaard et al (14) that EGF significantly inhibits the formation of cysteamine-induced duodenal ulcer in the rat. However, the mechanisms remain unclear. It has been reported that cysteamine causes a long-lasting hypersecretion of gastric acid (15) and inhibits secretion of alkaline mucus from duodenal Brunner's glands (16). Therefore, it is possible that HMGF III could have exerted its anti-ulcer effects in mice by altering gastric acid and/or duodenal mucus synthesis in a pattern opposite to that evoked by cysteamine. In order to clarify the mechanisms of the anti-ulcer actions of HMGF III, the present study examines its effects on gastric acid secretion and mucus formation in the mouse stomach and duodenum.

MATERIALS AND METHODS

Isolation of Human Milk Growth Factor III

Human milk was kindly provided by Dr. Cutberto Garza of the Baylor College of Medicine (Houston, TX). HMGF III was prepared as previously described (1). In brief, after defatting, acidification, and dialysis, the human milk samples were applied to a Sephadex G-100 column (5 X 90 cm) equilibrated with 0.1 M NaCl and 0.01 M sodium acetate (pH 4.3) at 4°C, and chromatographed at a flow rate of 40 ml/hour. Fractions were collected, measured for absorbance at 280 nm, dialyzed, and tested for ability to stimulate DNA synthesis in confluent quiescent monolayers of BALB/C 3T3 cells. The active fractions were then dialyzed, lyophilized and stored at -20°C. Appropriate quantities of HMGF III were prepared immediately before use by dissolving the polypeptide in saline.

Experimental Animals

Male ICR mice, weighing 20-25 g, were used. They were fasted for 16 hours before experiments, being allowed free access only to tap water.

Gastric Acid Secretion.

Shay's method of pyloric ligation under ether anesthesia (17) was employed to measure gastric acid secretion in the mice. The animals were sacrificed by rapid cervical dislocation after 3 hours following which their stomachs were removed and the gastric contents collected. Following centrifugation of each sample at 1200 X g for 15 min, the total acidity of the supernatant was determined by titration with 0.01 N NaOH to an end point of pH 7.4 using an autotitrator system (Model TTT80, Radiometer Copenhagen Ltd). Both the volume of gastric secretion (ml/100 g body wt/h) and total acid output (μEq HCl/100 g body wt/h) were recorded. Atropine sulfate (E. Merck) was used to demonstrate the responsiveness of the preparation to inhibition of gastric secretion. HMGF III 20, 40, 80 mg/kg or atropine sulfate 10 mg/kg was injected subcutaneously (s.c.) immediately after pyloric ligation. Similar volumes of saline were given by the same route and at the same time to the controls.

Gastric and Duodenal Mucus Synthesis in vivo and in vitro

The methods described by Corne et al (18) and Norris et al (19) with slight modifications, were employed for measuring mucosal wall mucus. Carbenoxolone sodium (a gift from Biorex Laboratories Ltd., London) was used to test the mucus-producing responsiveness of the mucosa.

For in vivo studies, the mice received s.c. injections of HMGF III 10, 20, 40, 80 mg/kg or saline 10 ml/kg in the case of the controls, or intragastric administration of carbenoxolone sodium 1200 mg/kg or its solvent (distilled water) 10 ml/kg for the controls, 1 hour before sacrifice. Following cervical dislocation, the glandular portion of stomachs and the proximal 2 cm of duodena were rapidly removed, opened along the line of greater curvature, and weighed. After rinsing in saline, the tissues were stained by immersion for 90 min in 0.1% w/v alcian blue (Sigma) solution in 50 mM sodium acetate (pH 5.8) containing 0.15 M sucrose. Excess dye was removed by washing the tissues for 2 X 15-min periods in 0.25 M sucrose. The alcian blue-stained mucus was then destained by soaking it, with periodic shaking, in 0.5 M magnesium chloride for 2 hours. The blue supernatant was decanted, clarified by shaking with diethyl ether, and the optical density of the aqueous layer read against a buffer blank at 600 nm, using a spectrophotometer (Cary 219, Varian). The amount of mucus synthesis was expressed by the alcian blue-binding capacity in terms of μg/g wet tissue.

For in vitro studies, untreated mice were sacrificed by cervical dislocation, following which the glandular portion of the stomach and the proximal 2 cm of duodenum were rapidly removed, opened and weighed. The isolated gastric and duodenal tissues were separately incubated at 37°C, in oxygenated Krebs solution containing 1% N-acetyl-l-cysteine (Sigma).

Table 1. Effects of HMGF III on Gastric Secretion in Mice

Treatment (s.c.)	Number of mice	Volume of gastric secretion (ml/100 g/h)	Total acid output (μEq HCl/100 g/h)
Saline			
10 ml/kg	13	1.02 ± 0.15	102.33 ± 15.80
Atropine sulfate			
10 mg/kg	9	0.18 ± 0.06[a]	18.03 ± 1.71[a]
HMGF III			
20 mg/kg	10	0.78 ± 0.11	84.45 ± 10.43
40 mg/kg	10	0.79 ± 0.16	73.56 ± 10.36
80 mg/kg	10	0.94 ± 0.11	108.20 ± 14.92

The values are the means ± SEM.
a, $P < 0.001$ when compared with the saline-treated controls.

After 30 min, the tissues were removed, thoroughly rinsed in fresh Krebs solution, and incubated in the same solution containing HMGF III 2,, 4, 8 μg/ml, or carbenoxolone sodium 0.3 mg/ml. Incubation was terminated after 1.5 hour by transferring the tissues to ice-cold 0.25 M sucrose, and immersion for 5 min. The tissues were then stained with alcian blue, destained with magnesium chloride, and the optical densities of the supernatants read at 600 nm as described above.

Statistical Analysis

The data were expressed as the means ± SEM, and analyzed for significance of differences by Student's t-test.

RESULTS

Gastric acid secretion

Table 1 shows the effects of HMGF III on gastric acid secretion in mice. Treatment with atropine sulfate 10 mg/kg s.c. significantly reduced both the volume of gastric secretion and total acid output ($P < 0.001$ for both). This shows that the present method of measuring gastric acid secretion is sufficiently sensitive to be able to measure the responses to a drug which is known to inhibit gastric secretion. However, s.c. injection of HMGF III did not significantly change either the volume of gastric secretion or total acid output at a dose of up to 80 mg/kg.

Gastric and Duodenal Mucus Synthesis

The effects of HMGF III on gastric and duodenal mucus synthesis in mice *in vivo* and *in vitro* are shown in Tables 2 and 3, respectively.

Table 2. Effects of HMGF III on Gastric and Duodenal Mucus Synthesis *in vivo* in Mice

Treatment	Number of mice	Alcian blue-binding capacity (μg/g wet tissue)	
		Stomach	Duodenum
Distilled Water			
10 ml/kg p.o.	8	459 ± 48	1035 ± 91
Carbenoxolone sodium			
1200 mg/kg p.o.	9	803 ± 43[b]	1155 ± 116
Saline			
10 ml/kg s.c.	8	664 ± 34	1038 ± 90
HMGF III, s.c.			
10 mg/kg	8	794 ± 73	1188 ± 87
20 mg/kg	9	856 ± 91	1223 ± 55
40 mg/kg	9	984 ± 48[b]	1425 ± 134[a]
80 mg/kg	9	971 ± 49[b]	1545 ± 74[b]

p.o. = intragastric administration.
s.c. = subcutaneous injection.
The values are the means ± SEM.
a, $P < 0.05$; b, $P < 0.001$ when compared with the appropriate controls.

Intragastric administration of carbenoxolone sodium 1200 mg/kg significantly enhanced gastric mucus synthesis ($P < 0.01$) (Table 2). Mucus production by the isolated stomach and duodenum incubated *in vitro* with carbenoxolone 0.3 mg/ml was significantly greater than those of the controls incubated without drug ($P< 0.01$ and $P < 0.05$, respectively) (Table 3). These results demonstrate the responsiveness of the present method, for quantitative measurement of mucosal wall mucus, to a drug known to stimulate mucus synthesis. The inability of carbenoxolone to increase significantly the duodenal mucus synthesis *in vivo* is expected because the drug is known to be largely, and rapidly, absorbed in the stomach following oral administration (20). In the *in vivo* study, s.c. injection of HMGF III significantly increased both gastric and duodenal mucus formation in a dose-dependent manner (Table 2); the greatest responses were seen with 40 mg/kg. The use of a larger dose, 80 mg/kg, caused either no further increase or only a slight elevation in the gastric and duodenal mucus synthesis, respectively. Similar findings were observed with the *in vitro* study. Both gastric and duodenal mucus production were stimulated in a dose-dependent fashion by HMGF III (Table 3).

Table 3. Effects of HMGF III on Gastric and Duodenal Mucus Synthesis *in vitro* in Mice

Treatment	Number of preparations	Alcian blue-binding capacity (μg/g wet tissue) Stomach	Duodenum
Saline			
1 ml	10	397 ± 42	1245 ± 93
Carbenoxolone sodium			
0.3 mg/ml	9	574 ± 37[c]	1583 ± 84[a]
HMGF III			
2 μg/ml	9	410 ± 33	1297 ± 91
4 μg/ml	10	477 ± 25	1635 ± 90[b]
8 μg/ml	10	538 ± 25[b]	1860 ± 161[c]

The values are the means ± SEM.
a, $P < 0.05$; b, $P < 0.02$; c, $P < 0.01$ when compared with the controls.

DISCUSSION

This investigation reveals that s.c. injection of HMGF III effectively increases gastric and duodenal mucus formation in mice. It is generally thought that increased mucus production tends to enhance mucosal resistance to peptic ulcer formation (20). Thus, it would appear that the anti-ulcer effect of HMGF III against cysteamine-induced duodenal ulcers in mice (13) could be due to increased duodenal mucus synthesis. As gastric mucus formation is also stimulated, it is likely that HMGF III may also be effective in preventing gastric ulceration induced by other methods. The present study also shows that the isolated stomach and duodenum when incubated *in vitro* with HMGF III are able to produce more mucus, thus, indicating that HMGF III can exert a direct action on the mouse stomach and duodenum to increase mucus synthesis. This finding, coupled with the ability of HMGF III to resist inactivation by HCl (3), suggests the likelihood that HMGF III could be more potent in stimulating mucus production and antagonizing ulcer formation if administered orally.

It has been well known that epidermal growth factors derived from the mouse submaxillary gland and from human urine (urogastrone) can inhibit gastric acid secretion in various species of experimental animals and in man (21). As HMGF III exhibits chromatographic and biological characteristics similar to those of EGF-urogastrone (1-3), it might be expected to have an inhibitory action on gastric acid secretion. However, the present study indicates that HMGF III does not influence gastric acid secretion with the HMGF III given in quantities which are sufficient to significantly increase gastric and duodenal mucus formation in mice.

This finding is in accordance with previous investigations that gastric erosions induced by aspirin were prevented by EGF given in quantities not sufficient to inhibit gastric acid secretion (22). Furthermore, it has recently been demonstrated that oral administration of the synthetic human EGF/urogastrone has a stimulating effect on healing of duodenal ulcers in rats (23). However, while this synthetic polypeptide is a potent inhibitor of gastric acid secretion when administered intravenously, it has no effect on acid secretion when given intraduodenally, which suggests that the effect of EGF/urogastrone is probably due to a direct action on the duodenal mucosa. In our experiments, only partially purified HMGF III was used because not enough pure growth factor was available to treat the large number of mice. Therefore, the exact quantities of active component in the HMGF III preparations administered to the mice could not be estimated accurately. However, since the equivalent doses which stimulate mucus formation but are without effect on gastric acid secretion are able to prevent the formation of cysteamine-induced duodenal ulcers (13), it is conceivable that the stimulatory effect of HMGF III on gastric and duodenal mucus synthesis may play an important role in its anti-ulcer activity.

REFERENCES

1. Y. W. Shing and M. Klagsbrun, Human and bovine milk contain different sets of growth factors, Endocrinology 115:273 (1984).
2. Y. W. Shing and M. Klagsbrun, Isolation of growth factors from human milk, in: "Methods in Molecular and Cellular Biology", vol. 1, pp. 159-179, G. Sato, D. Sirbasku, and D. Barnes, eds., Alan R. Liss, New York (1984).
3. M. Klagsbrun and Y. W. Shing, Growth promoting factors in human and bovine milk, in: "Growth and Maturation Factors", vol. 2, pp. 161-192, G. Guroff, ed., John Wiley & Sons, New York (1984).
4. G. Carpenter and S. Cohen, Epidermal growth factor, Ann Rev Biochem 48:139 (1979).
5. C. Malo and D. Ménard, Influence of epidermal growth factor on the development of suckling mouse intestinal mucosa, Gastroenterology 83:28 (1982).
6. R. Calvert, J. F. Beaulieu, and D. Ménard, Epidermal growth factor (EGF) accelerates the maturation of fetal mouse intestinal mucosa in utero, Experientia 38:1096 (1982).
7. R. H. Starkey and D. N. Orth, Radioimmunoassay of human epidermal growth factor (urogastrone), J Clin Endocrinol Metab 45:1144 (1977).
8. G. Carpenter, Epidermal growth factor is a major growth-promoting agent in human milk, Science 210:198 (1980).
9. P. E. Petrides, M. Hosang, E. Shooter, F. S. Esch, and P. Böhlen. Isolation and characterization of epidermal growth factor from human milk, FEBS Letters 187:89 (1985).
10. L.C. Read, L. Summer, S.M. Gale, C. George-Nascimento, F.J. Ballard, and J.C. Wallace, Properties of synthetic-gene recombinant human epidermal growth factor: comparison with the natural growth factor from human urine and milk, J. Endocrinology 109:245 (1986).
11. J. M. Bower, R. Camble, H. Gregory, E. L. Gerring, and I. R. Willshire, The inhibition of gastric acid secretion by epidermal growth factor, Experientia 31:825 (1975).
12. H. Gregory, J. M. Bower, and I. R. Willshire, Urogastrone and epidermal growth factor, in: "Growth Factors", pp. 75-84, K. W. Kistrup, J. H. Nielsen, eds., Pergamon, Oxford (1979).
13. S. Dai, M. Klagsbrun, and Y. W. Shing, Human milk-derived growth factor prevents duodenal ulcer formation, Pediatric Res 19:916 (1985).
14. P. Kirkegaard, P. Skov Olsen, S. S. Poulsen, and E. Nexø, Epidermal

growth factor inhibits cysteamine-induced duodenal ulcers, Gastroenterology 85:1277 (1983).
15. W. G. Groves, J. H. Schlosser, and F. D. Mead, Acid hypersecretion and duodenal ulcers produced by cysteamine in rats, Res Commun Chem Pathol Pharmacol 9:523 (1974).
16. P. Kirkegaard, S. S. Poulsen, C. Halse, F. B. Loud, P. Skov Olsen, and J. Christiansen, The effect of cysteamine on the Brunner gland secretion in the rat, Scand J Gastroenterol 16:93 (1981).
17. H. Shay, D. C. H. Sun, M. Gruenstein, A quantitative method for measuring spontaneous gastric secretion in the rat, Gastroenterology 26:906 (1954).
18. S. J. Corne, S. M. Morrissey, and R. J. Woods, A method for the quantitative estimation of gastric barrier mucus, J Physiol (Lond) 242:116 (1974).
19. D. B. Norris, T. J. Rising, and T. P. Wood, In vitro measurement of gastric mucus synthesis by a dye binding technique, Br J Pharmacol 80:581 (1983).
20. R. M. Pinder, R. N. Brogden, P. R. Sawyer, T. M. Speight, R. Spencer, and G. S. Avery, Carbenoxolone: A review of its pharmacological properties and therapeutic efficacy in peptic ulcer disease, Drugs 11:245 (1983).
21. M. D. Hollenberg, Epidermal growth factor-urogastrone, a polypeptide acquiring hormonal status, Vitam Horm 37:69 (1979).
22. S. J. Konturek, T. Brzozowski, I. Piastucki, A. Dembinski, T. Radecki, A. Dembinska-Kiec, A. Zmuda, and H. Gregory, Role of mucosal prostaglandins and DNA synthesis in gastric cytoprotection by luminal epidermal growth factor, Gut 22:927 (1981).
23. P. Skov Olsen, S. S. Poulsen, K. Therkelsen, and E. Nexø, Oral administration of synthetic human urogastrone promotes healing of chronic duodenal ulcers in rats, Gastroenterology 90:911 (1986).

ß-CASOMORPHINS:

DO THEY HAVE PHYSIOLOGICAL SIGNIFICANCE?

Hansjörg Teschemacher

Rudolf Buchheim-Institut für Pharmakologie
der Justus Liebig-Universität,
Frankfurter Str. 107, D-6300 Gießen

INTRODUCTION

Milk represents an essential food for the newborn mammal. In addition to nutrients milk contains a variety of important materials, e.g. substances with immunological relevance[1] and a considerable number of hormones whose physiological significance is not yet clearly understood[2].

Recently, various opioids have been demonstrated to occur in bovine milk or casein hydrolysates[3,4]; however, only opioids derived from α-casein[5,6] or from ß-casein[7,8], respectively, may be regarded to be opioids of mammary origin. For the ß-casein-derived opioids, various efforts have been undertaken to ascertain whether these opioids might be cleaved from ß-casein in the newborn mammal after milk intake or whether they have already been produced in the pregnant or nursing female. When released under such in vivo conditions, they might play any role in mammalian reproduction as suggested also for endogenous opioids such as ß-endorphin[9].

ORIGIN AND AMINO ACID SEQUENCE OF ß-CASOMORPHINS

Recently, a fragment of the bovine ß-casein amino acid sequence[10], ß-casein (60-66), has been shown to display opioid activity[7,11,12]. In view of its origin and pharmacological

Table 1. Amino Acid Sequences of Bovine and Human ß-Casein Containing Segments With Opioid Activity, Human or Bovine ß-Casomorphins, Respectively (Boxed)

BOVINE ß- CASEIN:

BOVINE ß-CASOMORPHINS

1	ARG-GLU-LEU-GLU-GLU-LEU-ASN-VAL-PRO-GLY-GLU-ILE-VAL-GLU-SER-LEU-SER-SER-SER-GLU-GLU-SER-ILE-IHR-ARG
26	-ILE-ASN-LYS-LYS-ILE-GLU-LYS-PHE-GLN-SER-GLU-GLU-GLN-GLN-GLN-THR-GLU-ASP-GLU-LEU-GLN-ASP-LYS-ILE-HIS
51	-PRO-PHE-ALA-GLN-IHR-GLN-SER-LEU-VAL [TYR-PRO-PHE-PRO-GLY-PRO-ILE-PRO] ASN-SER-LEU-PRO-GLN-ASN-ILE-PRO
76	-PRO-LEU-IHR-GLN-THR-PRO-VAL-VAL-VAL-PRO-PRO-PHE-LEU-GLN-PRO-GLU-VAL-MET-GLY-VAL-SER-LYS-VAL-LYS-GLU
101	-ALA-MET-ALA-PRO-LYS-HIS-LYS-GLU-MET-PRO-PHE-PRO-LYS-TYR-PRO-VAL-GLN-PRO-PHE-THR-GLU-SER-GLN-SER-LEU
126	-THR-LEU-THR-ASP-VAL-GLU-ASN-LEU-HIS-LEU-PRO-PRO-LEU-LEU-LEU-GLN-SER-TRP-MET-HIS-GLN-PRO-HIS-GLN-PRO
151	-LEU-PRO-PRO-THR-VAL-MET-PHE-PRO-PRO-GLN-SER-VAL-LEU-SER-LEU-SER-GLN-SER-LYS-VAL-LEU-PRO-VAL-PRO-GLU
176	-LYS-ALA-VAL-PRO-TYR-PRO-GLN-ARG-ASP-MET-PRO-ILE-GLN-ALA-PHE-LEU-LEU-TYR-GLN-GLN-PRO-VAL-LEU-GLY-PRO
201	-VAL-ARG-GLY-PRO-PHE-PRO-ILE-ILE-VAL

HUMAN ß-CASEIN:

HUMAN ß-CASOMORPHINS

1	ARG-GLU-THR-ILE-GLU-SER-LEU-SER-SER-SER-GLU-GLU-SER-ILE-PRO-GLU-TYR-LYS-GLN-LYS-VAL-GLU-LYS-VAL-LYS
26	-HIS-GLU-ASP-GLN-GLN-GLN-GLY-THR-ASP-GLN-HIS-GLN-ASP-LYS-ILE-TYR-PRO-SER-PHE-GLN-PRO-GLN-PRO-LEU-ILE
51	[TYR-PRO-PHE-VAL-GLU-PRO-ILE-PRO] TYR-GLY-PHE-LEU-PRO-GLN-ASN-ILE-LEU-PRO-LEU-ALA-GLN-PRO-ALA-VAL-VAL
76	-LEU-PRO-VAL-PRO-GLN-PRO-GLU-ILE-MET-GLU-VAL-PRO-LYS-ALA-LYS-ASP-THR-VAL-TYR-THR-LYS-GLY-ARG-VAL-MET
101	-PRO-VAL-LEU-LYS-GLN-PRO-THR-ILE-PRO-PHE-PHE-ASP-PRO-GLN-ILE-PRO-LYS-LEU-IHR-ASP-LEU-GLU-ASN-LEU-HIS
126	-LEU-PRO-LEU-PRO-LEU-LEU-GLN-PRO-SER-MET-GLN-GLN-VAL-PRO-GLN-PRO-ILE-PRO-GLN-THR-LEU-ALA-LEU-PRO-PRO
151	-GLN-PRO-LEU-TRP-SER-VAL-PRO-GLU-PRO-LYS-VAL-LEU-PRO-ILE-PRO-GLN-GLU-VAL-LEU-PRO-TYR-PRO-VAL-ARG-ALA
176	-VAL-PRO-VAL-GLN-ALA-LEU-LEU-LEU-ASN-GLN-GLU-LEU-LEU-LEU-ASN-PRO-PRO-HIS-GLN-ILE-TYR-PRO-VAL-PRO-GLU
201	-PRO-SER-THR-THR-GLX-ALA-ASX-HIS-PRO-ILE-SER-VAL

activity, this heptapeptide was named ß-casomorphin-7. When it was shortened at the C-terminus by one, two or three amino acid residues, the resultant molecules were found to display opioid activity[13,14]; they were named ß-casomorphin-4, -5 or -6, respectively (for amino acid sequences see table 1). The bovine ß-casomorphin amino acid sequence is also contained in the amino acid sequences of ovine[15] and buffalo[16] ß-casein.

In 1984, the primary structure of human ß-casein was determined[17]. A comparison of the sequence of human with bovine ß-casein revealed a 10-residue-shifted alignment relationship and 47% identity. Thus, by analogy, human ß-casein (51-57) could be supposed to correspond to bovine ß-casein (60-66). In fact, except for two amino acid residues, the human ß-casein segment proved identical with the bovine ß-casein segment. It also displayed opioid activity as did its C-terminally shortened fragments. Thus, these peptides were named human ß-casomorphin-4, -5, -6 and -7[8,18] (for amino acid sequences see table 1).

OPIOID AND NON-OPIOID ACTIVITIES OF ß-CASOMORPHINS

Bovine[13,14] as well as human[8,18,19] ß-casomorphins have been shown to interact with opioid receptors in rat brain membranes. Their affinities were reported to be highest for µ-receptors, lower for other receptors such as δ-, κ- or ε-receptors. Therefore, ß-casomorphins are regarded to be selective µ-receptor ligands. Opioid effects on guinea pig ileum preparations have been reported for both bovine[13,20] and human[8,18] ß-casomorphins. For bovine ß-casomorphins, further opioid effects in mammals have been demonstrated, e.g. on the central nervous system such as analgetic[13,14,21] or electrophysiological[22] effects or on the endocrine system such as modulation of insulin[23], somatostatin[24], pancreatic polypeptide[25], thyrotropin[26] or prolactin[27] release.

Non-opioid activities have also been reported for intact ß-casomorphins or its derivatives or fragments. Bovine ß-casomorphin-7 and bovine ß-casomorphin-4 amide were shown to bind to specific binding sites of Tyr-MIF-1 in rat brain plasma membranes[28]. A human ß-casomorphin fragment was reported to possess immunostimulatory properties[29].

These as well as other ß-casomorphin effects have been reviewed[3,4].

DO ß-CASOMORPHINS HAVE PHYSIOLOGICAL SIGNIFICANCE?

For reasons outlined in the introduction, we suspect that ß-casomorphins may have important physiological functions. The most interesting species to study would be, of course, the human one. Whereas, however, at least certain studies in pregnant women after parturition are possible, studies in newborn babies as required in this context in general are unethical or even impossible. Thus, at least for the newborn mammal the bovine species seems to be the better choice; bovine ß-casomorphins, after all, have been identified.

Attempts to elucidate a physiological role of ß-casomorphins should concentrate primarily on collection of information about the presence and, subsequently, any function of ß-casomorphins in the respective organism.

Valuable information about the eventual presence of ß-casomorphins under *in vivo* conditions may be obtained from *in vitro* studies about their release from ß-casein and about their degradation. Evidence for the presence of ß-casomorphins in the respective organism under study can only result, however, from *in vivo* studies.

When administered like drugs, ß-casomorphins can elicit a series of effects as reported in the previous section. However, these effects depend on the sites at which ß-casomorphins are released and subsequently degraded in the respective organism. Possible ways to test the effects would be to use specific antagonists which block physiological responses eventually elicited by ß-casomorphins or to search for correlations between ß-casomorphin concentrations at any sites in the organism and the degree of physiological responses.

Presence of ß-Casomorphins Under in Vivo Conditions: Evidence From In Vitro Studies

Recently, a number of *in vitro* studies have been performed, which provide indication for the release of ß-casomorphins under *in vivo* conditions. Upon incubation of buffalo ß-casein with proteolytic enzymes occurring in the gastrointestinal tract a ß-casomorphin precursor was released as well as "ß-casomorphin-8" and a ß-casomorphin fragment[30]. Upon digestion of bovine milk by processes that mimic the natural gastrointestinal digestion process, ß-casomorphin immunoreactive materials were demonstrated at the "pancreatic" and at the "intestinal mucosa digestion level", whereas no such material was found at the "stomach digestion level"[31]. By less well defined enzymatic digestion procedures, ß-casomorphins or derivatives thereof were released from bovine casein[7,32] and a ß-casomorphin fragment was produced from human casein[29].

Data on the degradation of ß-casomorphins are available

as well. Bovine β-casomorphins seem to be resistant to pronase, thermolysin and carboxypeptidases A and B[11]. However, they have been shown to be degraded by carboxypeptidase Y[11] and, in particular, by several enzymes cleaving amino acid sequences specifically after proline residues (as contained in the bovine β-casomorphin sequence in positions two, four and six and in human β-casomorphin in positions two and six), i.e. dipeptidylpeptidase IV, proline-specific endopeptidase and post-proline cleaving enzyme[33,34]. One or several of these enzymes might be present in bovine plasma, since bovine β-casomorphins have been shown to be rapidly degraded in bovine plasma[35].

Although the reports on release and degradation of β-casomorphins under _in vitro_ conditions are preliminary, a release of β-casomorphins, precursors or derivatives thereof in the small intestine of newborns after milk intake appears to be likely. β-casomorphins might elicit any effects there, but would be degraded in the blood after their absorption from the gastrointestinal tract.

Presence of β-Casomorphins Under In Vivo Conditions: Evidence From In Vivo Studies

In vivo experiments are required to accept that β-casomorphins are of physiological importance. However, only a few studies have been performed to determine, whether β-casomorphins occur in newborn mammals after milk intake or in pregnant or nursing females.

In newborn calves, blood samples were collected before and after their first milk intake after birth and plasma extracts were analysed for β-casomorphins by radioimmunoassay[36]. In 20 out of 24 calves a β-casomorphin-7 immunoreactive material was found after milk intake, whereas none was detected in the calves before milk intake. However, the chromatographical characterization of this material showed that it was not identical with β-casomorphin-7, but rather represented a precursor thereof. In addition, in contrast to β-casomorphin-7, the material proved very stable against enzymatic attack in the newborn calves' plasma. From such a precursor, ß-casomorphin-7 could be cleaved at any site of action in the newborn calf to elicit opioid or non-opioid effects.

In order to obtain information whether ß-casomorphins might already occur in the organism of the mother, blood was collected from pregnant women and from lactating or non-lactating women after parturition and plasma extracts were analysed for ß-casomorphins by radioimmunoassay[37]. In fact, ß-casomorphin-8 immunoreactive materials were found in most of these women, whereas no such material was found in men or in non-pregnant women included in the study as controls. The material apparently contained ß-casomorphins as well as precursors thereof, from which ß-casomorphins could be cleaved in the immediate neighbourhood of any site of action in the women's organism.

Finally, although not in a homologous system, degradation under in vivo conditions has been shown for ß-casomorphins as well; after intracerebroventricular injection in rats, bovine ß-casomorphin-5 was observed to be degraded rapidly[38]. Thus, as could be expected from the respective in vitro studies, ß-casomorphins seem to be degraded under in vivo conditions, which is a prerequisite for the flexibility of a signal transfer system; ß-casomorphins might play the role of transmitters or hormones in such a system.

Possible Functions of ß-Casomorphins

In newborn calves after milk intake, a ß-casomorphin precursor has been shown to be present in the blood. It may be assumed that this precursor is also present in the gastrointestinal tract. The demonstration of ß-casomorphins or ß-casomorphin precursors in the small intestine of adult humans or minipigs after bovine milk or casein administration, respectively, is compatible with this assumption[31,39]. ß-casomorphins eventually cleaved from a precursor in the small intestine could elicit opioid effects in the intestinal wall, e.g. influence gastrointestinal motility[20]. They might, thus, participate in the regulation of gastrointestinal functions as "food hormones"[40] in the newborn.

Whereas no β-casomorphin immunoreactive material has been found in the blood of adult humans after milk intake[41], in newborn calves a β-casomorphin precursor is apparently absorbed from the gastrointestinal tract after milk intake. Thus protected against enzymatic attack in the plasma[35,36], a β-casomorphin could be carried to any site in the recipient, where it might be cleaved from the precursor to elicit opioid or nonopioid effects. Candidates could be central effects in the newborn, e.g. triggering of certain behavioural phenomena[42]; entrance of β-casomorphins into the brain may be possible[43].

In pregnant women and lactating or non-lactating women after parturition, β-casomorphin immunoreactive materials probably identical with β-casomorphins as well as precursors thereof have been found[37]. Thus, β-casomorphins, or at least precursors thereof, may be present in the mother and might play a physiological role during pregnancy. An influence on the fetus is unlikely, since β-casomorphin degrading enzymatic activity has been demonstrated in the human placenta[44] and no β-casomorphin immunoreactive material has been found in plasma of newborn calves[36]. In contrast, β-casomorphins might participate in the endocrine regulation of pregnancy. In view of previously reported data[27,45,46], it seems to be likely that these peptides would modulate the release of prolactin or gonadotropins. An influence on the milk ejection reflex[47] may be considered in this context as well.

A material eventually related to β-casomorphins has been found in the plasma of some women suffering from post-partum psychosis[48]; the significance of this finding remains to be elucidated.

SUMMARY

Milk contains a considerable number of hormones, whose physiological significance is not yet clearly understood. Recently, bovine as well as human ß-casein have been demonstrated to contain amino acid sequences with opioid activity. In view of their origin and pharmacological activity these peptides have been named "ß-casomorphins". Several studies have been performed to determine whether these ß-casomorphin amino acid sequences might be cleaved from ß-casein in the pregnant or nursing female or after milk intake in the newborn and whether any opioid effects might be elicited in those situations.

In fact, a ß-casomorphin precursor has been found in the plasma of newborn calves after their first milk intake after birth, whereas this material was not present before milk intake. Further, ß-casomorphin immunoreactive materials , which may represent ß-casomorphins or precursors,have been found in the plasma of pregnant women and lactating or nonlactating women after parturition.

Thus, in newborn mammals as well as in pregnant and in lactating or non-lactating females, ß-casomorphin precursors may be present, from which at any site in the respective organism ß-casomorphins might be cleaved to elicit opioid or non-opioid effects. Attractive candidates would be the central nervous system in the newborn for triggering behavioural phenomena or the endocrine system in the mother for modulating the endocrine control of pregnancy.

Although a physiological significance of ß-casomorphins remains to be elucidated, there is growing evidence that these peptides or other casein fragments with opioid activity may have physiological effects upon mammalian reproduction.

REFERENCES

1. L. A. Hanson, The mammary gland as an immunological organ, Immunol. Today 3: 168 (1982).
2. O. Koldovsky, Hormones in milk, Life Sci. 26: 1833 (1980).
3. V. Brantl and H. Teschemacher, Opioids in milk, Trends Pharmacol. Sci 4: 193 (1983).

4. V. Brantl and K. Neubert, Opioid peptides derived from food proteins, Trends Pharmacol. Sci. 7: 6 (1986).
5. C. Zioudrou and W. A. Klee, Possible roles of peptides derived from food proteins in brain functions, in: "Nutrition and the Brain", R. J. Wurtman and J. J. Wurtman, eds., Raven Press, New York (1979).
6. S. Loukas, D. Varoucha, C. Zioudrou, R. A.Streaty and W. A. Klee, Opioid activities and structures of α-casein-derived exorphins, Biochemistry 22:4567 (1983).
7. V. Brantl, H. Teschemacher, A. Henschen and F. Lottspeich, Novel opioid peptides derived from casein (ß-casomorphins): I. Isolation from bovine casein peptone, Hoppe-Seyler's Z. Physiol. Chem. 360:1211 (1979).
8. V. Brantl, Novel opioid peptides derived from human ß-casein: human ß-casomorphins, Eur. J. Pharmacol. 106:213 (1984).
9. I. S. Zagon, P. J. McLaughlin, D. J. Weaver and E. Zagon, Opiates, endorphins and the developing organism: A comprehensive bibliography, Neurosci. Biobehav. Rev. 6:439 (1982).
10. B. Ribadeau-Dumas, G. Brignon, F. Grosclaude and J.-C. Mercier, Structure primaire de la caseine ß-bovine: Sequence complete, Eur. J. Biochem. 25:505 (1972).
11. A. Henschen, F. Lottspeich, V. Brantl and H. Teschemacher, Novel opioid peptides derived from casein (ß-casomorphins). II. Structure of active components from bovine casein peptone, Hoppe-Seyler's Z. Physiol. Chem. 360:1216 (1979).
12. F. Lottspeich, A. Henschen, V. Brantl and H. Teschemacher, Novel opioid peptides derived from casein (ß-casomorphins). III. Synthetic peptides corresponding to components from bovine casein peptone, Hoppe-Seyler's Z. Physiol. Chem. 361:1835 (1980).
13. V. Brantl, H. Teschemacher, J. Bläsig, A. Henschen and F. Lottspeich, Opioid activities of ß-casomorphins, Life Sci. 28:1903 (1981).
14. K.-J . Chang, P. Cuatrecasas, E. T. Wei and J.-K. Chang, Analgesic activity of intracerebroventicular

administration of morphiceptin and ß-casomorphins: Correlation with the morphine (μ)receptor binding affinity, Life Sci. 30:1547 (1982).

15. B. C. Richardson and J. C. Mercier, The primary structure of the ovine ß-caseins, Eur. J. Biochem. 99:285 (1979).

16. P. Petrilli, F. Addeo and L. Chianese, Primary structure of water buffalo ß-casein: Tryptic and CNBr peptides, Ital. J. Biochem. 32:336 (1983).

17. R. Greenberg, M. L. Groves and H.J. Dower, Human ß-casein. Amino acid sequence and identification of phosphorylation sites, J. Biol. Chem. 259:5132 (1984).

18. G. Koch, K. Wiedemann and H. Teschemacher, Opioid activities of human ß-casomorphins, Naunyn-Schmiedeberg's Arch. Pharmacol. 331:351 (1985).

19. M. Yoshikawa, T. Yoshimura and H. Chiba, Opioid peptides from human ß-casein, Agric. Biol. Chem. 48:3185 (1984).

20. W. Kromer, W. Pretzlaff and R. Woinoff, Regional Distribution of an opioid mechanism in the guinea-pig isolated intestine, J. Pharm. Pharmacol. 33:98 (1981).

21. G. Grecksch, C. Schweigert and H. Matthies, Evidence for analgesic activity of ß-casomorphin in rats, Neurosci. Lett. 27:325 (1981).

22. K. G. Reymann, A. N. Chepkova, K. Schulzeck and T. Ott, Effects of ß-casomorphin on dentate hippocampal field potentials in freely moving rats, Biomed. Biochim. Acta 44:749 (1985).

23. V. Schusdziarra, A. Holland, R. Schick, A. de la Fuente, M. Klier, V. Maier, V. Brantl and E. F. Pfeiffer, Modulation of post-prandial insulin release by ingested opiate-like substances in dogs, Diabetologia 24:113 (1983).

24. V. Schusdziarra, R. Schick, A. de la Fuente, A. Holland, V. Brantl and E. F. Pfeiffer, Effect of ß-casomorphins on somatostatin release in dogs, Endocrinologyy 112:1948 (1983).

25. V. Schusdziarra, R. Schick, A. Holland, A. de la Fuente, J. Specht, V. Maier, V. Brantl and E. F. Pfeiffer, Effect of opiate-active substances on pancreatic poly-

peptide levels in dogs, Peptides 4:205 (1983).
26. T. Mitsuma, T. Nogimori and M. Chaya, ß-Casomorphin inhibits thyrotropin secretion in rats, Exp. Clin. Endocrinol. 84:324 (1984).
27. J. Nedvidkova, E. Kasafirek, A. Dlabac and V. Felt, Effect of beta-casomorphin and its analogue on serum prolactin in the rat, Exp. Clin. Endocrinol. 85:249 (1985).
28. J. E. Zadina and A. J. Kastin, Interactions between the antiopiate Tyr-MIF-1 and the Mu opiate morphiceptin at their respective binding sites in brain, Peptides 6:965 (1985).
29. F. Parker, D. Migliore-Samour, F. Floch, A. Zerial, G. H. Werner, J. Jolles, M. Casaretto, H. Zahn and P. Jolles, Immunostimulating hexapeptide from human casein: amino acid sequence, synthesis and biological properties, Eur. J. Biochem. 145:677 (1984).
30. P. Petrilli, D. Picone, C. Caporale, F. Addeo, S. Auricchio and G. Marino, Does casomorphin have a functional role? FEBS Lett. 169:53 (1984).
31. J. Svedberg, J. de Haas, G. Leimenstoll, F. Paul and H. Teschemacher, Demonstration of ß-casomorphin immunoreactive materials in in vitro digests of bovine milk and in small intestine contents after bovine milk ingestion in adult humans, Peptides 6:825 (1985).
32. K.-J. Chang, Y. F. Su, D. A. Brent and J.-K. Chang, Isolation of a specific μ-opiate receptor peptide, morphiceptin, from an enzymatic digest of milk proteins, J. Biol. Chem. 260:9706 (1985).
33. B. Hartrodt, K. Neubert, G. Fischer, U. Demuth, T. Yoshimoto and A. Barth, Degradation of ß-casomorphin-5 by proline-specific endopeptidase (PSE) and post-proline cleaving enzyme (PPCE), Pharmazie 37:72 (1982).
34. B. Hartrodt, K. Neubert, G. Fischer, H. Schulz and A. Barth, Synthese und enzymatischer Abbau von ß-Casomorphin-5, Pharmazie 37:165 (1982).
35. G. Kreil, M. Umbach, V. Brantl and H. Teschemacher, Studies on the enzymatic degradation of ß-casomorphins, Life Sci. 33, Suppl. I:137 (1983).
36. M. Umbach, H. Teschemacher, K. Praetorius, R.

Hirschhäuser and H. Bostedt, Demonstration of a ß-casomorphin immunoreactive material in the plasma of newborn calves after milk intake, Reg. Peptides 12:223 (1985).

37. G. Koch, K. Wiedemann and W. Zimmermann, Human ß-casomorphin-8 immunoreactive materials in the plasma of nursing mothers, Naunyn-Schmiedeberg's Arch. Pharmacol. 332:R 85 (1986).

38. H. Stark, B. Lössner and H. Matthies, Degradation of ß-casomorphin in the rat brain in vivo, Biomed. Biochim. Acta 45:557 (1986).

39. H. Meisel, Chemical characterization and opioid activity of an exorphin isolated from in vivo digests of casein, FEBS Lett. 196:223 (1986).

40. J. E. Morley, Food peptides: A new class of hormones? J. Am. Med. Assoc. 247:2379 (1982).

41. H. Teschemacher, M. Umbach, U. Hamel, K. Praetorius, G. Ahnert-Hilger, V. Brantl, F. Lottspeich and A. Henschen, No evidence for the presence of ß-casomorphins in human plasma after ingestion of cows' milk or milk products, J. Dairy Res. 53:135 (1986).

42. J. Panksepp, L. Normansell, S. Siviy, J. Rossi III and A. J. Zolovick, Casomorphins reduce separation distress in chicks, Peptides 5:829 (1984).

43. A. Ermisch, H.-J. Rühle, K. Neubert, B. Hartrodt and R. Landgraf, On the blood-brain barrier to peptides: ^{3}H ß-casomorphin-5 uptake by eighteen brain regions in vivo, J. Neurochem. 41:1229 (1983).

44. G. Püschel, R. Mentlein and E. Heymann, Isolation and characterization of dipeptidyl peptidase IV from human placenta, Eur. J. Biochem. 126:359 (1982).

45. S. S. C. Yen, M. E. Quigley, R. L. Reid, J. F. Ropert and N. S. Cetel, Neuroendocrinology of opioid peptides and their role in the control of gonadotropin and prolactin secretion, Am. J. Obstet. Gynecol. 152:485 (1985).

46. C. A. Leadem and S. P. Kalra, Effects of endogenous opioid peptides and opiates on luteinizing hormone and prolactin secretion in ovariectomized rats, Neuroendocrinology 41:342 (1985).

47. D. M. Wright, Evidence for a spinal site at which opioids

may act to inhibit the milk-ejection reflex, J. Endocrinol. 106:401 (1985).

48. L. H. Lindström, F. Nyberg, L. Terenius, K. Bauer, G. Besev, L. M. Gunne, S. Lyrenäs, G. Willdeck-Lund and B. Lindberg, CSF and plasma ß-casomorphin-like opioid peptides in post-partum psychosis, Am. J. Psychiatry 141:1059 (1984).

PRODUCTION OF GROWTH FACTORS BY NORMAL HUMAN MAMMARY CELLS IN CULTURE

William R. Kidwell, David S. Salomon and S. Mohanam

National Cancer Institute, Bethesda, Maryland 20892

Graeme I. Bell

Chiron Labs, Emeryville, California 94608

INTRODUCTION

Recent evidence from several laboratories has demonstrated that human breast cancer cell lines produce a variety of polypeptide growth factors, including platelet derived growth factor (PDGF), insulin like growth factor (IGFI), transforming growth factors alpha and beta (TGFα and TGFβ). mammary-derived growth factor I (MDGFI), mammary-derived growth factor II (MDGFII), gastrin releasing peptide (GRP), human mammary tumor growth factor (h.MTGF) and epidermal growth factor (EGF). It has been assumed that production of the factors was a pathological phenomenon, one accompanying neoplastic transformation. However, we and others have shown that most of these are normal components of milk. Consequently, we postulated that the tumor cell-derived factors might be normal glandular secretory products.

To test this postulate, we have developed procedures for the isolation and cultivation of ducts and alveoli from normal human mammary tissue obtained from patients undergoing reduction mammoplasty for enlarged breasts. The glandular elements were cultured in serum-free medium for up to 30 days and underwent extensive cell proliferaton. Analyses of the medium in which the mammary cells were cultured clearly demonstrated that they released MDGFI, TGFα and TGFβ. Similar analyses of normal rat and mouse mammary epithelial cell-conditioned medium indicated that TGFα, TGFβ and EGF were synthesized. Experiments to determine whether PDGF and GRP are also made by normal mammary cells have not yet been performed.

Cell culture methods were also been utilized to assess the potential role of the various factors in the regulation of mammary epithelial cell proliferation. Thus, we could show that the division of normal mouse mammary cells was significantly slowed in culture by specific antibodies against mouse EGF. Furthermore we showed that mouse mammary cells expressed EGF mRNA as did numerous mouse mammary tumors and preneoplastic glandular lesions.

Information regarding production of each of the growth factors by cultured mammary cells (primary or established) will be considered here.

TRANSFORMING GROWTH FACTOR ALPHA (TGFα)

Reduction mammoplasty tissue was obtained fresh from surgery. The glandular elements, recognized by their yellowish color, were excised and placed in digestion buffer containing collagenase and hyaluronidase (1). After 16 hr incubation at 37° C, the ductal and alveolar structures were collected onto Nitex filters by filtration. The filters were inverted and the glandular epithelium was washed off into a collecting flask under sterile conditions. Ducts and alveoli were then pelleted by low speed centrifugation. Suitable dilutions of the preparations were added to culture dishes in I-MEM growth medium containing 0.2 mM $CaCl_2$, insulin (10 ng/ml), hydrocortisone (50 ng/ml), transferrin (5 ug/ml) and fetuin (Pedersen, 1 mg/ml). The cells were incubated for 10 days and the culture medium harvested for growth factor assays. For radioimmunoassays of TGF and IGFI, the culture medium was concentrated by lyophilization. The residue was resuspended in o.4 N HCl and applied to a Waters Sep-pak C_{18} reverse phase cartridge on which the growth factors bound tightly. Ninety percent of the proteins were eluted by washing extensively with 4% acetic acid. The growth factors were then recovered by washing the cartridges with 4-6 ml of methanol:0.2% trifluoroacetic acid. This eluate was evaporated to dryness under a stream of N_2 and the residue resuspended in a small amount of water for radioimmunoassay. TGFα was analyzed using anti-TGFα antibodies from Triton Biosciences (Houston, TX) or Biotope, Inc. (Seattle, WA) (2). The results are presented in Table I. TGFα was also analyzed by bioassays in which aliquots of concentrated conditioned medium from the cultures was tested for its ability to stimulate the growth of normal rat kidney cells in soft agar suspension cultures (3). IGFI activity was tested using the assay kit and materials provided by BTI, Inc., Cambridge, Mass. No IGFI was found in the normal cell-conditioned medium although mammary tumor cell lines make IGFI (4).

Table 1. Transforming growth factor alpha activity in human mammary cell-conditioned growth medium

Cell-Conditioned Medium Assayed	TGFα Activity	
	Radioimmunoassay (ng/ml)	Bioassay*
Normal Human Mammary Cells	0.85 ± 0.05	325 ± 58
MCF-7, Established Human Mammary cell Line	0.25 ± 0.03	127 ± 34
Control Medium (No Cells)	0.02 ± 0.01	4 ± 2

*The number of NRK cell colonies formed using 100 ul conditioned medium added to the test cultures: 14 days' growth in agar.

TGFα mRNA levels were also been determined in normal and transformed mouse mammary cells by Northern blot assays by us (5) and in normal and tumorigenic human mammary cell lines by Derynck et al (6). The latter group reported that TGFα message in HBL100 a normal human mammary cell line, was about half that seen in the tumorigenic MCF-7 cell line (6). Our assays of normal and Harvey ras oncogene-transformed MuMG cells derived from Namru mice indicated an mRNA abundance ratio of about 1:20 in favor of transformed cells. At this time, the limited number of samples analyzed is too small to conclude whether TGFα expression is greatly enhanced by transformation. Our data however, permit us to conclude that normal mammary tissue from both the mouse and human makes at least some TGFα . Findings consistent with this conclusion are the reports from our laboratory and Lippman's indicating that TGFα production is amplified by ovarian hormones particularly estrogens (7,8).

A potential role of TGFα in the neonate is suggested by the fact that human milk contains readily detectable levels of the growth factor (9). Activity was measured both by radioimmunoassay and by bioassay. The latter assay was possible only after milk proteins were separated by preparative isoelectric focusing. This is the case because TGFα and EGF are similarly efficient in the bioassay, stimulation of NRK cell growth in soft agar. Fractions focusing at a pH of about 6.5, the pI of TGFα , were positive in the radioimmunoassay and in the soft agar growth assay (9,10). EGF was also detected in fractions of human milk focusing at a pH of about 4.4 the region where authentic human EGF was found to focus (10). In most <u>in vivo</u> and <u>in vitro</u> assays, EGF and TGFα have been found to be equivalent (11). However, exceptions to this do exist. Angiogenesis (the formation of new blood vessels), bone resorption and re-epithelialization of burn wounds are three situations in which TGFα is more potent than EGF (12). Whether TGFα in milk is absorbed in an active form from the gut (as is EGF; Koldovsky and Read, this volume) is not known.

Transforming Growth Factor Beta (TGFβ)

While there are indications that TGFα might be higher in transformed than in normal mammary tissues, the same does not appear to be true for TGFβ . We analyzed the serum-free conditioned medium obtained from primary cultures of normal, benign and malignant human breast epithelium using the soft agar bioassay. To assay TGFβ in the bioassay one tests the effects of the conditioned medium on the soft agar growth of NRK cells in the presence of optimal levels of EGF or TGFα . The results of such an assay are shown in Table 2. Approximately the same amounts of TGF were produced per cell whether the cells were derived from normal or abnormal tissues.

Table 2. Production of TGFβ by normal, benign and malignant human mammary epithelium

Cell Source	NRK Cell Colonies*
Normal Mammary Glands	303 ± 33
Benign Mammary Tissue	250 ± 50
Malignant Mammary Tissue	246 ± 4

*Colonies formed per ml conditioned medium, normalized to 10^6 mammary cells per dish.

EPIDERMAL GROWTH FACTOR (EGF)

Epidermal growth factor in man and rodents is synthesized as a high molecular weight precursor (Mr = 128,000) from which a small internal acidic sequence is cleaved to yield the mature growth factor (Mr 5200, pI 4.4) (13). EGF is abundant in milk of a variety of species, including man, although the site of EGF production is not well defined. In man, not much data has been generated on this. In a preliminary report, Gol-Winkler et al (14) presented data suggesting that EGF mRNA is present in both normal and malignant human breast tissues. Their assay protocol entailed Northern hybridization of nick-translated human EGF cDNAs to slot blots of total RNA from the mammary tissues. In a well controlled study, Collette et al. (15) showed by radioimmunoassay that breast cyst fluids had high concentrations of EGF. In fact, these workers found that the mean values of EGF in breast fluids were higher than those in colostrum, eg 550 ng/ml vs 197 ng/ml (16). Although a direct demonstration that EGF mRNA is translated into protein by normal human mammary epithelium has not been described, it appears to be likely. MCF-7 cells, a human breast cancer cell line, makes the mRNA and produces a protein that competes with authentic human EGF in a radioimmunoassay (17).

More evidence for EGF production by normal mammary epithelium comes studies with experimental animals. We have shown, for example, that mouse and rat mammary tissues contain RNA that hybridizes with EGF cDNA (12). This RNA species has also been found to correspond in size to that of prepro-EGF, that is, 5 Kb (13). Relative amounts of the RNA in various physiological states of the mammary tissue of the two species is given in Table 3.

Table 3. Relative abundance of EGF mRNA in rat and mouse mammary glands

Species	Physiological State	mRNA (Relative Abundance)
Rat	Virgin Female (Intact)	+
Rat	Virgin Female (ovariectomized)	+
Rat	Lactating	++
Mouse	Virgin Female (Intact)	+
Mouse	Mid-pregnancy	++
Mouse	Tumor Tissue	+
Mouse	Kidney*	++++

*The kidney served as a positive control since this tissue is known to be a high EGF mRNA producer (18).

Direct evidence was obtained for the translation of EGF mRNA into a functional protein of importance to mouse mammary cell function, as presented in Table 4. Normal mouse mammary epithelium was isolated using porcedures essentially as described for human mammary ducts and alveoli. The cells were introduced into culture in a serum-free medium containing insulin, fetuin, transferrin and hydrocortisone, but no EGF. Antibodies against various segments of prepro-EGF, generated by Graeme Bell's group (18), was added to the cell cultures. Control serum was also tested. As shown in Table 4, there was an inhibition of cell growth by 5 different antibody preparations. At the maximum antiserum concentration, cell growth over a 4 day period was inhibited by about 60%. It would have been interesting to have evaluated the effects of TGFα and EGF antisera simultaneously. However, this experiment was not performed because of the low titer of the TGFα antisera currently available.

Table 4. Effects of anti-prepro-EGF peptide antibodies on mouse mammary cell growth in primary culture

Antiserum[a]	Amino Acid Sequence[b]	% Inhibition[c]
I	39-227	44
II	348-691	57
III	1096-1217	24
IV	845-1095	26
V	692-1095	26

[a] The antiserum was prepared against a gene fusion product of prepro-EGF gene fragment and the carboxy-terminal gene fragment of beta-galactosidase. The fusion product was expressed in E. coli and antibodies generated against the E. coli lysates. Control antiserum was generated against non-transfected E. coli.

[b] The amino acid sequence given is that of the corresponding fragment of prepro-EGF against which the antiserum was generated.

[c] Fifty ul of 1:100 dilute antiserum or control serum was added to the cultures which contained 3 ml of growth medium. Cells were counted 4 days later.

Interestingly, the most inhibitory antiserum was against amino acid sequence 348-691, which is the region of prepro-EGF containing 3 tandem EGF-like sequences. The EGF peptide is located in sequence 845-1095 and the corresponding antiserum is IV. Since the various precursor amino acid sequences have not been purified and used to determine antibody titers, no firm conclusions regarding the biological activity of various prepro-EGF segments can be drawn at this time.

Regarding the EGF in milk, Oka et al (19) have suggested that the salivary glands of mice are the likely source and moreover, that the salivary gland is essential for the development of mammary glands of mice into a fully functional, lactational status. Their conclusions are based on the following observations of their group. Sialadectomy of pregnant mice did not adversely affect the course of pregnancy but the nursing pups died in high numbers, unless foster fed with non-sialadectomized mothers. The adverse affects of sialadectomy were reversed by giving the mothers EGF during the latter stages of pregnancy. Milk yields of the sialadectomized mothers were reduced compared to intact animals and finally, explant cultures of mammary tissue from sialadectomized animals were less responsive to stimulation by mammotrophic hormones than tissues from intact animals.

These findings are very interesting from a glandular developmental standpoint. However, it is very clear from studies of other investigators that the EGF levels in milk are not dramatically lowered by sialadectomy (20). Since EGF and TGFα are biologically equivalent in a variety of assays, it will be interesting to assess the effects of sialadectomy on TGFα levels of milk and whether TGFα administration would also reverse the adverse affects of salivary gland ablation.

Mammary-derived Growth Factor I (MDGF I)

MDGF I is a Mr 62,000 acidic protein (pI 4.8) that we have purified from human milk and from human breast cancer tissues (21). The factor interacts with high affinity receptors on mammary and other cell types (Kd = 10^{-10}). A 10^{-11} M concentration of MDGF I stimulates normal mammary cell growth by about 30%. More dramatically, the same concentration of the factor differentially stimulates mammary cells to elaborate basement membrane collagen by as much as 10 fold (21). Studies of collagen biosynthesis have shown that the effects of MDGF I are mediated via an enhanced production of collagen mRNA. Based on extensive studies of collagen synthesis inhibitors (22, 23, 24), we believe that the growth stimulation of the mammary cells is effected by the enhanced production of basement membrane matrix, the scaffolding on which the cells normally rest in vivo. Also, MDGF I responsiveness is negated by plating the mammary cells on a basement membrane matrix instead of on plastic surfaces.

Indirect evidence suggests that MDGF I is made by mammary cells and functions as an autocrine or paracrine factor, that is, a factor that is produced by a cell and acts on the same cell (autocrine factor) or on neighboring cells (paracrine factor). The evidence is two-fold. First, an MDGF I-like factor has been detected in well differentiated rat mammary tumors that make a basement membrane (25). The rat factor is, however, absent or barely detectable in poorly differentiated tumors that do not make a basement membrane (25). This establishes a link between the concentration of the factor in mammary tumors and the presence of a product elaborated by the tumors in response to the factor. Second, we have examined the possibility that MDGF I is actually made by mammary cells in culture. Normal and transformed human mammary epithelia were recovered and placed in culture as described. After 5 days' growth, the cell conditioned medium was harvested and tested for its ability to inhibit the binding of ^{125}I-MDGF I to mammary cell membranes. The results of the radioreceptor competition assays are listed in Table 5. Conditioned medium from normal mammary cells contained lower levels of MDGF I receptor-competing activity than the medium recovered from the tumor cell cultures, but both were positive in the assay.

Table 5. Radioreceptor competition assays for MDGF I in growth medium conditioned by normal and malignant human mammary cells in primary culture

Mammary Cells	μl Medium	pMol ^{125}I-MDGF I Displaced
Normal	50	4.5
Normal	100	11.1
Normal	50	10.5
Normal	100	17.0
Malignant	50	20.2
Malignant	100	34.6

In the radioreceptor assay, 2.5 ug of A431 cell membrane protein was dried onto microtiter wells. The wells were washed and 5 ng ^{125}I-MDGF I was added in 1.6 ml binding buffer (DMEM containing 0.1% bovine serum albumin and 50 mM HEPES buffer, pH 6.8). After binding for 1 hr at room temperature, the wells were washed 4 times with DMEM and the wells were cut out and counted in a gamma counter. The dishes from which the conditioned medium was harvested contained about 10^5 cells each. Medium that had not been exposed to cells served as a control. The MDGF I labeled probe had a specific activity of about 10^5 cpm/ng.

The above results suggest that MDGF I is probably made by the mammary cells. However, we cannot rule out the possibility that the cells take up and store factors in vivo and release them in vitro. To eleminate this possibility will require metabolic labeling of the cultured cells and demonstration that labeled MDGF I is produced.

An obvious question regarding the growth factors that might be made by and act on mammary cells is why the same cells respond to exogenously added growth factor. From Table 5 it would appear that the cells have elaborated a sufficient amount of the factor for self-stimulation. However, the factor in the medium might not be in an active form, or it might be associated with a binding protein that attenuates biological activity. In the case of rat mammary epithelium, it was clearly shown that tumor epithelium, which makes more MDGF I-like activity than normal rat mammary epithelium, was less responsive to purified rat factor than were the normal mammary cells (25). There is evidence that some growth factors are elaborated as an inactive complex by normal cells, whereas the free, active factor is made by tumor cells. An example of this is TGFβ which is made by both transformed and normal cells, but in the latter case the factor becomes activated only after acid-dissociation (26).

Mammary-derived Growth Factor II (MDGF II)

MDGF II is an EGF and TGFα -like growth factor that was purified to near-homogeneity from human milk (27). An apparently identical activity has been detected in acid-ethanol extracts of human carcinoma *in situ*. No information regarding its abundance in normal human mammary cells is available at present. The protein has an apparent molecular weight of 6000 on gel permeation HPLC chromatography under low ionic strength eluting conditions and 17,000 in high ionic strength buffers. The activity was detected on the basis of growth promotion of NRK cells in soft agar. It also competes with EGF and TGFα for the same receptor on cell membranes. MDGF I can be distinguished from EGF and TGFα by its pI (4.0 vs 4.4 and 6.5, respectively) and by its different behavior on HPLC chromatography. The factor is not neutralized in the soft agar bioassay by antibodies against human EGF. Nor is there any activity in partially purified MDGF I preparations that competes with TGFα in a radioimmunoassay (27).

Other Growth Factors In Milk That Are Probably Made By Normal Human Mammary Cells

Because there is little or no information about the production of these factors in normal mammary cells, we have included them in a single category here. Included in the group are gastrin-releasing peptide (GRP), platelet-derived growth factor (PDGF), insulin, colony stimulating factor, human mammary tumor growth factor (hMTGF), three human milk activities that Shing and Klagsbrun have identified (27), and a mammary cell growth inhibitor (MCGI). Some properties and tissue distributions are given in Table 6.

Mammary Cell Growth Inhibitor (MCGI)

MCGI is an acidic protein that was first detected in extracts of lactating bovine mammary glands by Grosse's group (38,39). The factor is an acidic polypeptide (pI 5.0) that is highly active in inhibiting the growth of rat mammary ascites tumor (38) and normal mammary cells (our unpublished findings). Antibodies against MCGI have been used in enzyme-linked immunoassays of human milk and cross reacting material found in abundance in the milk fat globule fraction. Additionally, an inhibitory substance with the same size and pI of bovine MCGI was been purified from human milk. Human MCGI strongly inhibits mouse mammary epithelial cell growth but has no effect on the growth of human mammary fibroblasts.

Grosse's group has reported that MCGI varies depending on the physiological state of the bovine mammary gland. Thus the amount of MCGI is highest in non-dividing, lactating glands and lowest in pregnancy when cell division is highest (38). From these observations it is postulated that MCGI is a physiologically important regulator of mammary epithelial cell growth. It remains to be seen whether MCGI is an actual biosynthetic product of the mammary gland.

Table 6. Physical properties of various milk-derived growth factors and tissues where detected

Factor	Size[a]	pI	Detected In	Ref. No.
GRP	Multiple Sizes	?	Human and Bovine Milk	28,29
			Human Breast Tumors	30
			Normal Rat Mammary Cell Line (WRK 1)	31
			Mouse Mammary Tumor Cell Line (64/24)	31
PDGF			Human. Goat, Sheep and Bovine Milk	32
hMTGF	16	8.0	Human Mammary Tumor Extracts[b]	34
Insulin	6	5.6	Human Milk	35
MGFI	34-38	4.2-5.2	Human Milk	27
MGFII	10-15	3.2-4.8 7.4-8.5	Human Milk	27
MGFIII	5-6	4.4-4.7	Human Milk	27
CSF	250	4.4-4.9	Human Milk	36
IGFI	7	6.5	Human Milk	37
			Human Mammary Tumor Cell Lines	4
MCGI	13	5.0	Bovine Milk	38,39
			Human Milk[c]	-
52 K	52	?	Human Mammary Tumors[b]	40-44
MTGF	?	?	Human Mammary Tumor Cell Lines[b]	45

[a]Size in daltons x 10^{-3}.
[b]Human milk has not been analyzed for these factors, but because they are present in breast tumors or breast tumor cell lines, they have been included here.
[c]Using antibodies obtained from Grosse, we have shown that human milk contains factors that are immunologically related to MCGI.

SUMMARY

We have detailed the available, pertinent data on the types of growth factors that have been described in human milk and presented both direct and indirect evidence that some of these are products of the normal mammary gland. Additionally, we have reported that some of these putative glandular products can stimulate the growth of normal mammary cells in *in vitro* culture systems. Further, evidence that at least one of the factors made by the gland is under estrogenic hormone regulation was presented. What is not clear from our studies, or those of others, is whether any of the growth factors play a significant role in the neonate. We have postulated earlier that the milk growth factors might represent an "insurance policy" for the neonate (12). Indeed, there is evidence that at least one factor (EGF) can facilitate the repair of ulcerated lesions in the gastro-intestinal tract (reviewed in ref. 12). Moreover, very recent reports have indicated potential wound repair roles for PDGF, TGF , and TGF (46). One can envision the use of growth factor supplements in artifical formulas, and even in normal milk, to facilitate gastro-intestinal development of infants that are premature or that have gastrointestinal lesions.

REFERENCES

1. W. R. Kidwell, M. Bano and D. S. Salomon, Growth control of normal mammary cells on collagen in serum-free medium. *In*: "Methods for serum-Free Culture of Cells of the Endocrine System," D. Barnes, D. Sirbasku and G. H. Sato, eds., Alan R. Liss, Inc., New York, 1984, pp 105-125.

2. I. Perroteau, W. R. Kidwell, R. Pardue, M. Debertoli and D. S. Salomon, Immunoreactive alpha-transforming growth factor in human breast cancer extracts and cell lines, *Breast Cancer Res. Treat.* 7:201 (1986).

3. D. S. Salomon, J. A. Zwiebel, M. Bano, I. Losonczy, P. Fehnel and W. R. Kidwell, Presence of transforming growth factors in human breast cancer cells, *Cancer Res.* 44:4069 (1984).

4. K. K. Huff, D. Kaufman, K. H. Gabbay, E. M. Spencer, M. Lippman and R. B. Dickson, Secretion of an insulin-like growth factor-1-related protein by human breast cancer cells, *Cancer Res.* 46:4613 (1986).

5. D. S. Salomon, I. Perroteau, W. R. Kidwell, J. Tam and R. Derynck, Loss of growth responsiveness to EGF and enhanced production of alpha-transforming growth factor in *ras*-transformed mouse mammary cells, *J. Cell Phys.* In press.

6. R. Derynck, D. V. Goeddel, A. Ullrich, J. Gutterman, R. Williams, T. S. Bringman and W. H. Berger, Synthesis of mRNAs for transforming growth factors and EGF receptor by human tumors. Proc. Natl. Acad. Sci. USA. In press.

7. W. R. Kidwell S. Liu and D. S. Salomon, Ovariectomy reduces the level of transforming growth factor alpha and its mRNA in primary, 7, 12-DMBA-induced rat mammary tumors, Breast Cancer Res. Treat. In press.

8. R. B. Dickson, K. K. Huff, E. M. Spencer and M. E. Lippman, Induction of epidermal growth factor-related peptides by 17 beta-estradiol in MCF-7 cells, Endocrin. 118:138 (1986).

9. D. S. Salomon, W. R. Kidwell, G. S. Smith and G. I. Bell, Presence of alpha transforming growth factors in human breast tumors and milk, Cancer Res. 26:198 (1985).

10. J. A. Zwiebel, M. Bano, D. S. Salomon and W. R. Kidwell, Partial purification of transforming growth factors from human milk, Cancer Res. 46:933 (1986).

11. J. P. Tam, Physiological effects of TGF in the new-born mouse. Sci. 229:673 (1985).

12. W. R. Kidwell and D. S. Salomon, Growth factors in human milk, in: "Minor Protein Constituents in Human Milk," S. Atkinson and B. Lonnerdahl, eds., CRC Reviews, Boca Raton. In press.

13. J. Scott, M. Urdea, M. Quiroga, R. Sanchez-Pescador, N. Fong, M. Selby, W. Rutter, and G. I. Bell, Structure of a mouse submaxillary gland mRNA coding for EGF and seven related peptides. Sci. 221:236 (1983). 1983.

14. R. A. Gol-Winckler, A. Simenon, H. Dijkmas and J. A. Martial, Presence of mRNA encoding polypeptide growth factors and EGF receptor in primary human breast cancer, Second Annual Meeting on Oncogenes, Hood College, Frederick, Maryland, p. 93, July 1986.

15. J. Collette, J-C. Hendrick, and P. Franchimont, Presence of lact-albumin, epidermal growth factor, epithelial membrane antigen and gross cystic disease fluid protein in breast cyst fluid, Cancer Res. 46:3728 (1986).

16. J. M. Beadmore, D. I. Lewis-Jones and R. C. Richards, Urogastrone and lactose concentrations in precolostrum, colostrum and milk, Peadr. Res. 17:825 (1983).

17. K. Mori, M. Kurobe, S. Furukawa, K. Kubo and K. Hayaishi, Human breast cancer cells synthesize and secrete EGF-like immunoreactive factor in culture, Biochem. Biophys. Res. Commun. 136:300 (1986).

18. R. B. Rall, J. Scott and G. I. Bell, Mouse pre-proEGF synthesis by kidney and other tissues, Nature 313:228 (1985).

19. S. Okamoto and T. Oka, Evidence for a physiological function of epidermal growth factor: pregestational sialadectomy of mice decreases milk production and increases offspring mortality during the lactation period, Proc. Natl. Acad. Sci. USA 81: 6059 (1985).

20. E. W. Gresik, H. Van Der Noen and T. Barka, Transport of ^{125}I-EGF into milk and effect of sialadectomy on milk EGF in mice, Amer. J. Physiol. 247:E349 (1984).

21. M. Bano, D. S. Salomon and W. R. Kidwell, Purification of a mammary derived growth factor from human milk and human mammary tumors, J. Biol. Chem. 260:5745 (1985).

22. M. S. Wicha, L. A. Liotta, S. Garbisa and W. R. Kidwell, Basement membrane collagen requirements for growth and attachment of mammary epithelium, Exp. Cell Res. 124:181 (1979).

23. M. S. Wicha, L. A. Liotta, B. K. Vonderhaar and W. R. Kidwell, Effects of inhibition of basement membrane collagen deposition on rat mammary gland development, Dev. Biol. 80:253 (1980).

24. W. L. Lewko, L. A. Liotta, M. S. Wicha, B. K. Vonderhaar and W. R. Kidwell, Sensitivity of N-nitrosomethylurea-induced rat mammary tumors to cis-hydroxyproline, an inhibitor of collagen production. Cancer Res. 41:2855 (1981).

25. M. Bano, D. S. Salomon and W. R. Kidwell, Detection and partial characterization of collagen synthesis stimulating activities in mammary adenocarcinomas, J. Biol. Chem. 258:2729 (1983).

26. R. Pircher, D. Lawrence and P. Julien, Latent β-transforming growth factor in non-transformed and Kirsten sarcoma virus-transformed normal rat kidney cells, clone 49F, Cancer Res. 44:5538 (1984).

27. Y. Shing and M. Klagsbrun, Human and bovine milk contain two different sets of growth factors, Endocrin. 115:273 (1984).

28. G. D. Janke and L. H. Lazarus, Bombesin-like immunoreactive peptide in bovine milk, Proc. Natl. Acad. Sci. 81:578 (1984).

29. R. Ekman, S. Iversson and L. Jansson, Bombesin, neurotensin and proγ-melanotrophin immunoreactivity in human milk, Reg. Peptid. 10:99 (1985).

30. G. Guadino, M. E. DeBertoli and L. H. Lazarus, Bombesin-like immunoreactivity in human mammary tumors, 5th Internatl. George Washington Univ. Spring Symposium Proc., p 28, 1985.

31. G. Gaudino, M. E. DeBertoli and L. H. Lazarus, A bombesin-related peptide in experimental mammary tumors of rats, N.Y. Acad. Sci. In press.

32. K. D. Brown and D. M. Blakeley, Partial purification and characterization of a growth factor from goat colostrum, Biochem. J. 219:609 (1984).

33. E. Rosengurtz, J. Sinnett-Smith and J. Taylor-Papadimitriou, Production of PDGF-like growth factor by breast cancer cell lines, Int. J. Cancer 36: 247 (1985).

34. J. M. Rowe, S. Kasper, R. Shiu and H.G. Friesen, Purification and characterization of a human mammary tumor-derived growth factor, Cancer Res. 46:1408 (1986).

35. S. Cevreska, V. P. Kouacev, M. Stanovski and E. Kalamares, The presence of immunologically active insulin in milk of women during the first week of lactation and its relation to plasma insulin concentration. God. Zg. Med. Fak Skopje,21:35 (1975).

36. S. K. Sinha and A. A. Yunis, Isolation of a colony stimulating factor from human milk, Biochem. Biophys. Res. Commun. 114:797 (1983).

37. R. C. Baxter, Z. Zaltsman and J. Turtle, Immunoreactive somatomedin C/IGFI and its binding protein in human milk, Endocrin. 58:955 (1984).

38. F. D. Bohmer, W. Lehmann, H. Eberhardt, P. Langen and R. Grosse, Purification of a growth inhibitor for ehrlich ascites mammary carcinoma cells from bovine mammary glands, Exp. Cell Res. 150:466 (1984).

39. P. Langen W. Lehmann, H. Graetz, F. D. Bohmer and R. Grosse, Is ribonucleotide reductase in Ehrlich ascites mammary tumor cells the target of a growth inhibitor purified from bovine mammary gland?, Biochem. Biophys. Acta 43:1377 (1984).

40. B. Westley and H. Rochefort, A secreted glycoprotein induced by estrogen in human breast cancer cell lines, Cell 20:352 (1980).

41. M. Garcia, F. Capony, D. Derocq, D. Simon, B. Pau and H. Rochefort, Characterization of monoclonal antibodies to estrogen-regulated Mr 52,000 glycoprotein and their use in MCF-7 cells, Cancer Res. 45:709 (1985).

42. I. Touitou, M. Garcia, B. Westley, F. Capony and H. Rochefort, Effect of tunacamycin and endoglyciosidase H and F on the estrogen regulated 52,000 Mr protein seecreted by breast cancer cells, Biochemie 67:1257 (1985).

43. M. Morisset, F. Capony and H. Rochefort, The estrogen regulated 52,000 Mr protein is a cathepsin-like protease, Biochem. Biophys. Res. Commun. In press.

44. M. Garcia, G. Salazar-Retana, G. Richer, J. Domergue, F. Capony, H. Pujol, F. Laffargue, B. Pau and H. Rochefort, Immunohistichemical detection of estrogen regulated 52,000 mol wgt protein in primary breast cancer but not normal breast or uterus, J. Clin. Endocrin. Metab. 59: 564 (1984).

45. T. Ikeda, D. Danielpour, P. Galle and D. A. Sirbasku, General Methods for the isolation of acetic acid and heat-stable polypeptide growth factors for mammary and pituitary cells, in: "Methods for Serum-Free Culture of Cells of the Endocrine System," D. Barnes, D. Sirbasku and G. H. Sato, Alan R. Liss, New York, 1984, p.217.

SUMMARY OF WORKSHOP: HOST RESISTANCE

Lars A. Hanson and Randall M. Goldblum

Department of Immunology, University of Göteborg, Göteborg Sweden, Department of Pediatrics, University of Texas Medical Branch, Galveston, Texas

The workshop on host resistance was introduced by the review of epidemiologic studies by Michael S. Kramer from Canada. Kramer raised several questions about the validity of certain studies, but his review left little doubt that human milk feedings contribute to the protection of the infant against infections. With few exceptions, investigations of this protective capacity have focused upon the overall effect of breast feeding and thus do not allow assignment of specific protective function to individual factors in human milk. Kramer was less impressed with the studies reporting that breast-feeding protects against allergic disorders. The first two presentations of this workshop addressed the fate and function of human milk components of possible value in the infant's host defense.

Randall M. Goldblum and coworkers from the USA evaluated the effect of feeding maternal milk fortified with human milk fractions on the content of secretory IgA (SIgA), IgA, lactoferrin and lysozyme in the stool and urine of a group of premature infants. A matched group given cow's milk formulae served as controls. Human milk feeding increased not only the fecal content of each factor but also the urinary excretion of lactoferrin, IgA and specific SIgA antibodies to *Escherichia coli*. Serum levels of these factors were not altered by the type of feeding. Correlation analysis suggested that some of the increases, particularly stool antibodies and urinary IgA, were derived directly from transfer of milk factors, but that a portion of the increase in other stool factors and urinary lactoferrin was probably due to induction of the infant's own production. Further studies will be required to distinguish definitively between passive transfer and induction of the infant's production and to determine if these effects are restricted to the premature. The functions of the secreted factors also need to be investigated. However, the present studies suggest that induction of events in the infant's mucosa by mother's milk may be a further mechanism for protecting the young infant.

Barth from the Federal Republic of Germany presented in the discussion an interesting study of the uptake in pig ileum of bovine lactoferrin labeled by coupling of guanidino groups. This labeling technique permitted specific detection of the exogenous lactoferrin in the ileum. As much as 97% of the protein was absorbed by older piglets. The uptake was significantly less (89%) at 3 weeks. These results suggest efficient absorption of the heterologous, chemically modified lactoferrin. Although these results of lactoferrin absorption are consistent with the proportion of immunologically detectable lactoferrin recovered in the stools of human milk fed premature infants, the molecular form of the absorbed lactoferrin was not determined in these studies. Therefore, direct comparison with Goldblum's study was not possible.

Jan Holmgren from Sweden reviewed the studies of analogues for epithelial receptors of various bacterial pathogens in milk. Human milk contains factors which inhibit Vibrio cholerae agglutination of certain red cells. The factors are primarily of high molecular weight and seem to consist of the carbohydrate moiety of glycoproteins. Cow's milk has similar activity for the El Tor strains but no activity against classical V. cholerae.

The agglutination of red cells by E. coli carrying the virulence factors CFA/I or CFA/II is also inhibited by human milk. The inhibiting receptor analogues are again found to be high molecular weight glycoproteins. There are also gangliosides in human milk which block binding of V. cholerae and E. coli enterotoxins. Anne-Brit Kolsto from Norway suggested that from her studies of human milk using thin layer chromatography, very small amounts of GM1 gangliosides could explain these effects. Holmgren suggested that since the migration in thin layer chromatography of GM1 ganglioside isolated from human milk appeared somewhat atypical in some experiments, a part of the active principle could be fucosyl GM1, which also binds cholera toxin.

Holmgren also discussed studies with his colleagues evaluating the mechanism of protection of breast feeding against V. cholerae diarrhea. Using a family case-control study, it was possible to show that protection against symptomatic cholera was related to the SIgA antibody titres against V. cholerae lipopolysaccharide and enterotoxin in the mother's milk. There was no relation between degree of protection and the content in the mother's milk of lactoferrin or non-immunoglobulin factors which block hemagglutination by V. cholerae.

Stool cultures for that pathogen were positive in breast-fed infants with the same frequency as in non-breast-fed infants. This result may be parallel to the observations of Mata and Urrutia in Central America that breast-fed infants with Shigella in the stool did not show the severe symptoms seen in cow formula-fed infants infected with these bacteria. Recently, Duffy et al. from the USA also found that breast feeding does not prevent the occurrence of rotavirus in the gut, but decreases the symptoms of the infection.

The studies by Andersson et al. from Sweden showing the anti-adhesion effect of milk on Streptococcus pneumoniae and Haemophilus influenzae were also discussed. The blocking of adhesion of pneumococci and of H. influenzae to pharyngeal cells was mediated by a disaccharide and a high molecular weight substance, respectively. The importance of these anti-adhesion factors in preventing infection with such microorganisms is unclear. It is interesting, however, that cow's milk-based formulae also have a certain inhibitory activity on pneumococci, but actually increase the binding of H. influenzae to pharyngeal cells.

The last two papers in this workshop dealt with the effect of maternal antigen exposure and immune reactivity on two aspects of immunity in the infant: cell-mediated response to Mycobacteria and atopic hypersensitivity. Margaret Keller from the USA summarized her investigation of possible maternal transfer of tuberculin immunity. The study was designed to test the role of the placenta as well as the contribution of human milk in the transfer of tuberculin immunity. Four groups of infants were recruited, based on maternal delayed hypersensitivity skin test responsiveness to purified protein derivative (PPD) of tuberculin and the mode of infant feeding. The outcome variables, blastogenic responsiveness and lymphokine production by blood lymphocytes, were tested in infants at birth and at the 4 to 6th and 12th postnatal weeks. The results suggest that lymphocytes from about 15% of infants of PPD-positive mothers demonstrated a transient (4 to 6 weeks) blastogenic response to PPD. This effect appeared not to relate to breast feeding. The mechanism for induction of this responsiveness, including antigen exposure, transplacentally acquired anti-idiotype IgG, or more direct transfer of cellular immunity, as well as the importance of these responses, remain to be elucidated. Considering the high incidence of tuberculosis in many countries, this transfer of tuberculin reactivity could be important.

Ranjit K. Chandra from Canada presented his interesting studies on the effects of human milk on atopic eczema. As we had heard earlier, Kramer felt that there was no conclusive evidence that breast-fed babies were protected against atopic disease. Chandra contended, however, that such protection could be shown in certain groups of high-risk infants selected on the basis of high cord blood IgE levels, a family history of atopy and elevated serum IgE levels in parents. His data suggest that infants of mothers who avoided certain food antigens during their pregnancies and who breast fed those infants for four months had less atopic eczema than formula fed babies or infants breast-fed by mothers on unrestricted diets (p value close to 0.05). He suggested that avoidence of major food allergens was important during both pregnancy and lactation. A lively discussion followed the presentation, indicating the importance of the problem and the difficulties in designing definitive studies. Armond S. Goldman from the USA noted that eczema is a heterogenous disease and that the pathogenetic role of Type I hypersensitivity is not well established. It was also discussed that inclusion of other atopic diseases in studies of this kind would be of interest. Finally, it was pointed out that strict avoidance diets by breast feeding mothers could also have adverse effects, especially on the mother, in nutritionally marginal populations.

Several studies have illustrated that in developing countries, the heavy exposure of infants to microbes that cause repeated infections is the major cause of morbidity and mortality. One important consequence of the frequent infections is malnutrition. Clearly, breast feeding is important for protection of these infants. Detailed information about how this protection is mediated is only now becoming available.

THE EFFECT OF FEEDING HUMAN MILK ON THE DEVELOPMENT OF IMMUNITY IN LOW BIRTH WEIGHT INFANTS

R. M. Goldblum, R. Schanler, C. Garza, and A. S. Goldman

From the Departments of Pediatrics and Human Biological Chemistry and Genetics, University of Texas Medical Branch Galveston, Texas, and the Section of Neonatology and USDA/ARS Children's Nutritional Research Center, Department of Pediatrics, Baylor College of Medicine, Houston, Texas

INTRODUCTION

Epidemiologic evidence suggests that infants fed cow's milk rather than maternal milk have a higher frequency and severity of infectious, allergic, and possibly other inflammatory diseases of the gastrointestinal and respiratory tracts. While some of these illnesses may be attributed to adverse effects of ingesting heterologous milk, some reduction in illness in infants fed maternal milk may reflect alterations in the infants' immune function. The mechanisms for such protection might include passive transfer of the mucosal immune factors found in high concentration in human milk (1), interactions between the infant's immune factors and those in the milk (2), or possibly induction by human milk of synthesis or secretion of factors which contribute to the infant's host defense. However, evidence that any of these mechanisms is operative is limited.

Previous reports from our group have dealt with the effects of milk collection and storage methods (3) and donor selection (4-7) on the concentration of various immune factors in human milk. In addition, we have developed methods for reducing the infectious risk of banked human milk (8). Finally, in preparation for the present studies, we have evaluated the daily intake of immune factors in normal term infants (9). In this report, we will summarize preliminary studies quantifying immune factors in ingested milk and in the serum and certain external secretions of premature infants fed either fortified mother's milk (FMM) or cow's milk-based (CM) preparations. Quantitative association of daily intake and output of these immune factors provided an approach to distinguishing between direct and indirect contributions of human milk to the infants' mucosal immune factors.

MATERIALS AND METHODS

Patient population and feedings The selection of infants and criteria for entry in this study have been reported previously (10). Briefly, the infants were of very low birth weight but free of major clinical abnormalities and were able to tolerate total enteral feedings by 15 days of age. Assignment to the two feeding groups was based on parental desire to breast feed and the capability of the mother to deliver fresh breast milk to the nursery each day. Informed parental consent was obtained for all subjects. Fortified human milk was prepared daily by adding skim and cream fractions derived from pooled, heat-treated human donor milk to milk from the infant's mother (10). The CM formulae were typical of those used for premature infants (8).

Sample collection and measurements of immunologic factors All samples were collected during two 96 h balance periods, when the infants were approximately 2.5 and 5 weeks of age. During each balance period, an aliquot of each day's FMM and urine collection was frozen. The complete stool output for the 96 h period was pooled and an aqueous extract prepared and frozen as described previously (11). Total IgA and secretory IgA (SIgA) concentrations and specific SIgA antibodies to *Escherichea coli* O antigens were measured by quantitative immunoassays and enzyme-linked immunosorbent assays (ELISA) described previously (3). Lactoferrin and lysozyme were measured using sandwich ELISA with monospecific polyclonal antisera for capture and detector antibodies (11). None of these assays detected any activity in the CM, suggesting that they were specific for human immune factors or that related factors in the CM were so extensively denatured by processing that they did not react in our immunoassays.

Calculations and statistics The amount of each factor delivered in human milk during the balance period was determined by multiplying the concentration in the aliquot by the volume delivered each day. Similarly, total fecal and urinary outputs of the factors were calculated on the basis of the total volume of the sample from which the aliquot was derived. All data were expressed as mg of the measured factor per kg infant weight per 96 h balance period, to allow associations between inputs and outputs to be quantified. Amounts of some factors did not show characteristics of a normal distribution, so analysis of variance, Student's t-tests, and multiple linear regressions were performed on natural log transformed data. Tabular data are presented as median values with 0.25 and 0.75 quantiles given in parentheses.

RESULTS

Characteristics of study populations The results presented here were obtained from 18 infants fed FMM and 10 fed CM. The feeding groups were similar in gestational age, age at time of each balance period, birth weight, and total nitrogen and caloric intake (11). Since there were no differences in results between the two study periods, these data were pooled. Although IgA and SIgA were assayed separately, only total IgA data are presented, since IgA and SIgA correlated well, and the latter constituted 95% of the total.

Delivery of immune factors in milk When we compared the concentration of immune factors in maternal milk samples with the same samples after fortification, only the lysozyme content was significantly increased (median 69%) by supplementation. The factors delivered by the fortified milk preparation have been described and quantified previously (11) and are summarized in Table 1.

Table 1. Intake of Immune factors in FMM and CM Groups (mg/kg/96h)

Immune Factor	FMM (n = 33)	CM (n = 20)
Lactoferrin	1234 (794; 1504) [a]	Not detected
Lysozyme	80 (50; 140)	Not detected
Total IgA	654 (471; 918)	Not detected
SIgA Antibodies	Detected in all	Not detected

[a] Data are presented as the median (0.25 and 0.75 quantiles).

Stool output of immune factors The amounts of all factors were greater in the stools of infants fed FMM than in those fed CM (11) (Table 2).

Table 2. Fecal Output of Immunologic Factors in FMM and CM Groups (mg/kg/96h)

Immune Factor	FMM (n = 33)	CM (n = 20)
Lactoferrin	37 (13; 59) [a]	0.2 (0.1; 0.4)
Lysozyme	0.07 (0.02; 0.14)	0.009 (0.001; 0.018)
Total IgA	58 (18; 89)	2 (1; 5)
SIgA Antibodies	Detected in 28/31	Detected in none

[a] Data are presented as the median (0.25 and 0.75 quantiles). All differences between feeding groups were significant at $p < 0.01$.

Serum concentration of immune factors Total IgA, SIgA, lactoferrin and lysozyme concentrations in the infants' sera did not differ between the two feeding groups at either sampling period.

Urinary output of immune factors The amounts of lactoferrin, lysozyme, and IgA in the urine of infants fed FMM were higher than those in urines of infants fed CM (Table 3), but the differences were significant only for lactoferrin and IgA (12).

Table 3. Urinary Excretion of Immunologic Factors in FMM and CM Groups (mg/kg/96h)

Immune Factor	FMM (n = 33)	CM (n = 20)
Lactoferrin	4.66 (0.48; 7.09) [a] *	0.0226 (0.0092; 0.0328)
Lysozyme	0.114 (0.27; 0.143)	0.0631 (0.0301; 0.0921)
Total IgA	7.962 (0.719; 12.461) *	0.0817 (0.0; 0.127)
SIgA Antibodies	Detected in 8/33	Detected in none

[a] Data are presented as the median (0.25 and 0.75 quantiles).
* Significantly different from CM group at $p < 0.001$.

<u>Correlation analysis</u> Analyses were done in two ways. The amounts of different factors in a particular secretion were compared, and then the amount of each factor was compared with the amount of the same factor delivered to that infant by the FMM (Table 4).

Table 4. Correlations Between Amounts of Ingested and Secreted or Excreted Immunologic Factors in FMM Fed Infants

Fluids	Immune Factor	Correlation Coefficient
Comparisons Between FMM and Excreta		
FMM vs Stool	SIgA antibodies to E. coli	0.79 **
	SIgA, lactoferrin, lysozyme	NS
FMM vs Urine	Total IgA	0.72 **
	Lactoferrin	NS
Comparisons Within Milk or Excreta		
FMM	Lactoferrin, lysozyme, IgA, SIgA antibodies	NS
Stool	Lactoferrin, lysozyme,IgA, SIgA antibodies to E. coli	0.45-0.65 *
Urine	Lactoferrin, IgA	0.69 **

NS: Not significant
* $p < 0.01$
** $p < 0.001$

DISCUSSION

Several previous studies have evaluated the effects of human and cow's milk feeding on the fecal concentration of lysozyme (13), lactoferrin (14), and IgA (15) in term infants. However, few of these studies included timed collections to allow quantification of daily output, and in none was the intake of immunologic factors measured. To our knowledge, the effect of human milk on urinary immune factors has not been examined previously. In our studies, nutrient balance techniques allowed concurrent sampling of milk and collection of stool and urine output and serum samples, and thus provided an opportunity to correlate the intake of factors in milk with the output of those same factors in an infant's serum, urine and stools.

Our results indicate that human milk-fed premature infants have higher quantities of several immune factors in certain secretions than cow's milk-fed infants. The simplest explanation of this difference is that human milk delivers larger amounts of these substances to the infant than cow's milk. An additional or alternative explanation is that human milk may stimulate production of immune factors by the infant. These studies were designed to differentiate direct transfer by milk from host induction of IgA, SIgA, lactoferrin and lysozyme by allowing quantitative comparison of each factor in milk and at two mucosal sites in the infant: the gastrointestinal tract, a comparatively proximal site, and the urinary tract, a comparatively distal site. Parallel measurements in the serum were included to assess the most obvious route of potential transport of factors between these two sites. For the purpose of preliminary analysis, we interpreted positive correlations between the amount of any factor in milk and the amount of the same factor in stool, urine or serum to suggest that the factor was derived from the milk. Similarly, positive correlations between factors within stool or urine which were not correlated in milk were interpreted to suggest that those substances may be produced by the infant, and that the synthesis may be modulated by factors in the FMM.

Applying these experimental and analytical approaches, the results of regression analyses indicate that SIgA antibodies in the stool could have been derived wholly or predominantly from the human milk preparations (Table 4). For the other stool factors studied, however, direct transfer from milk was not indicated by this analysis. In addition, lactoferrin, lysozyme, total IgA, and SIgA antibodies were not correlated in the FMM, but were correlated in the stools of infants fed these preparations. Taken together, these observations are consistent with the hypothesis that coordinated induction of the production or release of these factors occurs in the gastrointestinal tract in response to some substance(s) in milk (11).

Two of the immune factors, lactoferrin and IgA, were also correlated in the urine. If the enhanced urinary output of these substances were due to absorption from the gut and clearance by the kidney, one might expect to see elevated levels in the serum, since the infants were receiving a continuous gastric infusion of FMM at the time of collection of the first serum sample. However, none of the immune factors which was elevated in the stools and urines of FMM infants was also elevated in serum samples. Furthermore, if the proteins in the urine were whole molecules or large fragments, given their molecular weights, they would be unlikely to be filtered by glomeruli. Thus, these findings suggest that FMM may not be the direct source of the enhanced urinary content of these factors.

Clearly more definitive experiments, including feeding of stable isotope-labeled human milk, are needed to identify the mechanisms which mediate the enhancement of immune factors in the secretions of these human milk-fed infants. Epidemiological studies designed to evaluate the possible benefits of this enhancement to the infant would seem indicated, especially if more mature infants show similar responses to human milk feeding, and if the structural and functional integrity of the immune factors in their secretions can be demonstrated.

ACKNOWLEDGEMENT

This work was supported by contracts from the National Institute of Child Health and Human Development (NO1-HD-2-2814 and NO1-HD-6-2918) and a grant from the National Institute of Allergy & Infectious Diseases (5 RO1 AI 21412-02).

REFERENCES

1. A. S. Goldman, A. J. Ham Pong, R. M. Goldblum. Host Defenses: development and maternal contributions in: "Advances in Pediatrics," L. A. Barness, ed., Year Book Medical Publishers, Inc. (1985).
2. L. Gothefors, and Marklund, S. Lactoperoxidase activity in human milk and in saliva of newborn infants. Infect. Immun. 11:1210 (1975).
3. R. M. Goldblum, C. Garza, C. A. Johnson, R. Harrist, B. L. Nichols, and A. S. Goldman, Human Milk Banking I. Effects of container upon immunologic factors in mature milk. Nutr. Res. 1:449 (1981).
4. A. S. Goldman, C. Garza, B. L. Nichols, and R. M. Goldblum, Immunologic factors in human milk during the first year of lactation. J. Pediatr. 100:563 (1982).
5. A. S. Goldman, R. M. Goldblum, and C. Garza, Immunologic components in human milk during the second year of lactation, Acta. Paediatr. Scand. 72:461 (1983).
6. A. S. Goldman, R. M. Goldblum, C. Garza, B. L. Nichols, and E. O. Smith, Immunologic components in human milk during weaning, Acta. Paediatr. Scand. 71:133 (1983).
7. A. S. Goldman, C. Garza, B. Nichols, C. A. Johnson, E. O. Smith, and R. M. Goldblum, Effects of prematurity on the immunologic system in human milk, J. Pediatr. 101:901 (1982).
8. R. M. Goldblum, C. W. Dill, T. B. Albrecht, E. S. Alford, C. Garza, and A. S. Goldman, Rapid high-temperature treatment of human milk, J. Pediatr. 104:380 (1984).
9. N. F. Butte, R. M. Goldblum, L. M. Fehl, K. Loftin, E. O. Smith, C. Garza, and A. S. Goldman, Daily ingestion of immunologic components in human milk during the first four months of life, Acta. Paediatr. Scand. 73:296 (1984).
10. R. J. Schanler, C. Garza, and E. O. Smith, Fortified mothers' milk for very low birth weight infants: results of macromineral balance studies, J. Pediatr. 107:767 (1985).
11. R. J. Schanler, R. M. Goldblum, C. Garza, and A. S. Goldman, Enhanced fecal excretion of selected immune factors in very low birth weight infants fed fortified human milk, Ped. Res. 20:711 (1986).
12. R. M. Goldblum, R. J. Schanler, C. Garza, and A. S. Goldman, Enhanced urinary lactoferrin excretion in premature infants fed human milk, Pediatr. Res. 19:342A (1985).
13. B. Haneberg, and P. Finne, Lysozymes in feces from infants and children, Acta. Paediatr. Scand. 63:588 (1974).
14. G. Spik, B. Brunet, C. Mazurier-Dehaine, G. Fontaine, and J. Montreuil. Characterization and properties of the human and bovine lactotransferrins extracted from the feces of newborn infants. Acta Paediatr. Scand. 71:979 (1982).
15. B. Haneberg, Immunoglobulins in feces from infants fed human or bovine milk, Scand. J. Immunol. 3:191 (1974).

INHIBITION OF BACTERIAL ADHESION AND TOXIN BINDING BY GLYCOCONJUGATE AND OLIGOSACCHARIDE RECEPTOR ANALOGUES IN HUMAN MILK

J. Holmgren, A.-M Svennerholm, M. Lindbald and G. Strecker[1]

Department of Medical Microbiology, University of Goteborg
S-413 46 Goteborg, Sweden and Laboratory of Biological
Chemistry, University of Lille, F-59655 Villeneuve-Cedex, France[1]

Following the discovery that the receptor for cholera toxin is a specific glycolipid, the ganglioside GM1 (see Holmgren, 1981), many other toxins and pathogenic bacteria have been found to attach to epithelial receptors being of a carbohydrate nature. Secretions, e.g. human milk, normally contain a variety of glycoproteins, glycolipids, and free oligosaccharides. At least in vitro the attachment of bacteria and toxins to epithelial cells can be completely inhibited by soluble receptor molecules. We assumed that some of the carbohydrate substances in secretions, including human milk, might be structure analogues of cell membrane receptors for bacterial pathogens or their toxins and could thus be able to function as competitive inhibitors. Such receptor analogues, if present in milk, could possibly be involved in the protection against enteric infections and respiratory infections ascribed to breast feeding. We have therefore tested the ability of glycoconjugates and oligosaccharides of human colostrum and milk to inhibit different Vibrio cholerae and enterotoxinogenic Escherichia coli (ETEC) bacteria from adhering to mammalian cells, and their enterotoxins from binding to GM1 receptor ganglioside (Holmgren et al. 1981; 1983). In this paper we will summarize the findings from these earlier studies and provide new, additional information on these systems. We will also briefly review the work by other investigators along the same lines, and discuss the possible significance of the findings.

MATERIALS AND METHODS

The following bacterial strains were used for our own work described in this paper: V. cholerae 1451 of classical biotype; V. cholerae T19479 of El Tor biotype; E. coli H10407 of CFA/I fimbrial type; E. coli E1392-75 or 411-5 of CFA/II fimbrial types. Cholera toxin was prepared from V. cholerae strain 569B, and E. coli LT from strain 286-C2.

Pooled mature milk from Swedish women was used. For the lipid extraction experiments non-centrifuged milk was used, whereas for other fractionations the clear middle layer after centrifugation was used. The various fractionation procedurès have been described in detail previously

(Holmgren et al., 1981; 1983). Purified oligosaccharides from human milk and from other sources were prepared as described (Ginsburg, 1978).

The assay methods used for determining the activity of various milk specimens, fractions and isolated oligosaccharides to inhibit bacterial attachment to erythrocytes of different species (hemagglutination; HA) or enterotoxin binding to GM1-ganglioside coated microtiter wells have been described (Holmgren et al., 1981; 1983).

VIBRIO CHOLERAE OF CLASSICAL AND EL TOR BIOTYPES

V. cholerae is the prototype for the several species of bacteria which can cause diarrhoeal disease by colonizing the small intestine and elaborating an enterotoxin (Holmgren, 1981). *V. cholerae* bacteria can agglutinate certain species of erythrocytes, and it has been suggested that the attachment of bacteria to erythrocytes mimicks the interactions that result in adherence of cholera vibrios to intestinal epithelium. At least two cell-associated hemagglutinins have been described on *V. cholerae*; one of these is inhibited by L-fucose, and the other is inhibited by D-mannose. We have found that the fucose-sensitive hemagglutinin is characteristically present on *V. cholerae* strains of the classical biotype, and that the mannose-sensitive hemagglutinin is associated with the El Tor biotype (Holmgren et al., 1983).

In our first study (Holmgren et al., 1981) we observed that both of these hemagglutinins were strongly inhibited by human milk specimens from not only Bangladeshi or Pakistani women who had substantial milk antibody titres to cholera antigens but also from Swedish mothers lacking such antibodies. This suggested that milk factors other than antibodies could act as inhibitors. We therefore separated milk specimens into non-immunoglobulin and immunoglobulin fractions by affinity chromatography on an immunosorbent column and tested their activity. This procedure removed >99% of both IgA and other immunoglobulins and all detectable antibody activity from the non-immunoglobulin fraction. Yet this fraction of Swedish milk and colostrum had the same inhibitory activity as that of the whole unfractionated samples, whereas the immunoglobulin fraction gave little if any inhibition. With milk from women living in cholera endemic areas, both the immunoglobulin and non-immunoglobulin fractions gave significant inhibition of *V. cholerae* hemagglutination.

Our further characterization (Holmgren et al., 1983) supported our proposal in the initial report that the hemagglutination inhibitory (HAI) activity of the non-immunoglobulin fraction for both classical and El Tor *V. cholerae* was mediated by carbohydrate structures.

We first separated pooled human milk from Swedish women into a high-molecular-weight fraction A (mw ≥10,000) and two low-molecular-weight fractions B (mw 1,000-10,000) and C (mw ≤1,000) by ultrafiltration through membrane filters with defined pore sizes, and tested the HAI activity of these fractions for the two different *V. cholerae* biotypes. For classical *V. cholerae*, the HAI activity was distributed about equally on fraction A and fraction B, with only about 5% of the activity recovered in fraction C (Fig 1). In contrast, about 80% of the HAI activity for the El Tor vibrios was found in fraction A, <5% in fraction B, and 15-20% in fraction C (Fig 1).

We showed that each of the HAI activities measured in these fractions were unrelated to immunoglobulin, resisted boiling, and were destroyed by periodate treatment. Gel filtration experiments of fraction A on Sephacryl

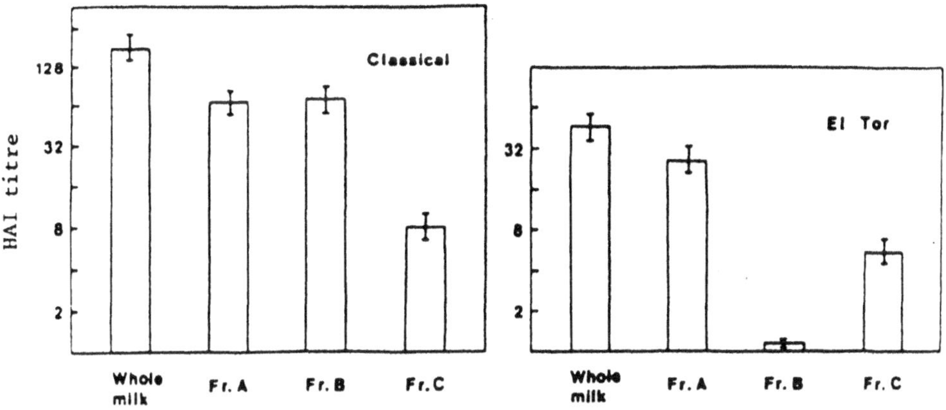

Fig 1 Molecular weight distribution of HAI titres (mean ± standard error of the mean) in human milk for classical and El Tor V. cholerae. Fraction A mw ≥10,000; fraction B mw 1,000-10,000; fraction C mw <1,000.

S-300 showed that the HAI activity for El Tor vibrios was of very high molecular weight (mw 150-500,000); in contrast, the predominating HAI activity in fraction A for classical V. cholerae had a molecular weight of approx. 10-20,000 although some activity was also found in the same high-molecular-weight fractions that were active for El Tor. Gel filtration of fraction B on Bio-Gel P-10 identified the HAI activity for classical V. cholerae in this fraction to have a molecular weight of approx. 1,000. The HAI activity of fraction A for both classical and El Tor vibrios was fully precipitated with 100% saturated ammonium sulphate, whereas the HAI activity of fraction B and C were not precipitated.

We assumed that the HAI activities in fraction A was due to high-molecular-weight glycoconjugates, the activity in fraction B mainly to free oligosaccharides, and that in fraction C mainly to mono- and disaccharides. We performed direct experiments to examine whether the high-molecular weight HAI activities were due to glycolipids or glycoproteins or both. Whole milk was lipid extracted with chloroform--methanol-water (4:8:3), and acid glycolipids (gangliosides) and neutral glycosphingolipids were isolated. None of the glycolipid preparations had any detectable HAI activity for either classical or El Tor V. cholerae; on the other hand, the delipidized milk residue retained HAI activity for both types of vibrios. The effect of pronase on the high-molecular-weight HAI activities of fraction A was also tested. The results showed, that the pronase treatment dramatically changed the molecular weight of the HAI activity for classical as well as El Tor vibrios, such that they completely passed through a membrane with exclusion limit of mw ~10,000 (Table 1). These results indicate that the HAI activities in fraction A for both classical and El Tor V. cholerae were of a glycoprotein nature rather than a glycolipid nature.

Since L-fucose is the only monosaccharide capable of inhibiting classical V. cholerae HA (Holmgren et al., 1983), we assumed that L-fucose-containing glycoproteins in fraction A and oligosaccharides in fraction B were responsible for the HAI activity against classical

Table 1. Effect of pronase digestion on high-molecular-weight HA inhibitory activities in human milk

Fraction	HAI titre	
	Classical	El Tor
Fraction A	32	32
FractionA, pronase-digested	16	32
Retentate of pronase-digested fraction A on PM-10 membrane	<2	<2
Filtrate of pronase-digested fraction A on PM-10 membrane	16	16

V. cholerae. It was therefore of interest to test the HAI activity of bovine milk against classical and, for comparison against El Tor V. cholerae since bovine milk lacks fucose. The bovine milk completely lacked HAI activity against classical V. cholerae, even though it had strong HAI activity against El Tor V. cholerae (again residing in the high molecular weight fraction A).

Many of the oligosaccharides isolated from human milk contain L-fucose, (Kobata et al., 1978) and we tested a series of such purified human milk fucosides. The results show that all the tested human milk fucose-containing oligosaccharides had strong HAI activity against classical V. cholerae, and that none of them had any HAI activity against El Tor V. cholerae (Table 2).

Mannose is not a constituent of milk oligosaccharides but is often present on high-molecular-weight glycoproteins of both human and bovine milk and may thus be of importance for the HAI activity against El Tor V.cholerae obtained with high-molecular-weight milk glycoproteins. We therefore tested a series of different mannose-containing oligosaccharides isolated from various sources (not milk). The results show that all mannose-containing oligosaccharides had demonstrable HAI activity for El Tor V. cholerae (Table 3); this was in contrast to oligosaccharides lacking mannose.

ENTEROTOXIGENIC ESCHERICHIA COLI

Some of the above fractions and treatments were also used to characterize the inhibitory activity of pooled Swedish human milk for cell attachment (HA) of ETEC bacteria carrying either CFA/I or CFA/II fimbriae; these HA tests were carried out in the presence of 1% D-mannose to avoid any contribution to the HA reactions from type 1 pili on the bacteria. The results are shown in table 4. The HAI activity for both CFA/I and CFA/II ETEC was of high molecular weight, resisted boiling and was destroyed by periodate treatment. In contrast to the results for V. cholerae the high-molecular-weight fraction seemed to contain HAI active glycolipids as well as glycoproteins against ETEC.

In this context it is also of interest that glycoproteins from the colostrum of cows or sows have been found to competitively inhibit cell attachment (HA) of K88- or K99-fimbriated ETEC which often cause diarrhoeal disease in calves and piglets (Lindahl et al., 1986). Various colostral sialoglycoproteins were active, consistent with reports that sialic acid may be part of the epithelial receptors for these colonization fimbriae.

Table 2. Effect of fucose-containing human milk oligosaccharides on classical but not El Tor V. cholerae cell attachment (hemagglutination)

<table>
<tr><th rowspan="2">Oligosaccharide</th><th colspan="2">HA inhibition</th></tr>
<tr><th>Classical</th><th>El Tor</th></tr>
<tr><td>Gal(β1-4)Glc
|(α1-2) |(α1-3)
Fuc Fuc</td><td>++</td><td>-</td></tr>
<tr><td>Gal(β1-3)GlcNAc(β1-3)Gal(β1-4)Glc
|(α1-2) |(α1-4)
Fuc Fuc</td><td>++</td><td>-</td></tr>
<tr><td>Gal(β1-3)GlcNAc(β1-3)Gal(β1-4)Glc
|(α1-4)
Fuc</td><td>++</td><td>-</td></tr>
<tr><td>Fuc
|(α1-4)
Gal(β1-3)GlcNAc(β1-3)
Gal(β1-4)Glc
Gal(β1-4)GlcNAc(β1-6)
|(α2-6)
NeuAc</td><td>++</td><td>-</td></tr>
<tr><td>Fuc
|(α1-4)
Gal(β1-3)GlcNAc(β1-3)
Gal(β1-4)Glc
Gal(β1-4)GlcNAc(β1-6)</td><td>++</td><td>-</td></tr>
</table>

CHOLERA TOXIN AND E. COLI HEAT-LABILE ENTEROTOXIN

In our initial study (Holmgren et al., 1981) we found that the nonimmunoglobulin fraction of human milk or colostrum could effectively inhibit the binding of both cholera toxin and E. coli LT to the receptor ganglioside GM1 as tested in vitro. Furthermore, Otnaess and Svennerholm (1982) observed a significant protective effect of a nonimmunoglobulin fraction of human milk against challenge with either of these toxins in the rabbit intestine. This suggested that GM1 ganglioside and/or a structurally closely related glycolipid or glycoprotein was the active component(s) in milk, since competitive inhibition studies with a variety of sugars have identified very precise structural requirements for inhibition of these high affinity toxin-receptor interactions (Holmgren et al., 1973).

Kolstõ-Otnaess et al. (1983) have characterized the inhibitory activity for E. coli LT and cholera toxin further and shown that it is associated with the monosialoganglioside fraction of human milk. A component identical or closely similar to GM1 ganglioside was described as the active factor, and the

Table 3 Effect of various mannose-containing oligosaccharides against classical and El Tor *V. cholerae* cell attachment (hemagglutination)

	HA Inhibition Classical	HA Inhibition El Tor
$Man_{2,3....9}$ — GlcNac	-	++
Gal(β1-4)GlcNAC(β1-2)Man(α1-3) Gal(β1-4)GlcNAC(β1-2)Man(α1-6) > Man(β1-4)GlcNAc	-	+
Gal(β1-4)GlcNAC(β1-4) Gal(β1-4)GlcNAC(β1-2)Man(α1-3) Gal(β1-4)GlcNAC(β1-2)Man(α1-6) > Man(β1-4)GlcNAc	-	+
NeuAc(α2-6)Gal(β1-4)GlcNAc(β1-2)Man(α1-3)Man(β1-4)GlcNAc	-	++
NeuAc(α2-6)Gal(β1-4)GlcNAc(β1-2)Man(α1-3) Gal(β1-4)GlcNAc(β1-2)Man(α1-6) > Man(β1-4)GlcNAc Gal(β1-4)GlcNAc(β1-2)Man(α1-3) NeuAc(α2-6)Gal(β1-4)GlcNAc(β1-2)Man(α1-6) > Man(β1-4)GlcNAc	+	++
NeuAc(α2-6)Gal(β1-4)GlcNAc(β1-2)Man(α1-3) Gal(β1-4)GlcNAc(β1-2)Man(α1-6) > Man(β1-4)GlcNAc	-	+

Table 4. Antiadhesive activity of human milk against CFA/I and CFA/II *E. coli*

Milk fractions	HAI titre CFA/I H10407	CFA/II 411-5
Whole milk	400	200
Molecular size fractions		
- A (mw ≥10,000)	800	400
- B & C (mw <10,000)	<5	<5
100°C for 15 min	400	200
0.1 M periodate	<20	<20
Lipid extraction fractions		
- Neutral glycolipids	32	(+)*
- Gangliosides	(+)*	(+)*
- Delipidized "glycoprotein" rest	128	32

*Uncertain result because of concomitant hemolysis

milk ganglioside fraction from 2 ml of human milk could completely inhibit 0.1 μg of cholera toxin in rabbit intestinal loop experiments.

We have not done any experiments of our own to determine whether any ganglioside in addition to GM1 in human milk can inhibit cholera toxin or *E. coli* LT. However, when testing gangliosides isolated from normal intestinal mucosa or small cell lung cancer tissue we have found that GM1 ganglioside with a fucose residue attached to the external galactose (fucosyl-GM1 ganglioside) has almost the same binding affinity and inhibitory activity for cholera toxin and *E. coli* LT as GM1 ganglioside (P. Fredman et al., to be published). Based on this we would like to draw attention to the possibility that fucosyl-GM1 ganglioside may be a toxin-inhibiting component of milk as well.

INHIBITION OF *STREPTOCOCCUS PNEUMONIAE* AND *HAEMOPHILUS INFLUENZAE*

The competitive inhibition of bacterial and toxin cell attachment by receptor-like glycoconjugates or oligosaccharides in human milk is not restricted to enteric pathogens. Andersson et al (1986) recently demonstrated an inhibitory effect against the attachment of *S. pneumoniae* to human pharyngeal or buccal epithelial cells in both high- and low-molecular-weight fractions of human milk. The inhibitory activity in the high-molecular-weight fraction was independent of specific antibody content; it was present after immunoadsorption and in the milk from IgA-deficient women. The inhibitory activity in the low-molecular-weight fraction was in part explained by the content of oligosaccharides containing the GlcNAc(β1-3)Gal-disaccharide unit that has been reported to act as attachment receptor structure for pneumococci. Andersson et al. (1986) also showed antiadhesive activity in milk against *H. influenzae*; this activity was restricted to the high-molecular-weight fraction of the milk and was also unaffected by immunoadsorption.

Table 5. Heat-stable cholera HAI factor levels (and for comparison cholera antibodies) in mothers' breast-milk in relation to cholera infection outcome for the breast-fed children. (In collaboration with R. Glass, International Centre for Diarrhoeal Disease Research, Bangladesh)

Milk factor	Median titres in breast-fed children with different outcome		
	Uninfected	Asymptomatic infection	Ill
Heat-stable HAI	N=20	N=20	N=20
Classical	128	128	128
El Tor	8	8	12
IgA antibodies	N=63	N=11	N=19
Anti-LPS	600	600	150*
Anti-CT	150	150	30**

*p<0.01 **p<0.02 for comparison with asymptomatic infection group

PERSPECTIVE

Breast-feeding of infants has been found to reduce the morbidity in mainly gastrointestinal infections but possibly also in upper respiratory tract infections. In view of the present findings it is tempting to propose a protective role for receptor-like glycocompounds in the breast-milk. However, human milk is known to contain several other anti-microbial components. These include e.g. immunoglobulins and lymphoid cells and macrophages, chemotactic factors, lactoferrin, lysozyme, lactoperoxidase, interferon, and bifidus factor. This complexity makes it very difficult indeed to assess the relative role of individual components.

In a prospective epidemiological study in Bangladesh (Glass et al., 1983), we were able to show that specific cholera antibody titres in breast milk are important for the outcome of cholera infection in the breast-fed children: infants and children who drank milk with high cholera antibody titres got ill less often than those who drank milk with lower antibody levels (p<0.01). We undertook similar analyses of heat-stable antiadhesive (HAI) activity for either classical or El Tor cholera vibrios and were unable to relate these HAI titres to the outcome for the breast-fed child (Table 5). However, similar analyses for lactoferrin and lysozyme also failed to show any correlation to protection (R Glass et al, unpublished data). Our interpretation is not that these factors are unimportant for the protective effect of breast-feeding, but rather, that different from antibodies, none of these factors singled out sufficiently from the others to correlate statistically significantly to protection against a specific infectious disease in an endemic setting.

Thus, we conclude that human milk contains a variety of receptor-like glycoproteins, glycolipids and oligosaccharides that through competitive inhibition have antiadhesive effects against several bacterial pathogens and toxins. A protective role for these soluble receptor analogues is likely but not yet proven.

ACKNOWLEDGEMENTS

We gratefully acknowledge the skilled technical assistance of Gunhild Jonson and Lena Ekman, and the financial support from the Swedish Medical Research Council (Project 16X-3382).

REFERENCES

ANDERSSON B, PORRAS O., HANSON L.Å., LAGERGÅRD T. and SVANBORG-EDÉN C. (1986): Inhibition of attachment of Streptococcus pneumoniae and Haemophilus influenzae by human milk and receptor oligosaccharides. J. Infect. Dis. 153, 232-237.

GINSBURG V. (1978): ed. Complex Carbohydrates. Methods in Enzymology, vol. 50, Academic Press, New York.

GLASS R., SVENNERHOLM A.-M., STOLL B.J., KHAN M.R., HOSSAIN K.M.B., HUQ M.I. and HOLMGREN J. (1983): Protection against cholera in breast-fed children by antibodies in breast-milk. N. Engl. J. Med. 308, 1389-1392.

HOLMGREN J. (1981): Action of cholera toxin and the prevention and treatment of cholera. Nature (London) 292, 413-417.

HOLMGREN J., LÖNNROTH I. and SVENNERHOLM L. (1973): Tissue receptor for cholera exotoxin: Postulated structure from studies with GM1 ganglioside and related glycolipids. Infect. Immun. 8, 208-214.

HOLMGREN J., SVENNERHOLM A.-M. and ÅHRÉN C. (1981): Nonimmunoglobulin fraction of human milk inhibits bacterial adesion (hemagglutination) and enterotoxin binding of Escherichia coli and Vibrio cholerae. Infect. Immun. 33, 136-141.

HOLMGREN J., SVENNERHOLM A.-M. and LINDBLAD, M. (1983): Receptor-like glycocompounds in human milk that inhibit classical and El Tor Vibrio cholerae cell adherence (hemagglutination). Infect. Immun. 39, 147-154.

KOBATA A., YAMASHITA, K. and TACHIBANA Y. (1978): Oligosaccharides from human milk. In Ginsburg, V. (ed.) Methods in Enzymology, vol. 50, p. 216-220, Academic Press, New York.

KOLSTÖ OTNAESS A.-B., LAEGREID A. and ERTRESVÅG, K. (1983): Inhibition of enterotoxin from Escherichia coli and Vibrio cholerae by gangliosides from human milk. Infect. Immun. 40, 563-569.

LINDAHL M., BROSSMER, R. and WADSTRÖM T. (1986): Sialic acid and N-acetylgalactosamine specific bacterial lectins of enterotoxigenic Escherichia coli (ETEC). In press.

OTNAESS A.-B. and SVENNERHOLM, A.-M. (1982): Non-immunoglobulin fraction of human milk protects rabbits against enterotoxin-induced intestinal fluid secretion. Infect. Immun. 35, 738-740.

TRANSFER OF TUBERCULIN IMMUNITY FROM MOTHER TO INFANT

Margaret A. Keller, Annette L. Rodriguez, Sarah Alvarez, *Noel C. Wheeler, and Diane Reisinger

Department of Pediatrics, UCLA School of Medicine, Harbor-UCLA Medical Center, Torrance California 90509; *Department of Biomathematics, UCLA School of Medicine

Previous investigators (1-3) have suggested that tuberculin immunity may be transferred from the mother to the infant via human milk, but other investigators (4-5) have found transplacental transfer of immunity. The purpose of our study was to examine transfer of tuberculin immunity by studying newborn lymphocyte blastogenesis induced by PPD antigen at 1-5 days of age, 4-6 weeks of age and 3 months of age. We also examined lymphokine production by neonatal lymphocytes as another index of tuberculin immunity. For this purpose we studied PPD-induced monocyte chemotactic factor production by lymphocytes from infants 4-6 weeks of age. In an attempt to define the mechanism of transfer of immunity, we studied male infants of tuberculin positive breast-feeding mothers at 4-6 weeks of age and examined the sex chromosomes of lymphocytes dividing in response to PPD-stimulation. This sex determination would identify blastogenic lymphocytes as either of maternal or newborn origin.

METHODS

Study Population

Healthy, postpartum adult females with known tuberculin status participated in the study with their healthy infants. Tuberculin status was determined by skin testing shortly after delivery, during pregnancy, or prior to pregnancy. Mothers with positive skin test ($\geq$ 12 mm) were not retested. However, all tuberculin negative mothers (no induration) were skin tested again 1 to 5 days postpartum and at 4 to 6 weeks postpartum. Our study consisted of four mother-infant groups: infants of tuberculin positive breast-feeding mothers, infants of tuberculin positive bottle-feeding mothers, infants of tuberculin negative breast-feeding mothers and infants of tuberculin negative bottle-feeding mothers. Breast-feeding infants were defined as those receiving at least 80% of their intake as human breast milk. Bottle-feeding infants never received any breast milk.

LYMPHOCYTE STUDIES

Ficoll-Hypaque differential centrifugation (6) was used to separate peripheral whole blood lymphocytes. Breast milk cells were first centrifuged at 1400 RPM (250xg) to remove fat and then washed twice in Hanks balanced salt solution, Ca^{++}, Mg^{++} free. When the cell yield permitted, milk cells were separated by Ficoll-Hypaque. Otherwise washed, unfractionated cells were used. Mononuclear cells were suspended in RPMI-1640 with gentamicin, antibiotic-antimycotic (Gibco-penicillin, streptomycin and fungizone) and additional L-glutamine. Purified protein derivative (Connaught) PPD, 2 mg/ml, was diluted to the following final concentrations for the lymphocyte cultures: 200, 100, 50, 25, 10, 5, 2.5, 0.5, 0.1, and .05 μg/ml. Lymphocytes ($1x10^5$) suspended in RPMI-1640 with 5% heat inactivated pooled human AB sera were added to each well of the microtiter plate with or without antigen. Cultures were incubated for 7 days at 37°C. Additional lymphocyte cultures were incubated for 3, 5 or 6 days when cell yields permitted. After incubation, 0.4 μCi of ^3H-thymidine was added to each well and incubated for 24 additional hours. Cells were harvested onto glass fiber filters and assayed in a liquid scintillation counter.

MONOCYTE CHEMOTACTIC FACTOR PRODUCTION AND ASSAY

Lymphocytes were incubated in serum free media for 72 hours with 10 μg/ml PPD or no antigen. Supernatants were harvested and frozen. Monocyte chemotaxis was assayed using the method of Altman et al. (7). Test supernatant (at 1:2 or 1:3 dilution) from PPD-stimulated or control cultures was placed in the lower portion of the blind well chamber separated from adult test monocytes by a 5.0 μm Nuclepore filter. The average number of monocytes per field migrating to the opposite side of the filter was determined. The number of monocytes migrating toward stimulated supernatants minus the number migrating toward control supernatants was used as Δ migrating monocytes, a chemotactic index.

STATISTICAL METHODS

Several approaches were used to analyze the blastogenesis and lymphokine data. Infants were classified as lymphocyte blastogenesis responders or Δ monocyte responders if they satisfied defined cutpoint (cut off point) criteria. These cutpoints were determined for stimulation index, Δ counts per minute (cpm), and Δ monocytes by determining with 99% confidence the tolerance limits between which will be 99% of infant responses from tuberculin negative mothers (8). The upper tolerance limit was used as a cutpoint to define positive and negative responses. These tolerance limits varied for the three time periods studied: Δ cpm $\geq$ 6325, S.I. $\geq$ 3.6 at entry; Δ cpm $\geq$ 3612, S.I. $\geq$ 3.1 at 4-6 weeks of age, Δ cpm $\geq$ 3261, S.I. $\geq$ 2.9 at 3 months. If tolerance limit criteria were met for either stimulation index or Δ cpm, an infant was considered to have a positive response. In addition, the counts per minute in stimulated wells had to be statistically different from counts per minute in nonstimulated wells using Student's t-test (9) on logarithmically transformed data. The same tolerance limit criteria were used to define Δ monocytes $\geq$ 18.9 as a positive response in the monocyte chemotactic assay. Fisher's exact test (10), two way analysis of variance(11) and two way analysis of covariance (11) with tests for equality of slopes were used to analyze data.

RESULTS

Table I summarizes the data in terms of traditional analysis by S.I. or Δ cpm using tolerance limit criteria for a positive response. At 4-6 weeks there were differences in the groups in incidence of positivity.

For breast-feeding infants of tuberculin positive mothers, 4/23 (17%) and for bottle-feeding infants of tuberculin positive mothers 2/15 (13%) had lymphocyte blastogenesis to PPD. None of the infants of tuberculin negative mothers had lymphocyte blastogenesis to PPD. However, no group was statistically different from any other group due to the sample sizes. However, if all infants of tuberculin positive mothers were compared with all infants of tuberculin negative mothers at 4-6 weeks regardless of the feeding method, 6/38 (16%) infants vs. 0/33 (0%) infants had lymphocyte blastogenesis to PPD. These results were statistically significant using Fisher's Exact test ($p<.05$).

TABLE I

Infants with Lymphocyte Blastogenesis to PPD

	Infants of PPD positive Mothers		Infants of PPD negative Mothers	
	Breast	Bottle	Breast	Bottle
Age				
1-5dy	2/33(6%)	0/26 (0%)	1/24 (4%)	0/14 (0%)
4-6wk	4/23(17%)	2/15(13%)	0/19 (0%)	0/14 (0%)
3 mn	0/12(0%)	1/9 (11%)	0/14 (0%)	0/9 (0%)

We then performed a two way analysis of variance on logarithmically transformed data. There was no effect of feeding method or PPD status on S.I. or Δ cpm at 1-5 days of age. At 4-6 weeks of age, there was a significant effect of both mother's PPD status and feeding method on S.I. and Δ cpm. The difference was between the infants of tuberculin positive breast-feeding mothers and infants of tuberculin negative bottle-feeding mothers. These results were not helpful in examining transfer via milk vs. transplacental transfer. At 3 months of age, there was no significant effect of PPD or feeding on Δ cpm, and for S.I. there was a significant interaction between feeding and tuberculin status. This interaction prevents any conclusion about effect of PPD or feeding method on S.I.

Our next approach was to use linear regression analysis for stimulated cpm vs. nonstimulated cpm for each group at each time. The correlation coefficients (12) ranged from 0.84 to 0.96 with $p < .0005$ for all groups. These results demonstrated the strong correlation between the stimulated cpm and nonstimulated cpm. Table II presents the mean nonstimulated counts per minute for each of the groups at the different times and shows the trend for bottle-feeding infants to have lower nonstimulated cpm than breast-feeding infants which would effect stimulated counts.

TABLE II

Infant Mean NS Counts Per Minute (± S.E.)

	1-5 days	4-6 weeks	3 months
PPD+ Breast-feeding	2738 ± 572 (N = 33)	2470 ± 489 (N = 23)	2678 ± 818 (N = 12)
PPD+ Bottle-feeding	2720 ± 497 (N = 26)	1953 ± 489 (N = 15)	545 ± 192 (N = 9)
PPD- Breast-feeding	3338 ± 824 (N = 24)	2288 ± 469 (N = 19)	1556 ± 479 (N = 14)
PPD- Bottle-feeding	1154 ± 275 (N = 14)	1749 ± 496 (N = 14)	510 ± 173 (N = 9)

Since our linear regression analysis had shown such a strong dependence of stimulated cpm on nonstimulated cpm, analysis of covariance was performed on the stimulated cpm as the dependent variable and the nonstimulated cpm as the covariate. The test for equality of slopes of the regression of stimulated cpm on nonstimulated cpm at both 1-5 days and 3 months of age showed no difference in the slopes for any PPD or feeding group. Analysis of covariance did not demonstrate any effect of maternal PPD status or feeding method on the blastogenesis results at 1-5 days of age and 3 months of age.

The results at 4-6 weeks of age are quite different. When the equality of slopes test was performed as a condition for the analysis of covariance, the assumption of equal slopes was rejected ($p<.05$). The slopes (stimulated cpm vs. nonstimulated cpm) for PPD positive breast-feeding and PPD positive bottle-feeding groups did not differ. Similarly the slopes for PPD negative breast-feeding and PPD negative bottle-feeding groups did not differ. However, each PPD positive group differed from each PPD negative group regardless of feeding method ($p<.0005$). The slopes and p values for these pairwise comparisons are summarized in Table III. These results strongly suggest transplacental transfer of immunity.

TABLE III

Equality of slopes for regression of stimulated cpm on nonstimulated cpm

	SLOPE	p value for pairwise comparison
PPD+ Breast-feeding	2.2	0.240
PPD+ Bottle-feeding	1.9	
PPD+ Breast-feeding	2.2	<0.0005
PPD- Breast-feeding	1.3	
PPD+ Breast-feeding	2.2	<0.0005
PPD- Bottle-feeding	1.1	
PPD+ Bottle-feeding	1.9	<0.0005
PPD- Breast-feeding	1.3	
PPD+ Bottle-feeding	1.9	<0.0005
PPD- Bottle-feeding	1.1	
PPD- Breast-feeding	1.3	0.078
PPD- Bottle-feeding	1.1	

MATERNAL LYMPHOCYTE BLASTOGENESIS

Maternal lymphocyte blastogenesis 4-6 weeks post-partum correlated well with the skin test classification. Mother's and baby's S.I. and Δ cpm were compared by Pearson's correlation coefficient (12). The only strong correlation was for baby S.I. at 4 weeks vs. mother's S.I. (4-6 weeks post-partum) in the PPD+ bottle-feeding group ($r=0.82$, $p<.005$). The similar correlation for the tuberculin positive breast-feeding group was only $r = .03$.

MILK DATA

The mean stimulation index (± S.E.) for milk lymphocytes from tuberculin positive mothers was 2.7 (± 1.1) and from tuberculin negative mothers was 1.8 (± 0.4). The wide variability in cpm precluded any demonstration of effect of maternal PPD on breast milk lymphocyte response.

MONOCYTE CHEMOTACTIC FACTOR

Supernatants from lymphocyte cultures obtained from 45 infants at 4-6 weeks of age were assayed in fourteen assays for monocyte chemotactic activity. 12% (2/17) of breast-feeding infants and 9% (1/11) of bottle-feeding infants of tuberculin positive mothers and none of the infants of tuberculin negative mothers produced lymphokine. However, numbers in each group were too small for any statistically significant differences. The results are consistent with an hypothesis of transplacental transfer of immunity.

CHROMOSOME STUDIES

Three male infants with PPD- responsive lymphocytes had metaphase spreads examined for the sex of the dividing cells. All metaphase spreads (179, 132, 50 from the three infants) were male XY.

DISCUSSION

The purpose of our study was to examine transfer of tuberculin immunity from mother to infant and to determine if breast-feeding had a role in this transfer. Possible results of our study were the following: no transfer of immunity, transfer solely via breast milk, transfer solely via placenta, or transfer via both milk and placenta.

At first we examined our data using traditional analyses, stimulation index and Δ cpm. However, since responses of neonates are much lower in general than adult responses, we used very strict tolerance limit criteria to define a positive response based on the distribution of responses of infants of tuberculin negative mothers. At 4-6 weeks of age, 17% (4/23) of breast-feeding and 13% (2/15) of bottle-feeding infants of tuberculin positive mothers had lymphocyte blastogenesis to PPD while none of the infants of tuberculin negative mothers had such responses. However, the number of subjects studied was not large enough for a statistically significant difference between breast-feeding and bottle-feeding tuberculin positive groups. However, if all infants of tuberculin positive mothers were compared with all infants of tuberculin negative mothers at 4-6 weeks, 6/38 (16%) infants vs. 0/33 (0%) infants had lymphocyte blastogenesis to PPD. These results were statistically significantly different using Fisher's Exact test ($p<.05$).

Our next approach was to use a two way analysis of variance to examine effect of PPD status or feeding method on the S.I. or Δ cpm. In addition we determined if there was any interaction between PPD status and feeding method on blastogenesis. These results were not helpful because there was an effect of PPD status and feeding method on S.I. and Δ cpm at 4-6 weeks of age. The difference was between the breast-feeding infants of tuberculin positive mothers and the bottle-feeding infants of tuberculin negative mothers which was not helpful in examining transplacental vs. breast milk transfer of immunity. Since in our correlation analysis we determined that there was a strong dependence of stimulated cpm on nonstimulated cpm, we used another statistical approach, the two-way analysis of covariance on stimulated counts with nonstimulated counts as the covariate. A condition required by this analysis is that all the slopes (stimulated cpm vs. nonstimulated cpm) for the four groups be equal. This assumption was satisfied at 1-5 days of age and three months of age. The analysis of covariance was then applied and showed no effect of PPD or feeding on results. The assumption of equal slopes at 4-6 weeks of age was rejected ($p<.05$) and further analysis of covariance could not be performed. However, there was no difference in the slopes for the breast-feeding vs. bottle-feeding infants of PPD positive mothers. Similarly, there was no difference between the slopes for breast-feeding and bottle-feeding infants of tuberculin negative mothers. The slope for each tuberculin positive group was significantly different from each tuberculin negative group regardless of feeding method. For a given nonstimulated cpm, the stimulated cpm was determined only by the tuberculin status of the mother. Breast-feeding or bottle-feeding had no effect. These results strongly suggest that transfer of tuberculin immunity from mother to infant evident at 4-6 weeks of age is transplacentally transferred. Our study also agrees with previous studies (2,4) in that the evidence of immunity had waned by 3 months of age. The lack of blastogenesis at 1-5 days may be the result of stress from delivery or immunoincompentence of the newborn.

Our study does disagree with the earlier study of Schlessinger and Covelli (3) in which 8/13 infants of tuberculin positive breast-feeding mothers, 1/13 infants of tuberculin positive bottle-feeding mothers, and 0/9 infants of tuberculin negative breast-feeding mothers had lymphocyte blastogenesis to PPD. This work suggested an effect of breast-feeding at age 4 weeks. In Schlessinger's study the nonstimulated cpm were very high in the breast-feeding tuberculin negative group. Differences in nonstimulated cpm may influence results. We feel very comfortable with our analysis of covariance since it takes into account the dependence of stimulated cpm on nonstimulated cpm. Another factor that may influence the result is the size of the control group. At 4-6 weeks of age, our control group was 33 infants vs. 9 infants in the Schlessinger and Covelli (3) study. Finally, another confounding variable would be ethnicity. 90% of our tuberculin positive mothers were Mexican-American while only 27% of the tuberculin negative mothers belonged to this group. However, ethnicity does not seem to cause any differences in the groups at 1-5 days or 3 months of age.

Unfortunately, our work does not answer the mechanism of transfer. Both transfer of subcellular factors (?transfer factor) and cells from the mother to the infant must be considered. Our study was not adequate to eliminate the possibility of cell transfer. It is possible that transferred maternal cells are quickly eliminated from the peripheral circulation but subcellular factors persist. It is clear that the cells dividing in response to PPD are not maternal cells. Another mechanism to be considered is <u>in utero</u> immunization. Since none of the mothers studied had active tuberculosis we feel that antigen exposure in utero was unlikely as a cause of transfer.

In summary, we feel that our data clearly supports the hypothesis that cellular immunity to tuberculin is transferred from the mother to her infant. We found no evidence to suggest a role for breast milk with this transfer, but clear evidence for transplacental transfer.

REFERENCES

1. Mohr, JA, 1973. The possible induction and/or acquisition of cellular hypersensitivity associated with ingestion of colostrum. J Peds 82:1062-1064.
2. Ogra, SS, D Weintraub, PL Ogra, 1977. Immunologic aspects of human colostrum and milk III. Fate and absorption of cellular and soluble components in the gastrointestinal tract of the newborn. J of Immunology 119:245-248.
3. Schlessinger, JJ, HD Covelli, 1977. Evidence for Transmission of Lymphocyte Responses to Tuberculin by Breast-Feeding. The Lancet 2, 529-532.
4. Pabst, H, J Crawford, M Grant, S Boyce, 1984. Transfer of Cell-Mediated Immunity (CMI) to the Fetus. Peds Res 18:262A.
5. Gallagher, MR, R Welliver, T Yamanaka, B Eisenberg, M Sun, PL Ogra, 1981. Cell-Mediated Immune Responsiveness to Measles AJDC 135:48-51.
6. Boyum,A, 1968. Separation of leukocytes from blood and bone marrow. Scand J Lab Invest 21:77.
7. Altman, LC, R Snyderman, JJ Oppenheim, SE Mergenhagen, 1973. A human mononuclear leukocyte chematactic factor: Characterization, specificity and kinetics of production of homologous leukocytes. J Immunol 110:801-810.
8. Introduction to statistical analysis, 1983. WJ Dixon and FJ Massey, Jr. 152, 583. Mc Graw-Hill Book Company.
9. Documenta Geigy. Scientific Tables, 6th ed., Ardsley, NY. Geigy Pharmaceuticals, 1968:32.
10. Documenta Geigy. Scientific Tables, 6th ed. Ardsley, NY, Geigy Pharmaceuticals 1962, pp 109-123.
11. Neter, J., Wasserman, W, 1974. Applied linear statistical models, regression, analyses of variance, and experimental designs. Irwin, Inc., Homewood, IL.
12. Beyer, WH (ed) CRC Handbook of Tables for Probability and Statistics, 2nd ed. The Chemical Rubber Co., Cleveland, OH, p 389.

PREVENTION OF ATOPIC DISEASE: ENVIRONMENTAL ENGINEERING UTILIZING ANTENATAL ANTIGEN AVOIDANCE AND BREAST FEEDING

Ranjit Kumar Chandra

Department of Pediatrics
Memorial University of Newfoundland
St. John's, Newfoundland, Canada A1B 3V6

INTRODUCTION

Atopic disease is a common cause of recurrent and chronic illness in young children. Its incidence appears to have increased in the last 15-20 years. The physical, psychological and health costs of atopic disease have prompted attempts at prevention. Much of the recent work has focussed on eczema, the most common manifestation of atopy. Heredity, immunoregulatory abnormalities, environmental factors and food hypersensitivity are considered to be important pathogenetic factors.[1-5] In this selective review, the concept of environmental engineering to overcome the handicap of hereditary predisposition is proposed and developed. Eczema is used as the prototype of atopic disease.

HEREDITY

The occurrence of atopic disease and the capacity to produce immunoglobulin E (IgE) are determined by inherited factors. Within families, there is a clustering of clinical symptoms. Thus, within one family the skin may be the predominant target organ resulting in eczema; in another family, it may be the bronchus, resulting in asthma. If one parent has history of atopic disease, the risk of the offspring developing it is about 45 percent. This risk increases to 70 percent in the event of a biparental history of atopy. It is clear that of all patients with atopic disease, as many as 30 percent will not volunteer a clear history of similar disease in first-degree relatives. This poses the dilemma of identification of high risk and emphasizes the need for developing tests that would identify such individuals without family history. Serum IgE levels reflect basal IgE production and appear to be controlled by a two-allele system, in which a low IgE level is inherited as an autosomal dominant trait and a high IgE level as autosomal recessive. Allergen-specific reagin production may be linked to several haplotypes of the HLA system. Other genetic influences include the number and mediator content of most cells, autonomic nervous system reactivity, and target organ sensitivity.

PREDISPOSING FACTORS AND SENSITIZATION

Sensitization to food protein antigens can be demonstrated in many young patients with atopic eczema. This may occur as early as prenatally, or soon after birth. Thus any attempts at reduction of atopic disease must begin very early.

Risk Factors

Besides hereditary predisposition, several factors increase susceptibility to atopic disease.[6] Deficiencies of IgA and IgG increase the risk of allergic disorders. It has been estimated that almost 20 percent of all individuals with IgA deficiency develop atopic disease at some point in their lives. Conversely, selective IgA deficiency, transient or permanent, is found 3-5 times more often in atopics than in the general population. Patients with common varied immunodeficiency also are at high risk. Some investigators found defect plasma opsonization in as many as 20 percent of all patients with atopic disorders. Others have failed to confirm such a high incidence. Low birth weight infants, particularly those born before 37 weeks of gestation, have a high incidence of atopic disease. This may be due in part to the enhanced absorption of macromolecules in the small intestine. Viral gastroenteritis, such as that caused by rotavirus, increases the risk of sensitization to food antigens for several weeks after the episode. Cystic fibrosis patients as well as obligate heterozygotes have a higher prevalence of exercise-induced bronchospasm, eczema and other atopic manifestations than matched controls. Recent data suggest that nasopharyngeal incoordination, so common in the first few weeks of life, is an etiologic antecedent of atopic respiratory disease.

Sensitization

The critical determinants of sensitization are listed in Table 1. The human fetus is capable of producing IgE beginning as early as 10 weeks gestation. However, the placenta serves as an effective barrier to the *in utero* transfer of IgE from the mother to the infant. The frequency of prenatal sensitization to food and inhalant antigens is not known. The absence of antigen-specific IgE in the majority of newborns may indicate that sensitization during pregnancy is an uncommon event. Alternatively, it may reflect the insensitive nature of the test; in other situations, e.g. tetanus, anamnestic response can be observed in the absence of detectable preimmunization antibody.

Table 1. Sensitization

Prenatal exposure
Mode of feeding
Solids
Macromolecular absorption
Gut microflora
Inhalants (dust, pets, tobacco)
Drugs
Infection

STRATEGIES FOR PREVENTION

Identification of Individuals at High Risk

Besides positive family history, other methods must be explored for identifying the newborn infant at high risk of developing atopic disease. Such screening techniques must satisfy certain basic criteria. One, the procedure must be sensitive and specific. Two, it should not involve unacceptable risk or discomfort. Three, the health infrastructure should be competent to deal with individuals identified to be at high risk. Four, there should be effective treatment or prevention available. Finally, the entire program should be cost effective. Based on current information, neonatal screening efforts using cord blood IgE levels and suppressor T cells, would meet these criteria. Croner and colleagues[7] found that over 70 percent of infants with positive family history and elevated cord blood IgE developed atopic eczema by their second birthday. Follow up studies have shown an even greater predictive value. Others have found 5-10 fold increase in incidence of atopic disease if cord blood IgE is elevated. Most of these studies have methodological problems in that data on the type of feeding, breast or formula, and other confounding variables, e.g. cigarette smoke, pets, dust, drugs, infections, were not reported. These are critical factors determining the outcome. In order to examine this issue further, we conducted a prospective study of cord blood IgE levels and followed the infants for 36 months.[8] Our data are shown in Fig. 1. Those with elevated cord blood IgE are at high risk of developing eczema and it is this group which appears to benefit most from breast feeding. Those with normal concentration of IgE in cord blood derive a marginal benefit, if that, from exclusive breast feeding.

The recognition of the immunoregulatory role of T cells on IgE synthesis has led to studies of lymphocyte subset number and function in patients with atopic disease including infantile eczema. Most of the reports document a reduction in the number of suppressor T cells. The discrepancy between the results of some studies may be related to

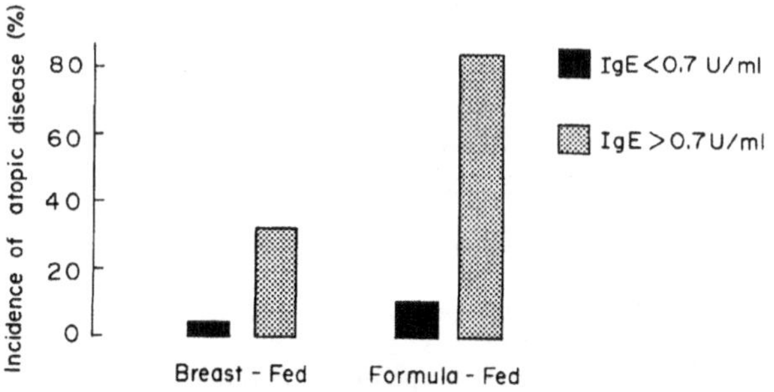

Fig. 1. Incidence of atopic eczema related to cord blood IgE concentration and mode of infant feeding. The threshold for definition of "high" IgE was 0.7 units/ml.

methodologic differences. Suppressor cell abnormalities are seen less often in adults with allergic disease. These differences may depend upon the duration of disease. One of the objections to the biological significance of immunoregulatory defects in atopic eczema is the possibility that changes in suppressor T cells are the result rather than the cause of atopy. Histamine and other potent mediators have marked effects on inflammatory cells and lymphocytes. However, a prospective study of infants 4-6 weeks old without history of overt symptoms also showed similar abnormalities in the number (Table 2) and function of suppressor T8 cells.[2] Since these findings preceded the development of atopic eczema, it appears that they may have a primary role in the development of this disorder.

Table 2. T-Cell Subpopulations in Infants With and Without Family History of Atopy

Group	% of cells		T4/T8
	T4-positive	T8-positive	
Negative history			
Healthy	51±3.2*	28±1.4[a]	1.9±0.21[c]
Symptomatic	47±3.7	20±1.9[b]	2.6±0.33[d]
Positive history			
Healthy	45±4.1	25±2.0[a]	2.0±0.25[c]
Symptomatic	49±3.5	17±2.5[b]	2.8±0.51[d]

* Mean ± S.E.M.
Values with different superscript letters are significantly different.

Prevention

The synergistic steps of environmental control for prevention of atopic disease are listed in Table 3. Reports of the association between breast-feeding and atopic disease have been critically reviewed in recent publications.[4,5,9-11] The scientific validity and generalizability of many studies are questionable. This is because the investigations have not given attention to one or more of the following critical factors (Table 4).

Table 3. Environmental Engineering

Pregnancy	Avoid allergenic foods
Infancy	Prolonged breast feeding Delay solid foods Reduce exposure to inhalants Reduce risk of infection Eliminate parental smoking

Table 4. Confounding Variables Influencing the Scientific Validity of Studies of the Association Between Breast-Feeding and Atopic Disease

Prospective versus retrospective
Duration and exclusive nature of breast-feeding
Definition of disease
Time of introduction of solids
Small number of subjects
Drop out rate
Social demographic characteristics
Attendance at day care facility
Environmental control (smoking, dust, pets)
Compliance
Blinding
Lack of supportive immunology data
Statistical analysis

Our experience[12] with the influence of breast-feeding and maternal food antigen avoidance during pregnancy and lactation on incidence of atopic eczema in infants is shown in Tables 5 and 6. Of 71 breast-fed infants, 16 developed eczema compared with 25 out of the 38 formula-fed. Group 1 mothers avoided cow's milk and other dairy products, egg, fish, peanut, and beef. This together with exclusive breast-feeding for about six months resulted in a significant reduction in frequency of eczema in infants. The expected benefit from maternal antigen avoidance is almost entirely offset by exposure of the infant to these substances after birth.

The mechanisms contributing to a protective effect of human milk are not clear. Among others, reduced exposure to food allergens, early maturation of natural intestinal barrier to macromolecular uptake, and reduced frequency of infection may be important.

Table 5. Influence of Mode of Feeding on Incidence of Atopic Eczema in High Risk Infants

Group	n	Eczema present	No eczema
Breast-fed	71	16*	55
Formula-fed	38	25*	13

*Significantly different on Chi square analysis, $X^2 = 8.6$, $P<0.05$.

Table 6. Influence of Maternal Antigen Avoidance and Mode of Feeding on Incidence of Atopic Eczema in Infants

Groups	n	Eczema present	No Eczema
Group 1. Maternal Food antigen avoidance	55	17	38
1 (BF) Breast-fed	35	5	30
1 (FF) Formula-fed	20	12	8
Group 2. Control	54	24	30
2 (BF) Breast-fed	36	11	25
2 (FF) Formula-fed	18	13	5

Statistical analysis using Fisher's exact probability test: 1 (BF) versus 1 (FF) P = 0.0046, 2 (BF) versus 2 (FF) P = 0.0509, 1 (BF) versus 2 (BF) P = 0.0503, 1 (FF) versus 2 (FF) P = 0.7864, Group 1 versus Group 2 P = 0.6933.

REFERENCES

1. M-LE. Lubs, Empiric risks for genetic counselling in families with allergy, J. Pediat. 80:26-31 (1972).
2. R. K. Chandra and M. Baker, Numerical and functional deficiency of suppressor T cells precedes development of atopic eczema. Lancet ii:1393-4 (1983).
3. D. J. Atherton, Breast feeding and atopic eczema, Br. Med. J. 287:775-6 (1983).
4. R. S. Zeiger, S. Heller, M. Mellon, R. O'Connor, and R. N. Hamburger, Effectiveness of dietary manipulation in the prevention of food allergy in infants, J. Allergy Clin. Immunol. 78:224-238 (1986).
5. R. K. Chandra, "Food Allergy," Nutrition Research Education Foundation, St. John's (1986).
6. J. F. Soothill, C. E. Stokes, M. W. Turner, A. P. Normam, and B. Taylor. Predisposing factors and the development of reaginic allergy in infancy, Clin. Allergy 6:305-12 (1976).
7. S. Croner, N-IM. Kjellman, B. Eriksson, and A. Roth. IgE screening in 1701 newborn infants and the development of atopic disease during infancy, Arch. Dis. Child 57:364-369 (1982).
8. R. K. Chandra, S. Puri, and P. S. Cheema, Predictive value of cord blood IgE in the development of atopic disease and role of breast feeding in its prevention, Clin. Allergy 15:517-22 (1985).
9. B. Björksten, Does breast feeding prevent the development of allergy?, Immunol. Today 4:215-17 (1983).
10. M. Burr, Does infant feeding affect the risk of allergy?, Arch. Dis. Childh. 58:561-5 (1983).
11. U. M. Saarinen, Prophylaxis for atopic disease. Role of infant feeding, Clin. Rev. Allergy 2:151-67 (1984).
12. R. K. Chandra, S. Puri, C. Suraiya, and P. S. Cheema, Influence of maternal food antigen avoidance during pregnancy and lactation on incidence of atopic eczema in infants, Clin. Allergy Vol. 16 (in press) (1986).

SUMMARY OF WORKSHOP: POTENTIALLY HARMFUL EFFECTS OF HUMAN MILK UPON THE RECIPIENT INFANT

William C. Heird

Department of Pediatrics
College of Physicians and Surgeons of Columbia University
630 West 168th Street
New York, NY 10032

The presentations in this section addressed some of the potentially harmful effects of human milk upon the recipient infant. These included: transmission of viral infection via milk; transmission of drugs via milk; deposition in developing tissues of the trans-fatty acids present in most milks; sensitization of the infant secondary to transfer of foreign antigens via milk; vitamin K deficiency and hemorrhagic disease of the newborn secondary to the low vitamin K content of human milk.

Robert Pass from the United States presented data showing that human milk can be a vector for transmission of viruses to the receipient infant. Cytomegalovirus, for example, is reported to be present in the milk of at least 70% of sero-positive women. However, it is not clear that many other viruses with which the lactating woman may be infected find their way to her milk. Thus, much further research is required to identify the viruses that can be transmitted by milk to the receipient infant as well as those that are most likely to be transmitted. It is even more important, perhaps, to determine if specific maternal factors are involved in the transmission of some viruses but not others and, if so, to identify these maternal factors.

Much further research also is required to determine the clinical significance of the presence of viruses in milk. It appears, for example, that ingestion of milk contaninated with cytomegalovirus and rubella is much more likely to result in development of immunity to the contaminating virus than inactive disease. Further insight into this general area has important implications with respect to advising mothers concerning continuation of breast feeding during viremia, assessing the safety of milk donated by a mother with viremia for use in infants other than her own and establishing rational criteria for screening milk donated to human milk banks. Definition of maternal vs infant factors responsible for transmission of immunity vs infection via milk may lead to different criteria for feeding donor milk to term vs premature infants. Since the latter infants have a less mature immune system, they may be more susceptible to infection with the contaminating virus than the term infant whose immune mechanisms are more mature.

The dismal state of knowledge concerning the potentially harmful effects of drugs in milk upon the recipient infant was reviewed by John

Wilson from the United States. Since the specific effects of most drugs are known prior to their general availability, the effect of any drug taken by a lactating mother upon the recipient infant, theoretically, could be predicted if the likelihood of that drug's being excreted into milk were known. What is needed, in other words, is a more complete definition of the pharmacokinetics of drug distribution and excretion into human milk. The mathematical models proposed by Dr. Wilson represent a reasonable approach to defining the pharmacokinetics of classes of drugs in lactating women. With complete pharmacokinetic data available for typical examples of each class of drugs, the potentially harmful effects of other drugs of each class which might be prescribed for the mother should be predictable, as Dr. Wilson points out, using methods available in most laboratories. Such models also may provide much needed information concerning the transfer of ingested nutrients, toxins, and foreign antigens from mother to infant.

The theoretical reasons for concern regarding the presence of the trans-fatty acids in human milk were summarized by Berthold Koletzko from the Federal Republic of Germany. It is clear that these fatty acids are present in the lipid fraction of human milk and that the amount present reflects maternal intake (and indirectly dietary practices of the community). Although it is clear that dietary trans-fatty acids are incorporated into tissue lipids, it remains to be shown whether trans-fatty acids in tissues are harmful. Such information has important implications concerning the optimal diet for lactating women and also for nutritional standards for the general population.

Several issues concerning the presence of foreign food antigens in natural milk and transfer of these to the receipient infant were presented by Paul Harmatz from the United States. Using a rat model to study this overall process, Dr. Harmatz and colleagues have demonstrated that antigen absorption via the gastrointestinal tract is greater during lactation. This appears to be related to the greater enteric mucosal mass during lactation. Their data also suggest that the likelihood of the absorbed antigen's being transmitted into the milk is dependent upon the concentration of the antigen in plasma. In the rat model, it is clear that a specific antigen ingested by a lactating animal appears both in the animal's milk and in peripheral blood of young animals fed this milk. However, much more information is needed concerning the transfer of antigens ingested by the mother to her infant by way of milk. Perhaps the greatest need in this area is information concerning whether or not such antigen transfer gives rise to clinical problems and, if so, the magnitude of the problem. For example: Is this a potential problem for all infants or only for infants with a family history of food and/or other allergies? Is the problem of significant magnitude to warrant advising lactating women to avoid certain foods and, if so, which foods should be avoided? This issue was also addressed by Ranjit Chandra from Canada in the companion workshop on Host Resistance.

Rudiger von Kries from the Federal Republic of Germany summarized studies indicating that the concentration of vitamin K is somewhat lower in human milk than most infant formulas and that, in Germany, vitamin K deficiency is much more common in breast-fed infants than in formula-fed infants. Moreover, the incidence of vitamin K deficiency, including hemorrhagic disease of the newborn, has increased in developing countries that have been successful in efforts to encourage breast-feeding. On the other hand, routine neonatal vitamin K phophylaxis is not common in developing countries, where most births occur outside the hospital, or in West Germany. Thus, the question of whether breast-fed infants who receive prophylactic vitamin K at birth are more susceptible to vitamin K deficiency than formula-fed infants has not been addressed. The other interesting question raised by Dr. Von Kries concerns the possibility that the lower vitamin K content of human milk might be particularly harmful for infants with cholestatic liver disease. Answers to these questions are crucial for development of guidelines concerning vitamin K supplementation of either lactating women or breast-fed infants.

VIRAL CONTAMINATION OF MILK

Robert F. Pass

Department of Pediatrics
University of Alabama at Birmingham
Birmingham, Alabama 35294

There is a substantial body of evidence indicating that milk plays an important role in protecting the suckling animal from infections due to viruses and other agents. The presence of antibodies, phagocytes and immunocompetent cells in milk is often cited in support of both nursing and use of banked human milk for nourishment of hospitalized premature newborns. The fact that a variety of animal and human viruses can be transmitted from mother to offspring through milk is not widely recognized. The purpose of this report will be to review the range of viral agents that have been recovered from animal and human milk, to consider the potential importance of milk in transmission of these agents and to assess the implications of this information for nursing mothers and milk banks.

ANIMAL STUDIES

Viruses that have been isolated from milk of lactating dams during either experimental or natural infection are listed in Table I. The variety of viral families represented is noteworthy, but there are undoubtedly other viral agents that appear in and can be transmitted through milk. Unfortunately, studies of viral infection in laboratory animal models and in domestic animals often fail to determine whether virolactia is a feature of infection. Proving that transmission of virus occurs through milk requires careful foster mother experiments. In spite of these problems, it is clear that milk is an important means of spread for several viruses of economic importance in agriculture. Milk also appears to be important in the maintenance of infection in animal populations by certain retroviruses, a finding which may have parallels in human retrovirus infection.

Transmission of both murine mammary tumor virus and murine leukemia virus to suckling mouse pups through milk has been proven by carefully controlled foster mother studies, though other routes of vertical transmission are also operative and genetic factors greatly influence both transmission and subsequent occurrence of disease [7-12]. Bovine leukemia virus, the cause of bovine lymphosarcoma, has been found in over half of dairy herds. The virus is commonly present in milk from infected cows, and milk transmission appears to contribute to maintenance of infection in a herd [1-2]. Feline leukemia virus infection of cats produces a range of illnesses reminiscent of the clinical spectrum of acquired immunodeficiency syndrome in humans [4,5]. Feline leukemia virus has been detected in milk

Table I. Viral agents excreted in and transmitted through milk of naturally or experimentally infected animals

Virus [Reference]	Family	Host	Disease
Bovine leukemia virus [1,2]	Retroviridae	Cattle	Lymphosarcoma, lymphocytosis
Caprine arthritis-encephalitis virus [3]	Retroviridae	Goats	Arthritis, encephalitis, pneumonia
Feline leukemia virus [4-6]	Retroviridae	Cats	Lymphosarcoma, anemia, panleukopenia, immune deficiency
Murine leukemia virus [7-9]	Retroviridae	Mice	Lymphomas, leukemia
Murine mammary tumor virus [10-12]	Retroviridae	Mice	Mammary carcinoma
Foot-and-mouth disease virus [13,14]	Picornaviridae	Cattle	Vesicular eruption
Junin virus [15]	Arenaviridae	Rodents	None in rodents; Argentine hemorrhagic fever in humans
Tick born encephalitis viruses [16,17]	Togaviridae	Sheep, goats, humans	Progressive neurologic disease
Transmissible gastroenteritis virus [18,19]	Coronaviridae	Swine	Gastroenteritis

Table II. Viruses that have been found in or appear to be transmitted through human milk

Virus [Reference]	Family	Detected in Milk	Transmission through Milk	Associated Disease*
Cytomegalovirus [20-23]	Herpetoviridae	Frequently	Proven	None
Herpes Simplex Virus [24]	Herpetoviridae	1 case	Questionable	Disseminated neonatal herpes
Rubella Virus [25-28]	Togaviridae	With acute infection or immunization	Proven	None
Hepatitis B Virus [29-31]	Hepadnaviridae	From carrier mothers	†Unknown	Unknown
Human T Cell Lymphotrophic Viruses	Retroviridae			
I [32,33]		Yes	Very Likely	Unknown
HIV [34]§		Unknown	1 case	AIDS

*Refers to disease occurring as a result of acquisition of virus from mother's milk.

†Perinatal HBV infection is known to result in chronic antigenemia and chronic liver disease in children and undoubtedly contributes to later development of hepatoma.

§Now referred to as human immunodeficiency virus.

of both acutely ill, viremic cats and a nonviremic latently infected queen [4-6]. Although vertical transmission through milk occurs in both situations, its importance relative to other possible routes of transmission has not been defined.

Both foot-and-mouth disease virus and transmissible gastroenteritis virus have considerable economic importance. The former produces a vesicular eruption in cattle, and the latter results in severe, often fatal gastroenteritis in young pigs [13,14,18,19]. Since both viruses appear in respiratory secretions and feces as well as milk, the relative importance of the latter route has not been established.

The tick-borne encephalitis viruses are a group of antigenically related agents, several of which can cause human disease. Virus is excreted in milk of sheep and goats and can be transmitted to humans through this route, though ticks are the principal source of human infection. Louping-ill virus, a member of this group, produces neurological disease of sheep and other animals with progressive paralysis. The virus is excreted in milk and transmission through this route resulting in disease in offspring has been demonstrated [15,16].

Junin virus, the cause of Argentine hemorrhagic fever, is a rodent parasite which, like other members of the Arenavirus family, results in chronic viremia and viruria in infected animals. Virus has been recovered from mammary glands of experimentally infected guinea pigs [17]; milk-borne virus could contribute to maintenance of chronic infection in wild rodents.

Although most of the viruses listed in Table I are not known to cause human disease, it is important to note that every virus family listed includes members that are human pathogens. It is likely that some of these are also excreted in human milk.

VIRUSES IN HUMAN MILK

The number of viruses that have been isolated from human milk is limited, but so is the number of studies that have attempted to detect viral agents in human milk. As can be seen from Table II, agents that are major causes of fetal or neonatal infection have received the most interest. In humans, breast milk transmission has been proven to be a common means for spread of virus from mother to offspring only for cytomegalovirus.

CYTOMEGALOVIRUS

Review of the epidemiology of CMV infection reveals that in populations where there is a high rate of maternal seropositivity and breast feeding is widely used, a relatively high proportion of infants acquire CMV during the first year of life [21,35]. Further evidence for the role of milk in spread of CMV has been provided by virologic studies. Hayes, et al. isolated CMV from milk of 17 of 63 seropositive women [20]. Stagno, et al. recovered CMV from 38/278 women [21]. Dworsky and his co-workers studied 41 seropositive lactating women serially; CMV was isolated from at least one milk specimen from 13 (32%) [22]. Both Hayes and Dworsky found milk collected from seropositive women after the first week postpartum was more likely to yield CMV. Ahlfors and Ivarsson recently confirmed this pattern; between nine days and three months postpartum, CMV was isolated from milk of 16 of 23 (70%) antibody positive women studied from one to three times whereas none of the 18 milks collected before or eight samples collected after this interval were positive [23].

Table III. Virolactia, duration of breast feeding and transmission of CMV from seropositive mothers to offspring [22].

Category	N	Infected Infants
Breast Fed		
< 1 month	10	0
≥ 1 month	31	12 (39%)
CMV Isolated from Milk		
Yes	13	9 (69%)
No	28	3 (10%)

The importance of milk as a means of transmission of CMV from mother to offspring has been proven by examining sites of maternal excretion in relation to acquisition of CMV by the infant. Stagno, et al. found that infection rates were similar for babies born to mothers who shed CMV only in milk (58%) or only from the cervix (57%), but neither bottle fed babies nor those born to mothers who shed virus only in urine or saliva became infected [21]. Dworsky reported that CMV infection in infants nursed by seropositive mothers was related to the duration of breast feeding and the presence of detectable virolactia, Table III [22].

RUBELLA VIRUS

Both wild and vaccine strains of rubella have been isolated from human milk and apparently transmitted to offspring through this route [25-28]. Lasonsky, et al. demonstrated that postpartum rubella immunization with either the HPV-77 or RA 27/3 vaccine resulted in virolactia [27]. Rubella virus was recovered from milk of 9/13 (69%) vaccinees. Although a similar proportion of these women also shed virus from nasopharynx, infection in their infants was clearly related to breast feeding [28]. No significant illnesses have been described in infants who acquired rubella from their mother's milk.

HERPES SIMPLEX VIRUS

Isolation of herpes simplex virus (HSV) from milk has been reported [24]. However, in this case report virus was first recovered from milk three days after onset of illness in the newborn. Acquisition of virus from the maternal genital tract with spread to the breast from the infant's saliva or infection of the breast by oral contact with some other family member are both plausible explanations. In an additional case, fatal HSV-1 infection occurred in an infant whose mother had herpetic lesions on the surface of the breast [36]. Excretion of HSV in milk in the absence of breast lesions must be distinctly unusual. In studies of CMV, over 500 milks have been examined by tissue culture techniques that would also detect HSV, but no HSV isolations have been reported [20-23].

HEPATITIS B VIRUS (HBV)

Hepatitis B surface antigen has been detected in the milk of carrier mothers, with 72% in Taiwan found to be positive [29-31]. However, there is no evidence that human milk is a significant means of HBV transmission.

Studies in Taiwan found that breast feeding did not affect the rate of antigenemia in infants born to carrier mothers [37].

HUMAN IMMUNODEFICIENCY VIRUS (HIV)

In view of the consistency with which virolactia has been documented during animal retrovirus infection and the fact that lymphocytes are regularly present in milk, one would expect that the HIV would appear in human milk. Acquired immune deficiency syndrome (AIDS) has been reported in a breast fed infant whose mother apparently acquired virus through a postnatal blood transfusion [34], strong evidence for transmission of virus from mother to infant through milk. Human T cell lymphotrophic virus type 1 (HTLV-1), a related agent that has been linked with adult onset T cell leukemia/lymphoma, has been found in milk of carrier mothers in Japan and transferred to a marmoset by oral inoculation of human milk [32,33]. Since 20% of children born to HTLV-I seropositive mothers are seropositive by the age of one year, milk could be a major means of mother to child transmission of this virus. In view of the projected increases in prevalence of HIV infection and steady rise in infection rates in heterosexual women, careful studies of mechanisms of perinatal transmission are needed.

IMPLICATIONS FOR NURSING MOTHERS

Breast milk transmission of CMV appears to be a common event. Since there is no evidence that CMV acquired from mother's milk produces any acute or delayed untoward effects, breast feeding of a normal infant should never be stopped because of maternal CMV infection. Transmission of rubella through milk is likely to occur after postpartum maternal immunization, but there is no evidence that rubella acquired in this way is harmful to the infant. Although HSV infection, as indicated by prevalence of serum antibody, is common in most populations, silent shedding of virus in milk appears to be extremely rare. In the absence of breast lesions it does not seem reasonable to interfere with nursing by a mother with either antibody to HSV or a history of genital lesions. When any family member has oral or genital lesions, then steps should be taken to prevent transfer of virus from infected material to the baby [38]. Obviously, a mother with herpetic breast lesions should not nurse. Although breast feeding by hepatitis B carrier mothers did not increase the risk of transmission of HBV to the newborn in Taiwan where there may be multiple potential sources of virus, different results might be expected in areas where carrier rates are very low, such as Europe and the U.S. Where formula is readily available and can be safely prepared, breast feeding of infants by recently infected or carrier mothers is not recommended. Children born to mothers with AIDS frequently develop the disease. Although in utero transmission may be the route of most of these infections, mothers with AIDS should not be allowed to nurse their babies. More information on both the course of asymptomatic HTLV-III infection in women and risk of perinatal transmission is needed; because of the potential consequences of neonatal HTLV-III infection, mothers who are seropositive but do not have AIDS or related syndromes should be advised not to nurse their babies if formula is available.

IMPLICATIONS FOR HUMAN MILK BANKING

The infant being suckled by his own mother benefits by receiving antibodies specifically directed against viruses that have previously infected the mother and other antiviral activities found in human milk. In addition, the infant will have circulating antibodies acquired through transplacental transfer. Thus, when virolactia occurs in a chronic infection such as CMV, the newborn is protected by a considerable array of passively acquired immune factors. With CMV, neither systemic nor milk immunity effectively prevents transfer of virus to the newborn, but it is likely that they prevent disease, as has been demonstrated for transfusion acquired CMV in the newborn [39]. Feeding newborns unpasteurized banked human milk could potentially lead to situations in which an infant with no serum antibody to a specific agent (seronegative mother) is fed virus contaminated milk. Although transmission of CMV through banked human milk has not been described, its possible occurrence has not been sought among recipients of banked milk.

Again drawing on the parallel with blood transfusion, as well as the studies of mother to infant transmission of CMV, it seems highly likely that banked human milk could transfer virus and lead to disease in small, seronegative premature newborns. Reviewing the list of virus families with at least one virus (Tables I and II) known to be excreted in milk, there are undoubtedly human pathogens other than those in Table II that can cause virolactia. In addition to CMV, other agents that result in asymptomatic infection with chronic viremia or infection of lymphocytes, such as HIV, hepatitis B and the HTLV-I could pose major problems for milk banking in areas that have significant rates of infection in mothers.

Pasteurization of human milk at 62.5°C for 30 minutes has been shown to eliminate CMV infectivity as has rapid high-temperature treatment for 15 seconds at 72°C or 87°C [41,43]. However, storage at 4°C, freezing at -15°C or -20°C, and heating to 56°C have all been shown to be unreliable means of inactivating CMV in milk [40-43]. One could also ask whether a single laboratory inoculation of a milliliter or less of milk into tissue culture for viral isolation is a valid means of proving milk to be virus free. After all, the infant will likely receive a liter or more daily for weeks to months. In view of the uncertainties and potential risks, infants born to mothers with no antibody to CMV should not be fed banked human milk unless it is known to be from a seronegative donor. Whether milk donors should be screened for other agents, such as hepatitis B virus or HIV depends upon the prevalence of the infection in the donor population. The safest approach would be to use the same methods for selecting milk donors and screening them for viral infection that are used with blood banking.

ACKNOWLEDGEMENTS

This work was supported in part by a Program Project grant from the National Institute of Child Health and Human Development (HD 10699).

REFERENCES

1. J. F. Ferrer, Bovine leukosis: natural transmission and principles of control, J. Am. Vet. Med. Assoc. 175:1281 (1979).
2. J. F. Ferrer and C. E. Piper, Role of colostrum and milk in the natural transmission of the bovine leukemia virus, Cancer Res. 41:4906 (1981).
3. R. Oliver, A. Cathcart, R. McNiven, W. Poole, and G. Robati, Infection of lambs with caprine arthritis encephalitis virus by feeding milk from infected goats, Vet. Rec. 116:83 (1985).
4. D. W. Hardy, Jr., P. W. Hess, E. G. MacEwen, A. J. McClelland, E. E. Zuckerman, M. Essex, S. M. Cotter, and O. Jarrett, Biology of feline leukemia virus in the natural environment, Cancer Res. 36:582 (1976).

5. D. Hardy, The virology, immunology and epidemiology of the feline leukemia virus, in: "Feline Leukemia Virus," W. D. Essex, Jr., A. J. McClelland, eds., Elsevier, New York (1980).
6. M. Pacitti, O. Jarrett, and D. Hay, Transmission of feline leukaemia virus in the milk of a non-viraemic cat, Vet. Rec. 118:381 (1986).
7. A. B. Jenson, D. E. Groff, P. J. McConahey, and F. J. Dixon, Transmission of murine leukemia virus (Scripps) from parent to progeny mice as determined by p30 antigenemia, Cancer Res. 36:1228 (1976).
8. R. Jaenisch, Germ line integration and mendelian transmission of exogenous type C viruses, in: "Molecular Biology of RNA Tumor Viruses," Academic Press, New York (1980).
9. M. Zijlstra, R. E. Y. DeGoede, H. J. Schoenmakers, A. H. Schinkel, W. G. Hesselink, J. L. Portis, and C. J. M. Melief, Naturally occurring leukemia viruses in H-2 congenic C57BL mice, III. characterization of C-type viruses isolated from lymphomas induced by milk transmission of B-ecotropic virus, Virology 125:47 (1983).
10. J. J. Bittner, Some possible effects of nursing on the mammary gland tumor incidence in mice, Science 84:162 (1936).
11. J. Hilgers and P. Bentvelzen, Interaction between viral and genetic factors in murine mammary cancer, in: "Advances in Cancer Research, Vol. 26," Academic Press, New York (1978).
12. P. Hainaut, D. Vaira, C. Francois, C. M. Calberg-Bacq, and P. M. Osterrieth, Natural infection of Swiss mice with mouse mammary tumor virus (MMTV): viral expression in milk and transmission of infection, Arch. Virol. 83:195 (1985).
13. H. L. Backrach, Foot-and-mouth disease, Ann. Rev. Microbiol. 22:201 (1968).
14. H. Blackwell, P. D. McKercher, F. V. Kosikowski, L. E. Carmichael, and R. C. Gorewit, Physicochemical transformation of milk components and release of foot-and-mouth disease virus, J. Dairy Res. 50:17 (1983).
15. P. Sangiorgio and M. C. Weissenbacher, Congenital and perinatal infection with Junin virus in guinea pigs, J. Med. Virol. 11:161 (1983).
16. J. Nosek, O. Kozuch, E. Ernek, and M. Lichard, The importance of goats in the maintenance of tick-borne encephalitis virus in nature, Acta. Virol. (Praha) 11:470 (1967).
17. H. W. Reid, D. Buxton, I. Pow, and J. Finlayson, Transmission of louping-ill virus in goat milk, Vet. Rec. 114:163 (1984).
18. J. Kemeny and R. D. Woods, Quantitative transmissible gastroenteritis virus shedding patterns in lactating sows, Am. J. Vet. Res. 38:307 (1977).
19. L. J. Saif and E. H. Bohj, Passive immunity to transmissible gastroenteritis virus: intramammary viral inoculation of sows, Ann. NY Acad. Sci. 409:708 (1983).
20. K. Hayes, D. M. Danks, H. Givas, and I. Jack, Cytomegalovirus in human milk, N. Engl. J. Med. 287:177 (1972).
21. S. Stagno, D. W. Reynolds, R. F. Pass, and C. A. Alford, Breast milk and the risk of cytomegalovirus infection, N. Engl. J. Med. 302:1073 (1980).
22. M. Dworsky, M. Yow, S. Stagno, R. F. Pass, and C. A. Alford, Cytomegalovirus infection of breast milk and transmission in infancy, Pediatrics 72:295 (1983).
23. K. Ahlfors and and S. A. Ivarsson, Cytomegalovirus in breast milk of Swedish milk donors, Scand. J. Infect. Dis. 17:11 (1985).
24. M. Dunkel, R. R. Schmidt, and D. M. O'Connor, Neonatal herpes simplex infection possibly acquired via maternal breast milk, Pediatrics 63:250 (1979).
25. E. Buimovici-Klein, R. J. Hite, T. Byrne, and L. Z. Cooper, Isolation of rubella virus in milk after postpartum immunization, J. Pediatr. 91:939 (1977).

26. E. B. Klein, T. Byrne, and L. Z. Cooper, Neonatal rubella in a breast-fed infant after postpartum maternal infection, J. Pediatr. 97:774 (1980).
27. G. A. Losonsky, J. M. Fishaut, J. Strussenberg, and P. L. Ogra, Effect of immunization against rubella on lactation products. I. development and characterization of specific immunologic reactivity in breast milk, J. Infect. Dis. 145:654 (1982).
28. G. A. Losonsky, J. M. Fishaut, J. Strussenberg, and P. L. Ogra, Effect of immunization against rubella on lactation products. II. Maternal-neonatal interactions, J. Infect. Dis. 145:661 (1982).
29. C. C. Linnemann, Jr. and S. Goldberg, HBAg in breast milk, Lancet 2:155 (1974).
30. E. H. Boxall, T. H. Flewett, D. S. Dane, C. H. Cameron, F. O. MacCallum, and T. W. Lee, Hepatitis-B surface antigen in breast milk, Lancet 2:1007 (1974).
31. A. K. Y. Lee, H. M. H. Ip, and V. C. W. Wong, Mechanisms of maternal-fetal transmission of hepatitis B virus, J. Infect. Dis. 138:668 (1978).
32. K. Kinoshita, S. Hino, T. Amagasaki, S. Ikeda, Y. Yamada, J. Suzuyama, S. Momita, K. Toriya, S. Kamihara, and M. Ichimaru, Demonstration of adult T-cell leukemia virus antigen in milk from three seropositive mothers, Gann. 75:103 (1984).
33. K. Kinoshita, K. Yamanouchi, S. Ikeda, S. Momita, T. Amagasaki, H. Soda, M. Ichimaru, R. Moriuchi, S. Katamine, T. Miyamoto, and S. Hino, Oral infection of a common marmoset with human T-cell leukemia virus type-I (HTLV-I) by inoculating fresh human milk of HTLV-I carrier mothers, Jpn. J. Cancer Res. 76:1147 (1985).
34. J. B. Ziegler, R. O. Johnson, D. A. Cooper, and J. Gold, Postnatal transmission of AIDS-associated retrovirus from mother to infant, Lancet 1:896 (1985).
35. R. F. Pass, Transmission of viruses through human milk, in: "Role of Human Milk in Infant Nutrition and Health," L. K. Pickering, F. H. Morriss, eds., Charles C. Thomas Publishers, Springfield (1986).
36. J. Z. Sullivan-Bolyai, K. H. Fife, R. F. Jacobs, Z. Miller, and L. Corey, Disseminated neonatal herpes simplex virus type 1 from a maternal breast lesion, Pediatrics 71:455 (1983).
37. R. P. Beasley, C. E. Stevens, I-S Shiao, and H-C Meng, Evidence against breast feeding as a mechanism for vertical transmission of hepatitis B, Lancet 2:740 (1975).
38. Committee on Fetus and Newborn and Committee on Infectious Diseases, Perinatal herpes simplex virus infections, Pediatrics 66:147 (1980).
39. A. S. Yeager, F. C. Grumet, E. B. Hafleigh, A. M. Arvin, J. S. Bradley, and C. G. Prober, Prevention of transfusion-acquired cytomegalovirus infection in newborn infants, J. Pediatr. 98:281 (1981).
40. J. K. Welsch, M. Arsenakis, R. J. Coelen, and J. T. May, Effect of antiviral lipids, heat, and freezing on the activity of viruses in human milk, J. Infect. Dis. 140:322 (1979).
41. M. Dworsky, S. Stagno, R. F. Pass, G. Cassady, and C. A. Alford, Persistence of cytomegalovirus in human milk after storage, J. Pediatr. 101:440 (1982).
42. H. Friss and H. K. Andersen, Rate of inactivation of cytomegalovirus in raw banked milk during storage at -20°C and pasteurisation, Br. Med. J. 285:1604 (1982).
43. R. M. Goldblum, C. W. Dill, T. B. Albrecht, E. S. Alford, C. Garza, and A. S. Goldman, Rapid high-temperature treatment of human milk, J. Pediatr. 104:380 (1984).

TRANSFER OF MATERNAL FOOD PROTEINS IN MILK

Paul R. Harmatz, Donald G. Hanson, Marc Brown,
Ronald E. Kleinman, W. Allan Walker and Kurt J. Bloch

Departments of Pediatrics and Medicine, Harvard Medical School, and Combined Program in Pediatric Gastroenterology and Nutrition, and Clinical Immunology and Allergy Units Massachusetts General Hospital, Boston, Massachusetts

For more than 60 years, it has been recognized that nursing infants may develop allergic reactions to substances in the maternal diet.[1,2,3] Recent studies demonstrate that, colic,[4,5] atopic dermatitis,[6,7] and colitis[7,8] can be related directly to the presence of a specific protein in the mother's diet. To understand the basis for these disease processes, investigators have determined dose and molecular character of antigens in milk,[6,9,10] probed for a history of atopic disease in the family,[10] examined the relationship of antigen-induced disease to skin-test results obtained with the antigen,[6] and measured total and cow's milk-specific IgA antibody titers in breast milk of mothers with symptomatic infants.[10] In the latter study, Machtinger and Moss[10] found that all breast-fed infants with allergic symptoms had consumed maternal milk with low total and cow's milk-specific IgA antibody.

A knowledge of the physiologic influence of dietary antigen received via the milk on the developing neonatal mucosal and systemic immune systems is of fundamental importance in understanding the pathophysiology of disease induced by protein transferred in breast milk. In a recent study in human infants, a transient low level of food protein (cow's milk, egg white, or soy) specific IgE antibody was noted in 27 percent of an unselected population of infants at 8 months of age. These low levels of specific IgE antibody had largely disappeared (<2%) by 4 yr,[11] suggesting the development of systems to control this antibody response. The appearance of IgE antibodies specific for cow's milk proteins or egg white in nursing infants before any known feeding of these proteins[11] supports the hypothesis that the small quantities of dietary protein antigen transferred in breast milk are immunologically relevant. The induction of dietary protein-specific IgE antibodies, in fact, may be favored by prolonged breast feeding.[12,13] An enhanced IgE response or failure to eliminate the IgE response may be related to the low dose of dietary antigen presented in the milk. In animal studies,[14,15,16] the dose of antigen influenced the development of an IgE response and induction of systemic tolerance. High doses of antigen will generally suppress the IgE response and induce systemic tolerance, while low doses favor production of a specific IgE response and induce systemic tolerance. The neonate may constitute an exception; doses which normally induce tolerance in the adult animal favor priming for both antibody and cell-mediated immune responses in the newborn.[17,18] Other mechanisms offered to explain the increased immune response in the neonate include imbalance of suppressor/helper lymphocyte function in neonates and the transfer via the milk of maternal anti-idiotypic antibodies that might stimulate antibody synthesis in the neonate (cited in reference 11).

Because maternal dietary antigens appear to be involved in the induction and persistence of certain diseases of infants and because of the potential importance of these antigens for the development of the neonate's normal immune response, we have examined in detail the process of antigen transfer from lactating mother to infant. This transfer process requires that ingested proteins penetrate the gastrointestinal barrier of the lactating mother, reach the systemic circulation, avoid clearance from the systemic circulation before reaching the mammary gland, and be transferred from the circulation into the milk. The development of an animal model to examine this process of antigen transfer from mother to infant has allowed us to control for the prior antigen exposure of the animal, to utilize radiolabelled tracers, and to examine the systemic processing of antigen within the lactating mouse.[19,20]

Effect of Lactation on Dietary Antigen Uptake from the Intestine

To examine the capacity of lactating mice to absorb dietary antigen, lactating BD-Fl mice (6-10 days post-partum, no prior exposure to ovalbumin) and age-matched nulliparous controls were given 10 mg ovalbumin (OVA) in 0.5 ml saline, via a soft silastic tube. Blood was drawn after 15, 30, 60, and 120 min, and serum immunoreactive OVA (iOVA) concentration was measured by enzyme immunoassay (EIA).[20] After 30 and 60 min, the concentration of OVA in the serum of lactators was 3- to 4-fold greater than that of controls (Fig. 1).

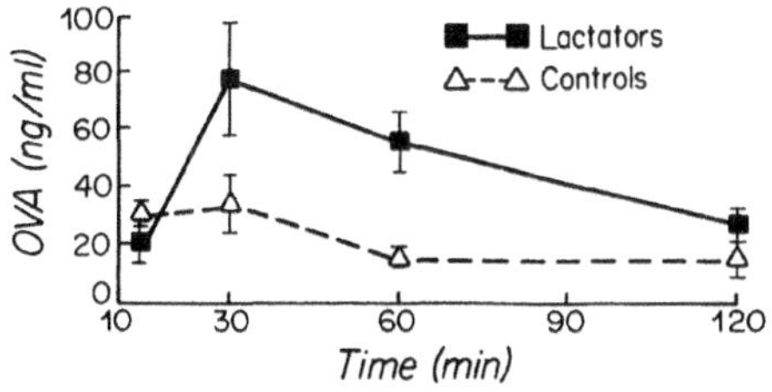

Fig. 1. Mean OVA concentration (+S.E.M.) in the serum of lactating and age-matched nulliparous controls.

Since plasma volume and clearance of OVA from the circulation may influence serum OVA concentration, these were measured after iv injection of ^{125}I-BSA or OVA, respectively. Plasma volume of lactators, measured with the use of ^{125}I-BSA, was 2-fold greater than that of controls, an observation consistent with a previously reported study in mice (21). After iv injection, the rate of clearance of unlabelled OVA was not different in the 2 groups. Because plasma volume was greater in the lactators and clearance of OVA from the circulation was not different in the two groups, we conclude that the differences in serum concentration of iOVA observed in lactating and control mice were associated with greater protein uptake from the gut in lactators.

To examine the mechanism of this enhanced uptake, the pattern of OVA movement within the intestine after feeding was determined. A solution containing 10 mg of OVA and small amounts (1 ug) of ^{125}I-OVA was fed to non-fasted lactating and control mice via a silastic tube. At 15, 30, and 60 min. after feeding, the small intestine was removed, divided into 5 equal length segments, and radioactivity measured within each segment. The peak of radioactivity had moved a smaller percentage of the distance from pylorus to cecum in the lactating mouse, consistent with the increased length of the small intestine in the lactating mouse. It suggests that, in lactating mice, antigen is localized for a longer time in areas of the intestine which are active in protein antigen uptake.

Antigen interacting with the intestinal surface may also contribute to enhanced transport of OVA from the intestinal lumen into the circulation. To assess association of OVA with the intestinal surface, the small intestine was removed from lactating and control mice, divided into 5 equal length segments from pylorus to cecum, and incubated at 37°C for 1 hr in an oxygenated physiologic buffer containing ^{125}I-OVA. After incubation, the individual segments were washed and associated radioactivity was measured. The specific location of the associated radioactivity within the intestinal wall was not determined in these studies. Significantly more radioactivity per total intestine (summation of the 5 individual segments) was found in lactating mice. The radioactivity obtained from mucosal scrapings was shown to be greater than 72% co-precipitable by rabbit anti-OVA-OVA complexes.

These data suggest that the greater concentration of specific antigen in the serum of lactating mice, 1 hr after feeding, resulted in part from the prolonged contact of antigen with the absorptive surfaces of the small intestine and from an increase in total OVA associated with the intestine of the lactating mouse.

Transfer of protein antigens into milk after intravenous injection

Although several different dietary protein antigens have been demonstrated in human milk,[4,9,10,22] the concentrations detected have varied widely (100pg - 1ug). It has been suggested that these differences may result from artifacts in analysis.[9,10] We have posed an alternate hypothesis, which holds that inherent differences in the protein molecules or in their mechanism(s) of transport, may account for the different concentrations found in milk.

Equal amounts (by weight) of ^{125}I-bovine serum albumin (^{125}I-BSA), ^{125}I-bovine gamma globulin (^{125}I-BGG), ^{125}I-OVA, and ^{125}I-bovine beta-lactoglobulin (^{125}I-BLG) were injected intravenously into CD-1 lactating mice 1 week post-partum. Immunoprecipitable radioactivity was measured in serum from time 1 min to 4 hr (Fig. 2).

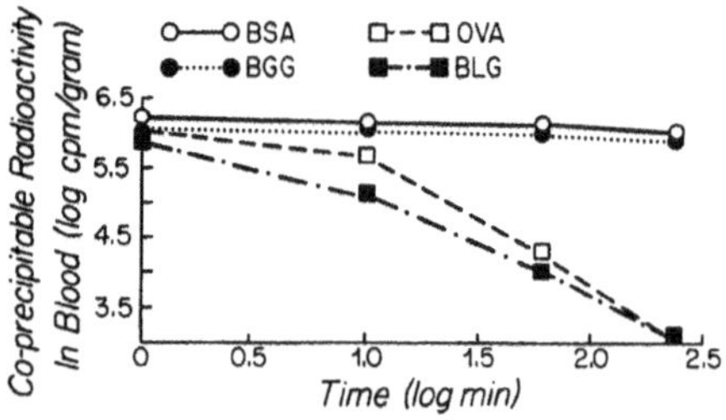

Fig. 2. Clearance of co-precipitable radioactivity from the circulation of lactating mice injected with ^{125}I-labelled antigens. Each data point represents the mean of three animals.

After injection of ^{125}I-BSA and ^{125}I-BGG, co-precipitable radioactivity decreased by less than 40% and 25%, respectively, over 4 hr. In contrast, the co-precipitable radioactivity in blood decreased to less than 0.1% of initial values of ^{125}I-BLG and ^{125}I-OVA in the same time period.

Milk was obtained from the same lactating females by manual expression 4 hr after injection of each of the 4 test proteins (Table 1). Radioactivity was greatest in the milk of animals injected with ^{125}I-BLG. The milk of animals injected with the other 3 antigens all had a lower level. Of more significance, the radioactivity in milk from mothers receiving ^{125}I-BSA and ^{125}I-BGG was 10-fold more immunoprecipitable than the radioactivity in milk from mothers receiving ^{125}I-OVA or ^{125}I-BLG. Persistence of antigen in the maternal circulation appeared to be positively associated with transfer of immunoreactive antigen into milk.

Table 1. Characterization of radioactivity in milk 4 hr after injection of ^{125}I-labelled antigens

Antigen	cpm/ml[1]	% Co-Precipitable Radioactivity[2]
^{125}I-BSA	16.7	46
^{125}I-BGG	14.7	56
^{125}I-OVA	17.4	5
^{125}I-BLG	25.4	3

[1]Mean of total radioactivity (cpm x 10^{-4}) in milk samples from 3 animals.

[2]Mean of radioactivity co-precipitated by specific rabbit antibody-antigen complexes from milk of 3 animals.

We next determined whether transfer of iOVA into milk might also be related to the dose administered. OVA (50, 200, or 800 ug) was injected iv into lactating mice; serum was obtained at 10 min and 4 hr, and milk was expressed at 4 hr. The iOVA concentration in serum and milk was measured by EIA. A positive correlation was noted between administered dose, serum concentration at both 10 and 4 hr, and concentration of OVA in milk at 4 hr (Fig. 3).

At present, it is unknown whether transfer is primarily dependent on the concentration of antigen in maternal blood or on the length of time that mammary tissue is exposed to circulating antigen, or to both, and whether other properties of the molecules may also contribute. Halsey et al[23] have also studied transfer of proteins into milk after iv injection of radiolabelled tracers. They demonstrated greater transfer of radioactivity after injection of IgA or IgM in comparison to IgG and suggested that a transport system for polymeric immunoglobulins might be involved. It is not known whether the mammary gland possesses specific transport mechanisms for the four proteins involved in our studies.

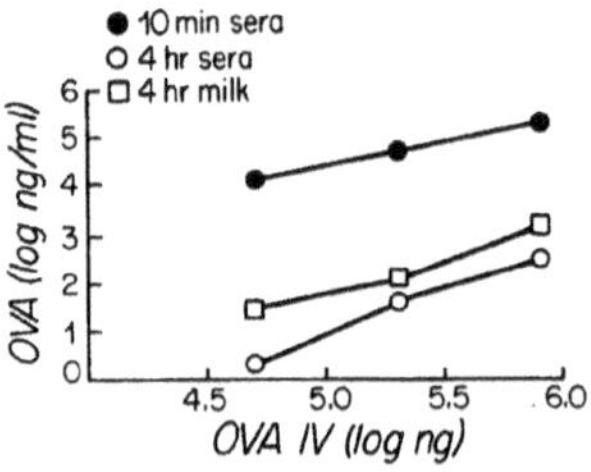

Fig. 3. Effect of selected doses of OVA administered iv on serum and milk iOVA concentration. Each data point represents mean of 2 animals.

Influence of Circulating Maternal Antibody on the Transfer of Dietary Antigen via Milk.

Circulating antibody of both IgG and IgA isotypes has been shown to influence systemic processing of circulating antigen.[24,25] We, therefore, examined the effect of circulating antibody in the lactating mouse on the transfer of antigen to the nursing neonate. Lactating mice were injected with either anti-BSA or antibodies of another specificity followed by ^{125}I-BSA. On injection of ^{125}I-BSA into mice prepared with antibodies of another specificity, immunoreactive radioactivity persisted in the maternal circulation and immunoreactive BSA was transferred into the milk (Fig. 4).

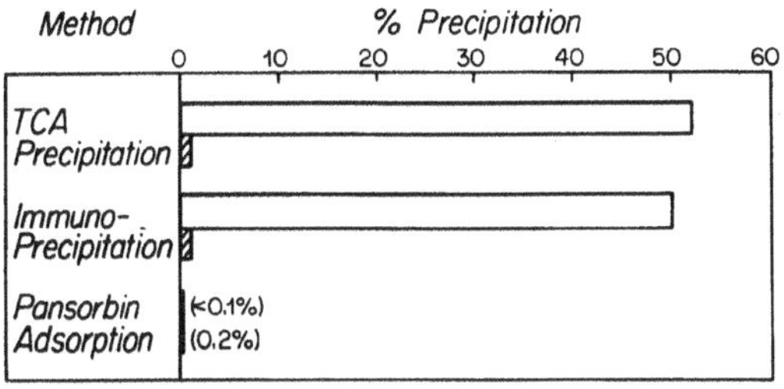

Fig. 4. Characterization of radioactivity in milk from lactating mice injected with either anti-BSA and ^{125}I-BSA (hatched bar) or antibodies of another specificity and ^{125}I-BSA (open bar). Milk was analyzed for protein-bound radioactivity by precipitation with an equal volume of cold 20% trichloroacetic acid. Immunoreactive protein was sought by co-precipitation with rabbit anti-BSA-BSA complexes. A 10% suspension of Staphylococcus aureus cells bearing protein A (Pansorbin) was used to test for IgG antibody-antigen complexes in milk.

On injection with anti-BSA, ^{125}I-BSA was cleared rapidly from the maternal circulation, and although two-fold more radioactivity was transferred into the milk in this group, it was not immunoprecipitable and did not bind to Staphylococcal Protein A (Figure 4).

Discussion

We have summarized experiments that examined the uptake of dietary antigen from the gut of the lactating mouse, the transfer of dietary antigens from the maternal circulation into milk after iv administration, and the effect of circulating maternal antibody on transfer of dietary antigen into milk. Although enhanced uptake of OVA was documented in mid-lactation (6-10 days), we have not as yet examined whether the changes responsible for enhanced uptake develop during pregnancy and thereby increase the potential for fetal exposure to circulating dietary antigen. Dahl et al[26] measured the serum concentrations in pregnant rats of the same four antigens used in our studies. Although they found no difference in serum antigen concentration between pregnant and control rats, it is important to note that they measured the serum concentration at a single time point 4 hr after feeding the antigen and, thus, may have missed differences that might have been present at other times.

The changes in intestinal structure, in particular in length of the intestine, that we believe are important in producing the increased uptake of OVA, develop primarily from early to mid-lactation.[27] Falth-Magnusson et al[28] have tested lactating women to determine whether altered intestinal permeability is associated with lactation. Utilizing a mixture of polyethylene glycols (PEG 400 and PEG 1000), they found no difference in maximum recovery or recovery ratio (1074 M.W./370 M.W.) in lactating or non-lactating mothers. Possibly the lactating woman does not undergo the changes in GI structure and function noted in the mouse. Alternatively, the use of a different probe (PEG) to measure macromolecular uptake may produce different results. Thus far, we have assessed intestinal uptake of dietary protein in lactating mice utilizing only one dietary protein. Other proteins also should be examined.

We also examined the transfer of four dietary proteins into milk after iv injection into the lactating mouse. As noted, the transfer of protein antigens into milk after iv injection is dependent on the specific protein injected. The transfer appears to correlate inversely with its rate of clearance from the maternal circulation and directly with the dose administered. Telemo et al[29] examined transfer of the same four proteins into milk of lactating rats after iv injection and obtained similar results.

We observed that circulating maternal antibody limited the transfer of antigen into milk, and suggest that this effect is related to enhanced rate of antigen clearance from the maternal circulation, most likely by the liver.[19] Skogh[25] also observed that the clearance of intravenously injected antigen was enhanced by the formation of an IgG antibody-antigen complex. IgG antibodies to dietary antigens are likely to be present in the circulation of lactating women.[30] Antibodies of other isotypes may also be important. Polymeric serum IgA antibodies mediate transfer of antigens from the circulation into bile;[24] such antibodies might contribute to limitation of antigen transfer from the lactating mother to the neonate via the milk.

References

1. F. B. Talbot, Eczema in childhood, Med Clin No Am 1:985 (1918).
2. W. R. Shannon, Demonstration of food proteins in human breast milk by anaphylactic experiments on guinea pigs, Am J Dis Child 22:223 (1921).
3. J. W. Gerrard, Allergy in breast fed babies to ingredients in breast milk, Annals of Allergy 42:69 (1979).
4. I. Jacobsson, T. Lindberg, Cow's milk proteins as a cause of infantile colic in breast-fed infants, Lancet ii:437 (1978).
5. I. Jakobsson and T. Lindberg, Cow's milk proteins cause infantile colic in breast-fed infants: A double-blind crossover study, Pediatrics 71:268 (1983).
6. A. Cant, R. A. Marsden, P. J. Kilshaw, Egg and cows' milk hypersensitivity in exclusively breast fed infants with eczema, and detection of egg protein in breast milk, Br Med J 291:932 (1985).
7. A. J. Cant, J. A. Bailes, R. A. Marsden, and D. Hewitt, Effect of maternal dietary exclusion on breast fed infants with eczema: two controlled studies, Br Med J 293:231 (1986).
8. A. M. Lake, P. F. Whitington and S. R. Hamilton, Dietary protein-induced colitis in breast-fed infants, Pediatrics 101:906 (1982).
9. P. J. Kilshaw and A. J. Cant, The passage of maternal dietary proteins in human breast milk, Int Archs Allergy appl Immun 75:8 (1984).
10. S. Machtinger and R. Moss, Cows milk allergy in breast-fed infants: The role of allergen and maternal secretory IgA antibody, J Allergy Clin Immunol 77:341 (1986).
11. G. Hattevig, B. Kjellman, S. G. O. Johansson, and B. Bjorksten, Clinical symptoms and IgE responses to common food proteins in atopic and healthy children, Clinical Allergy 14:551 (1984).
12. M. S. Kaplan and N. J. Solli, Immunoglobulin E to cow's milk protein in breast-fed atopic children, J Allergy Clin Immunol 64:122 (1979).
13. F. Bjorksten and U. M. Saarinen, IgE antibodies to cow's milk in infants fed breast milk and milk formulae, Lancet ii:624 (1978).
14. E. E. E. Jarrett, D. M. Haig, W. McDougall, E. McNulty, Rat IgE production. II. Primary and booster reaginic antibody responses following intradermal or oral immunization, Immunology 30:671 (1976).
15. E. E. E. Jarrett, Activation of IgE regulatory mechanisms by trans-mucosal absorption of antigen, Lancet ii:223 (1977).

16. D. G. Hanson, N. M. Vaz, L. C. S. Maia, M. M. Hornbrook, J. M. Lynch, C.A. Roy, Inhibition of specific immune responses by feeding protein antigens, Int Arch Allergy Appl Immun 55:526 (1977).
17. D. G. Hanson, Ontogeny of orally induced tolerance to soluble proteins in mice 1. Priming and tolerance in newborns, J Immunol 127:1518 (1981).
18. S. Strobel and A. Ferguson, Immune responses to fed protein antigens in mice. 3. Systemic tolerance or priming is related to age at which antigen is first encountered, Pediatr Res 18:588 (1984).
19. P. R. Harmatz, K. J. Bloch, R. E. Kleinman, M. K. Walsh, and W. A. Walker, Influence of circulating maternal antibody on the transfer of dietary antigen to neonatal mice via milk, Immunology 57:43 (1986).
20. P. R. Harmatz, D. G. Hanson, M. K. Walsh, R. E. Kleinman, K. J. Bloch and W. Allan Walker, Transfer of protein antigens into milk after intravenous injection into lactating mice, Am J Physiol 251:E227 (1986).
21. J. Jepson and L. Lowenstein, Erythropoiesis during pregnancy and lactation in the mouse. II. Role of erythroprotein, Proc Soc Biol Med 121:1077 (1966).
22. A. C. Kulangara, The demonstration of ingested wheat antigens in human breast milk, Med Sci Libr Compend 8:19 (1980).
23. J. F. Halsey, B. H. Johnson and J. J. Cebra, Transport of immunoglobulins from serum into colostrum, J Exp Med 151:767 (1980).
24. P. R. Harmatz, R. E. Kleinman, B. W. Bunnell, K. J. Bloch, W. A. Walker, Hepatobiliary clearance of IgA immune complexes formed in the circulation, Hepatology 2:328 (1982).
25. T. Skogh, Tissue distribution of intravenously injected dinitrophenylated human serum albumin. Effect of specific IgG and IgA antibodies, Scand J Immunol 16:465 (1982).
26. G. M. K. Dahl, E. Telemo, B. R. Westrom, I. Jacobsson, T. Lindberg, B.W. Karlsson, The passage of orally fed proteins from mother to foetus in the rat, Comp Biochem Physiol 77A:199 (1984).
27. A. W. Cripps and V. J. Williams, The effect of pregnancy and lacta0-tion on food intake, gastrointestinal anatomy and the absorptive capacity of the small intestine in the albino rat, Br J Nutr 33:17 (1975).
28. K. Falth-Magnusson, N. I. M. Kjellman, T. Sundqvist and K. E. Magnusson, Gastrointestinal permeability in atopic and non-atopic mothers, assessed with different-sized polyethyleneglycols (PEG 400 and PEG 1000) Clin Allergy 15:565 (1985).

29. E. Telemo, B. Westrom, G. Dahl, and B. Karlsson, Transfer of orally or intravenously administered proteins to the milk of the lactating rat, J Pedriatr Gastroenterol and Nutr 5:305 (1986).

30. R. M. Rothberg and R. S. Farr, Anti-bovine serum albumin and anti-alpha lactalbumin in the serum of children and adults, Pediatrics 35:571 (1965).

POTENTIALLY TOXIC EFFECTS OF DRUG AND TOXINS IN HUMAN BREAST MILK

John T. Wilson, R. Don Brown, Iain J. Smith, and
James L. Hinson

Section on Clinical Pharmacology
Departments of Pharmacology and Therapeutics and
Pediatrics
Louisiana State University Medical Center
1501 Kings Highway
Shreveport, Louisiana 71130

INTRODUCTION

The lactating human breast is an ORPHAN ORGAN. Too little is known about mechanisms for the transfer of xenobiotics into milk. This review highlights a lack of data on drugs and chemical toxins in milk and proposes a remedial approach.

The following criteria must be met if an effect in the nursing infant is due to a drug or chemical toxin in breast milk.

1. Maternal exposure. This requires documentation by a careful history and appropriate laboratory confirmation of drug or metabolite in maternal plasma.

2. Infant effect. This effect must be compatible with the known pharmacological effect of the drug or toxin, e.g., diazepam and drowsiness, and with the maturational susceptibility of the infant.

3. Known amount of drug or toxin in milk. The traditional estimate of this has used the milk/plasma concentration ratio. This ratio is not invariant and is, therefore, not without error in estimating drug or toxin dose in milk (1-10). Measurement of the amount of drug in milk at all feeding periods is the ideal method of assessing drug dose in milk. This is clearly impractical and probably impossible. A third approach utilizes predictive kinetics. This requires a properly defined pharmacokinetic model for drug disposition from plasma into milk. If a model can be described for each drug then this predictive kinetic approach can be used to calculate drug or toxin amount in milk for all desired periods after dosing.

4. Identification of drug or chemical toxin in the nursing infant. This includes drug or toxin in plasma, urine or other appropriate body fluid. Indirect laboratory measurements of an effect are also needed, e.g. thyroid function studies after in-

gestion of antithyroid drugs. The demonstration of drug levels and pharmacodynamics in the nursing infant may obviate the need for assessment of cause and effect by the preceding criteria. However, in cases of severe toxicity these criteria must still be satisfied. For items 3 and 4 the method of drug or chemical toxin quantitation must have appropriate sensitivity and specificity. This is imperative if their concentrations in breast milk and infant plasma or urine are to be detected. A method of insufficient sensitivity may lead to false assumptions about safety of the dose in milk or body burden in the infant.

Review of Infant Effects After Exposure to Drugs in Breast Milk

A recent survey of post-partum drug consumption by Scandinavian women has described the frequency with which drugs (by class) are ingested (11). We have reviewed the literature on infant effect after drug ingestion via breast milk. This review is summarized in Table I and is restricted to the drugs reported in the above study. Also included are the anticoagulants, antithyroid drugs, drugs of abuse and drugs of high intrinsic toxicity. Our reference sources were 1) the breast milk literature data base provided to one of us (JTW) by the World Health Organization, 2) the advisory statement on drug excretion in breast milk published by the American Academy of Pediatrics (12) and 3) Index Medicus. From these sources we found 306 reports about a drug in milk.

It is clear (Table I) that only 57 of the 306 (18.6%) reports mentioned an infant effect. Of these 57 reports, only 20 (37.7%) documented

Table I. Numbers of reports on infant effects and drug concentration in milk or infant plasma/urine

Drug Class	A	B	C	D	E
Analgesic/Antiinflammatory	27	7	3	3	2
Dermatological	3	3	0	1	0
Anti-asthmatic	15	4	4	1	1
Anti-infective	72	3	0	0	0
Sex hormones	71	4*	0	0	0
Antihistamines	5	2	1	1	1
Gastrointestinal	11	2	1	1	1
Benzodiazepines	22	6	6	5	5
Neuroleptics	8	2	2	1	1
Tricyclic antidepressants	7	4	3	3	2
Anticoagulants	7	4	2	4**	2
Antithyroid	11	3	1	3**	1**
Drugs of abuse or habit	32	9	8	5	4
Intrinsic toxicity	15	4	0	0	0
Total	306	57 18.6%†	31 58.5%††	28 52.8%††	20 37.7%††

A = Total number of papers reviewed.
B = Total number of reports which mention effect on the infant.
C = "Item B" and measured milk concentration.
D = "Item B" and infant drug concentration.
E = "Item B" and both infant and milk drug concentrations.
* = Breast hypertrophy (3 cases).
** = Includes indirect drug effect (laboratory tests).
† = B/A X 100 = % for column B.
†† = C/B X 100 or D/B X 100 or E/B X 100 = % for each respective column.

an infant effect and included measured drug concentrations in both milk and infant plasma or urine. Approximately half of the 57 reports contained quantitative information about drug in breast milk. Clearly the question about dose of drug in milk and infant effect needs attention to describe dose-safety estimates and especially to corroborate dose-effect relationships. Accordingly, we describe a predictive pharmacokinetic approach for estimating the dose of drug in breast milk. A study design that incorporates this approach, combined with an objective recording of infant effect and measurements of infant drug levels, will most likely substantiate the etiological role of drug (or chemical toxin) in breast milk with regard to its proposed adverse effect on the nursing infant.

PREDICTIVE KINETIC APPROACH

Milk/plasma ratios often do not accurately predict amounts of drug delivered to the infant via milk. These ratios may vary with time after dose and also with the number of doses. This variance is a function of drug distribution according to multicompartment pharmacokinetic models.

Pharmacokinetic Models

The following discussion of pharmacokinetic models is restricted to the simpler linear or first order models in which elimination of the drug occurs directly from the central or plasma compartment, the only exception to this being, of course, additional elimination of the drug through the milk. In order to simplify the discussion non-linear or zero order kinetics and mixed order kinetics are not discussed.

A simple one compartment model, is illustrated below (Fig. 1).

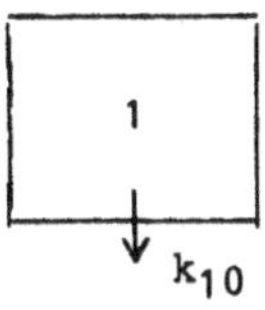

Figure 1. One compartment model.

In this model, the first order rate constant k_{10} regulates the rate of elimination of the drug from the central (plasma) compartment, including the egress of drug, via milk, caused by suckling. If the drug is given orally or by any mode of administration other than intravenously, another first order rate constant, k_a, regulates the absorption rate of the drug into the central compartment.

In order for drug distribution into milk to be described by this simple model, the entire body is considered as one compartment; i.e., distribution of the drug into all compartments of the body (milk, interstitial fluid, cerebrospinal fluid, etc.) must occur simultaneously and to the same extent. In other words, the M/P ratio = 1 at all times because milk concentrations and plasma concentrations are the same. One drug that may behave in this manner is antipyrine, the distribution of which appears to follow that of body water. This behavior is unlikely with all but a very few of the therapeutic agents normally employed. Let us, therefore, discuss more common compartmental behavior of drugs, and examine the consequences in terms of the distribution of drugs into milk.

The most common 'multicompartment' model of drug pharmacokinetics is that of the two compartment open model (Fig. 2).

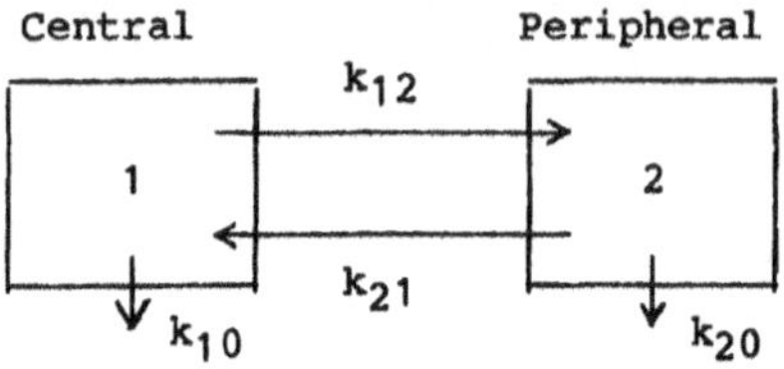

Figure 2. Two compartment open model.

Compartment one represents the central or plasma compartment and compartment two, the peripheral compartment, represents all other compartments of the body (i.e., interstitial fluid, liver, kidney, mammary tissue, milk itself, saliva, etc.). The rate constants k_{12} and k_{21} regulate the transfer of drug from compartment one to compartment two (k_{12}) and from compartment two to compartment one (k_{21}). In this model, k_{10} regulates the rate of elimination of the drug from the central compartment. As with the previous model, if the drug is given orally or by any route of administration other than intravenously, another rate constant, k_a, regulates the absorption rate of the drug into the central compartment. k_{20} represents the egress of drug (via milk) from the peripheral compartment caused by suckling. All the rate constants, with the exception of k_{20}, are first-order. The k_{20} is an on-demand rate "constant", hence it cannot be calculated as a simple first or zero order rate constant.

One must realize, however, that the two compartment model is a very general model for drug transfer into milk. It requires that all compartments other than the plasma compartment behave in a similar fashion to that of the milk compartment, i.e., the drug distributes into all extravascular tissues and spaces in an identical manner. An example of a drug that may fit a two compartment open model is theophylline. However, this is probably not true for most drugs. Certainly one would not expect a drug such as gentamicin, which is sequestered in renal tissue, to be completely characterized with this model.

More complicated theoretical models have been proposed to characterize drug transfer into milk (6). In the sequential three compartment open model (Fig. 3), the milk compartment is designated as compartment three.

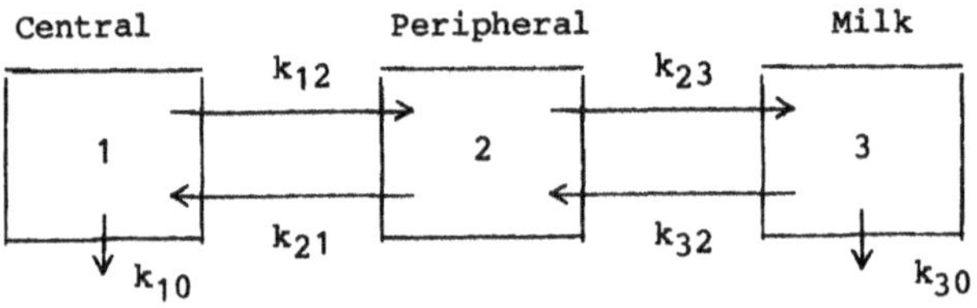

Figure 3. Sequential three compartment open model.

Compartment 2 represents drug distribution to all tissues and spaces other than milk and plasma. The drug gains access to milk from compartment two. A drug that appears to fit this model reasonably well is acetaminophen.

Other relatively simple potential compartmental models that can be used to describe the transfer of drugs into milk are the mamillary three compartment open model (Fig. 4) and the mixed four compartment open model (Fig. 5).

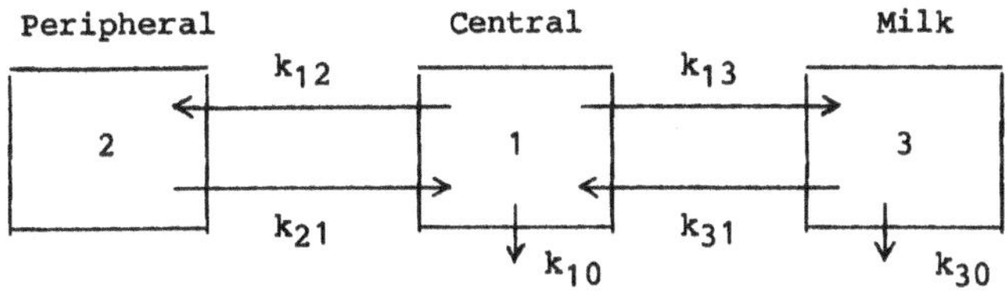

Figure 4. Mamillary three compartment open model.

In the mamillary three compartment open model, the drug would gain entrance into milk via direct, separate equilibrium with plasma.

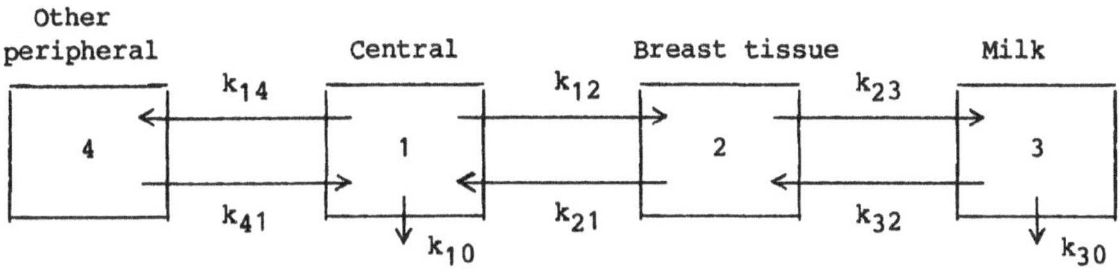

Figure 5. Mixed four compartment open model.

The four compartment model involves a separate compartment interacting with plasma and, in turn, with milk (other tissues are represented by the fourth compartment). This model is attractive since the second compartment could represent the mammary tissue itself, while the third compartment would represent milk.

Model Discrimination

Adequate characterization of drug distribution into milk must be accomplished so that predictions about drug amount being delivered to the infant can be made with a reasonable degree of accuracy. It is not sufficient to consider that the simplest model which seems to fit the data of a single dose experiment will describe the data adequately after repeated multiple doses. The M/P ratio is a function of transfer kinetics operative for all the compartments of a pharmacokinetic model (8). In a one compartment model, M/P ratio = 1. However, even in the simplest of the multicompartment models, the two compartment open model, the M/P ratio is likely to vary at the different times after a dose of drug is given. In addition, accumulation of drug in milk (deep compartment behavior) can occur even with the two compartment model, although it is more likely to occur with the other multicompartment models (6). Therefore, one must develop an approach to the analysis of data which will allow discrimination of the proper compartmental model.

The overall approach should consist of the following steps: (1) Computation of all polyexponential equations that might pertain to the models; (2) Calculation of the appropriate microscopic rate constants and compartmental volumes (e.g., k_{10}, k_{12}, k_{21}, and the two compartmental volumes in the i.v. two compartment open model); (3) Simultaneous fitting of milk and plasma drug concentrations for each possible model; (4) Model dependent and independent calculations of the same pharmacokinetic parameters, which can be used to assist in model discrimination; and (5) Model discrimination. It should be realized that model discrimination can occur during the performance of any of the first four steps, e.g. if the plasma-concentration time profile of an i.v. drug does not fit anything but equations consisting of one or two exponents, it is unlikely that models containing polyexponential equations of more than two exponents apply to this drug.

The plasma concentration, time profile of all drugs which follow first order or linear kinetics can be described by one or more exponential equations. The simplest equation is that of the one compartment i.v. model

$$C(t) = A \exp(-B*t)$$

where C(t) is the drug concentration at a given time, t, and A and B are the coefficient (intercept) and exponent of the equation (in this model, $B = \text{beta} = k_{10}$). This and other compartmental models can be represented by the generalized equation

$$C(t) = \sum_{i=1}^{m} A_i \exp(-B_i*t)$$

where m is the number of exponents involved in the equation (e.g., m = 2 in the i.v. two compartment open model). Initial estimates of these equations can be derived using ESTRIP (13) as well as other back projection or feathering techniques for stripping the components of a polyexponential equation.

Once polyexponential parameter estimates are obtained, the microscopic rate constants and volumes for the models in question are calculated. Consider, for example, the calculations for the i.v. two compartment open model. Drug concentrations in the plasma compartment of an i.v. two compartment pharmacokinetic model are described by the following exponential equation:

$$C(t) = A1 \exp(-B1*t) + A2 \exp(-B2*t)$$

where C(t) is the drug concentration at a given time t and A1, A2 and B1, B2 are the coefficients (intercepts) and exponents of the equation (in this model, B1 = alpha and B2 = beta). An estimate of this equation is derived using ESTRIP. The volume of the compartment (V_p) can then be calculated from the exponential equation as follows:

$$\text{Volume (plasma)} = \text{Dose} / (A1+A2)$$

Amounts of drug in the plasma compartment can then be calculated simply by:

$$\text{Amount in Plasma (t)} = \text{Conc (plasma)} * \text{Volume (plasma)}$$

However, this information is of little value if one is interested in the amount of drug delivered to an infant in breast milk. In order to obtain this information, the volume of the peripheral compartment must also be derived. This is done by first solving the plasma bi-exponential equation for k_{10}, k_{12} and k_{21}:

$$k_{21} = (A1*B2 + A2*B1) / (A1 + A2)$$

$$k_{10} = (B1*B2)/k_{21}$$

$$k_{12} = B1 + B2 - k_{21} - k_{10}$$

Once values are known for the above micro-rate constants, a bi-exponential equation for drug amount in the peripheral compartment (Xp) can be derived:

$$Xp = \frac{(k_{12} * Dose)}{(B1 - B2)} [exp(-B2*t) - exp(-B1*t)]$$

If X' is set equal to the fractional value, then X1' = -X' and X2' = X'and this equation resolves into the familiar bi-exponential equation:

$$Xp = X1' \exp(-B1*t) + X2' \exp(-B2*t)$$

However, this equation does not allow us to determine whether drug concentrations obtained from milk samples relate to those that would be calculated for the peripheral compartment in this two compartment model. In other words, this equation does not allow determination of whether the two compartment model is a reasonable model for the particular drug in question.

In order to continue the steps necessary in model discrimination, one must convert the amount coefficients into concentration coefficients by dividing by a milk volume (V_m). Estimation of the milk volume involves calculating the volume of milk necessary to fit the amount equation to each of the milk concentration samples, then selecting an initial estimate of milk volume from these values and iteratively fitting all drug milk concentrations to the resulting equation:

$$C\ (milk,t) = (X1'/V_m) \exp(-B1*t) + (X2'/V_m) \exp(-B2*t)$$

Iterative fitting is performed until the sum of the squared deviations

$$SS = \sum_{i=1}^{n} (\text{Predicted Conc}_i - \text{Observed Conc}_i)^2$$

is minimized. Thus are the rate constants and volumes calculated for the i.v. two compartment open model. (Solutions have also been developed for the more complicated theoretical models presented in this discussion (unpublished results)).

Once the microscopic rate constants and the initial estimates of plasma and milk volumes have been obtained for a particular model, simultaneous fitting is employed to resolve that particular model. Non-linear regression analysis using differential weighting should be employed during this step. The preferred method of weighting is through the use of the power variance model of Sheiner and colleagues as applied to Fletcher's modification of the Marquardt Algorithm (14, 15).

The key to simultaneous fitting is to understand that milk drug concentration data is fit in an inter-related way with that of the plasma drug concentration data. In other words, the equation that describes milk concentrations and the equation that describes plasma concentrations are both derived from the microscopic rate constants and compartmental volumes, as was the case with the data in the two compartment example given above.

In essence, the microscopic rate constants and the plasma and milk volumes are the parameters being resolved by non-linear regression analysis. This is done by utilizing these parameters to calculate the polyexponential equations for plasma and milk drug concentration as well as the time profiles every time that one or more of these parmeters are re-estimated during the program operation. Every time these equations are calculated, an objective function is also calculated. This objective function is a modified, weighted sum of squared deviations which incorporates the variance contributed by error. Different combinations of parameters are tested until one is arrived at that yields an objective function that is not significantly different from the previous smallest objective function. The significant difference which is set by the investigator to be an acceptable difference between these two objective functions is called the convergence criterion. The resulting set of parameters is then taken to represent the best estimate of the parameters. If convergence does not occur, this is a case of failed convergence and can be due to model misspecification (i.e., to that particular model being an incorrect model for the drug in question).

The calculation of model dependent and model independent parameters is another important step in model discrimination. Parameters such as area under the curve, clearance, mean residence time and volume of distribution at steady state can be calculated from the polyexponential equations of the models as well as by evaluating the drug concentration data by analyzing the moments of the data points (16). If a model applies to the drug in question, model independent calculations will agree reasonably well with the calculations based on that model.

Overall discrimination of the pharmacokinetic model begins with the measure of goodness of fit of the different polyexponential equations that can be derived from the curve stripping technique used, namely the sum of the squared deviations, $(Cpred_i - Cobs_i)^2$. It continues with a comparison of the objective functions obtained by non-linear regression analysis of the simultaneous fitting of the parameters of the different models and with the comparisons between model dependent and model independent calculations.

The final and crucial test of model discrimination is comparison of the predictive power of the models for repeated doses. The familiar sum of squared deviations is again used as the test criterion. In the event that two or more models remain after chronic dose evaluation, the simplest model which fits the single dose and chronic dosing data reasonably well is selected. Simulations of repeated dose profiles are described elsewhere (6).

The predictive kinetic approach has been tested in one experiment. Acetaminophen (82.5 mg/kg) was given intravenously for 63 minutes to a 45 kg lactating goat. Serial blood and milk samples were taken for acetaminophen assay both during and after the infusion. The resultant values are shown in Figure 6. Using the techniques described above, the simultaneous fitting of the milk and plasma profiles was performed. The simulated profiles fit the observed milk and plasma drug concentrations very closely ($r^2 = 0.983$).

Predictive Aspects

Model discrimination of plasma and milk drug concentration data allows reasonably accurate predictions to be made about the amounts of a particular drug delivered to the infant. These predictions can be made by simply multiplying the average predicted concentration in milk that occurs during the nursing interval by the volume of milk ingested during that interval.

However, in order to apply predictive pharmacokinetics to other mothers and infants, one must investigate the pharmacokinetics of that drug in a sufficient number of mothers and infants. Once this has been accomplished, then two approximations can be used to predict amounts of drug delivered in the milk of other lactating females.

The first approximation is to use the means and variances of the derived pharmacokinetic parameters to define the transfer of a particular

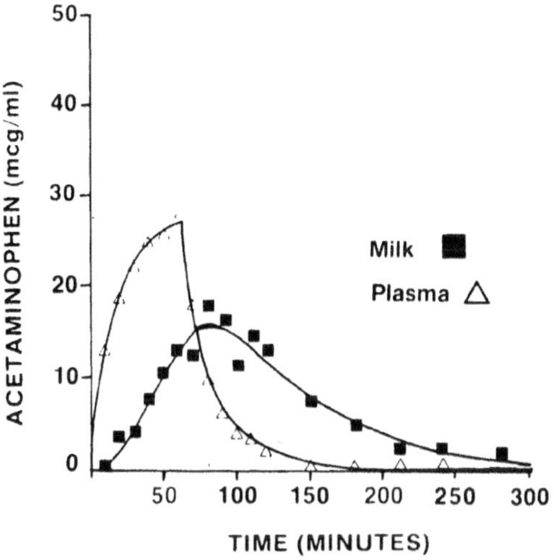

Figure 6. Resolution of the plasma and milk drug concentration profiles for a 63 minute infusion of acetaminophen (82.5 mg/kg) to a lactating goat. The estimates are based upon a sequential three compartment open model with plasma and milk represented by the first and third compartments respectively. The parameters are: $k_{10} = 0.089\ min^{-1}$, $k_{12} = 0.080\ min^{-1}$, $k_{21} = 0.085\ min^{-1}$, $k_{23} = 0.0051\ min^{-1}$, $k_{32} = 0.020\ min^{-1}$, $r^2 = 0.983$, $\alpha = 0.220\ min^{-1}$, $\beta = 0.041\ min^{-1}$, $\gamma = 0.016\ min^{-1}$, V_1 = l/kg 0.488, V_3 = l/kg 0.115, $t_{\frac{1}{2}\gamma}$ = 42.95 min. These values were obtained by the simultaneous fitting of both plasma and milk drug concentration profiles by an unweighted simplex optimization technique.

$$t_{\frac{1}{2}\gamma} = \text{elimination half-life} = \frac{\ln 2}{\gamma}$$

drug into milk for most lactating females. One would first define the composite milk drug concentration versus time profile for a woman receiving that drug (e.g., means with 95% confidence limits). Drug amounts delivered to the infant at different nursing intervals would be calculated, using the values for volume of milk obtained during nursing and the average duration of nursing as derived in the detailed evaluation of this particular drug. This information would then be used to determine whether sufficient drug may be delivered to the infant to cause toxicity and, if toxic effects were likely, to determine the time period after drug administration when the infant would be most at risk. Assessment of drug accumulation in milk after repeated drug doses is implicit to these applications.

The second predictive pharmacokinetic approximation is applied to a lactating woman for whom a limited number of plasma and milk samples are obtained after administration of a drug which has been evaluated previously in other lactating women. Non-linear regression analysis can be used to fit the limited data to a model by employing the means of the parameters estimated in the previous complete study of that drug, i.e., such study giving initial estimates of the parameters required by the analysis. If convergence occurs, one can then proceed to predict amounts of drug delivered to the infant. If convergence does not occur, then estimates based directly on the plasma and milk drug concentration data in that lactating woman cannot be made. Lack of convergence may be due to effect of confounding lactation or mammary component variables operative in that particular woman (8).

Practical Applications of Predictive Pharmacokinetics

The above approach may be illustrated by simulations of hypothetical drugs and by examples of calculations used to estimate drug dose to the nursing infant. The objective is to use a known milk concentration profile (i.e., from population kinetics or fitted from data) and the volume of milk the infant consumes to estimate the amount of drug delivered to the infant. This is the basis for potential toxicity prediction or confirmation of observed toxicity. If the amount of drug delivered to the infant is estimated below the toxic level, then alternative reasons for observed toxicity must be sought.

Let us first examine drug concentrations in milk for 24 hours after a single dose of a drug. The feeding times which correspond to the time of day consistent with the "normal" feeding times are 6:00 a.m. 10:00 a.m. 2:00 p.m. 6:00 p.m. and 10:00 p.m. and the drug is administered once at 6:00 a.m. During each 20 minute feed the infant (assumed to be 5 kg in weight) consumes 30 ml of breast milk per kg of infant body weight. This corresponds to a total volume consumed of 150 ml milk or 7.5 ml/min per feeding. The amount of drug that is delivered to the infant equals the product of the instantaneous drug concentration of the milk and the volume of consumed milk. From a practical consideration, the drug concentration in milk varies over a 20 min. feed. For this reason it is more appropriate to consider 1 minute intervals and to calculate the amount from the predicted drug concentration as well as the volume consumed during each interval, such that:

$$\text{Total drug per feed} = \sum_{n=0}^{19} C_{milk}(t+n) \times V_{milk}$$

These amounts per minute are essentially summated to arrive at total drug delivered to the infant during the feeding. In Figure 7 a single maternal dose of 10 mg/kg was administered at 6:00 a.m. The amount of drug received by the infant at each feeding was: 6:00 a.m. (<0.01 mg), 10:00 a.m. (0.08 mg), 2:00 p.m. (0.11 mg), 6:00 p.m. (0.09 mg) and 10:00 p.m. (0.07 mg) for a total over the first 24 hours of 0.35 mg (0.07 mg/kg infant). The amount of drug was calculated by an integral of milk drug concentration midpoint at each minute for 20 minutes as shown by the shaded bars in Figure 7. These bars represent periods of breast feeding. Such an approach applies regardless of the actual shape of the milk curve, infant feeding times or number of doses per day.

From examples shown in Figure 7 the infant dose can be compared to the maternal dose by the following calculation.

$$\%\ \text{maternal dose} = \frac{\text{infant dose (mg/kg)}}{\text{maternal dose (mg/kg)}} \times 100\%$$

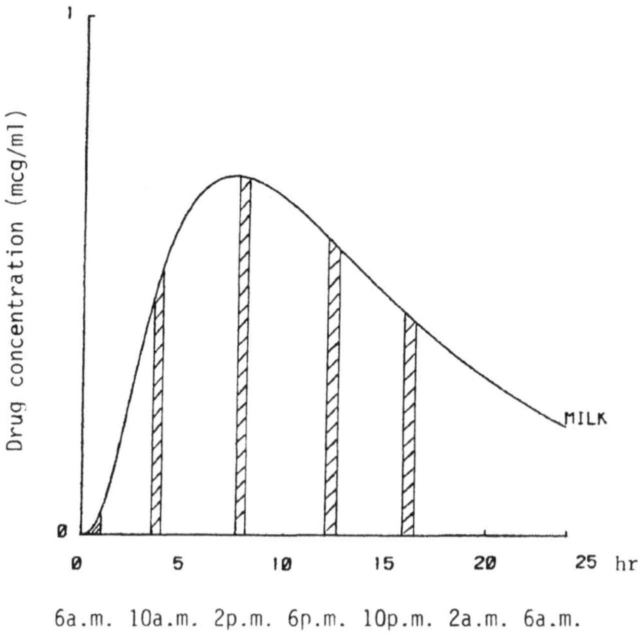

Figure 7. Concentration of drug in milk ingested by a 5 kg infant at each 20 minute feeding (5 feeds per day). A sequential three compartment open model with oral absorption is assumed for this simulation and the following parameters are used to prepare the simulation. $k_a = 1.50\ hr^{-1}$, $k_{10} = 0.80\ hr^{-1}$, $k_{12} = 0.25\ hr^{-1}$, $k_{21} = 0.22\ hr^{-1}$, $k_{23} = 0.06\ hr^{-1}$, $k_{32} = 0.18\ hr^{-1}$, $f = 0.80$, $X_o = 10.0$ mg/kg, $V_p = 1.5$ l/kg, $V_m = 0.3$ l/kg, $t_{\frac{1}{2}\gamma} = 7.3$ hr. Shaded bars represent feeding times.

The infant dose is that summated for all feedings during a day. The maternal dose is that administered during a day. In Figure 7 the % maternal bodyweight adjusted dose is 0.7%. A linear relationship of dose to body weight for infant and mother is assumed, but we acknowledge that this does not precisely occur. An upper limit for % maternal dose is arbitrarily placed at 10%. An amount below this percentile is considered as unlikely to harm the infant if toxicity is a function of dose.

These evaluations of drug toxicity are not limited to a single dose. They apply to repeated doses given for several days. Figure 8 shows a simulation for an accumulating system with repeated daily doses. In this figure, the bar graph shows the amount of drug an infant receives for 24 hours (5 feeds) and assumes once a day dosing. A maternal dose of 10 mg/kg administered at 6:00 a.m. each day for 10 days produced the milk and plasma drug profiles shown. In this simulation the drug accumulated in milk. The lower part of this figure shows the daily amount of drug summated for five feedings per day. In Figure 8, milk drug concentrations continue to rise into days 5-6 and steady state is not achieved until that time ($M/P_{ss} = 5$). The amount of drug received by the infant parallels the drug in milk pattern and percent of maternal dose rises to 9.06% at 10 days of dosing.

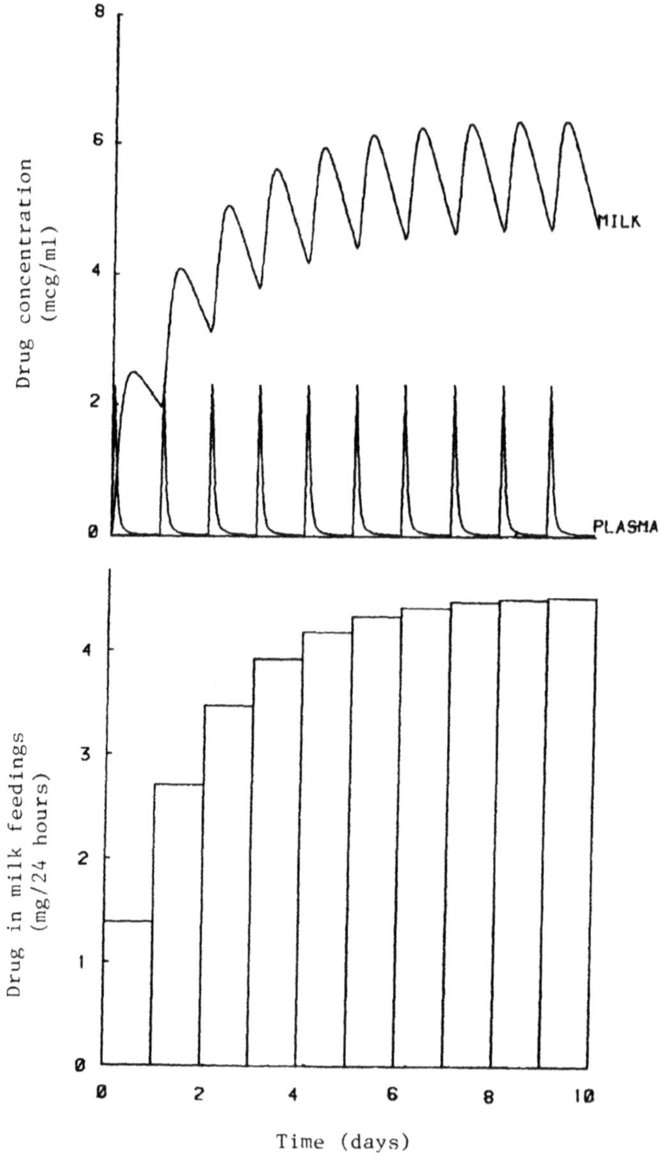

Figure 8. Amount of drug received by a 5 kg infant over each 24 hour period (5 feeds per day) under repeated dose conditions of one dose per day. A sequential three compartment open model with oral absorption is assumed for this simulation. The parameters are: $k_a = 1.50$ hr^{-1}, $k_{10} = 0.80$ hr^{-1}, $k_{12} = 0.25$ hr^{-1}, $k_{21} = 0.22$ hr^{-1}, $k_{23} = 0.18$ hr^{-1}, $k_{32} = 0.05$ hr^{-1}, $f = 0.80$, $X_o = 10.0$ mg/kg, $V_p = 1.5$ l/kg, $V_m = 0.3$ l/kg, $t_{\frac{1}{2}\gamma} = 31.2 \cdot$hr.

For the example shown in Figure 8, we considered that less than 10% of the maternal weight adjusted dose would be non toxic to the infant. The percent of a maternal (or pediatric) drug dose considered as safe for the nursing infant depends upon its intrinsic pharmacologic action and the rate of clearance by the infant (i.e., the extent of accumulation). Risk assessment also depends upon our ability to define the milk drug concentration (either by predictive population kinetics or by fitting of individual data) and knowledge of the volume of milk consumed. Once known, the amount of drug an infant receives can be calculated and potential toxicity assessed in relation to these amounts. Very low amounts of drug in milk (either predicted or observed) render suspect an alleged toxicity in the infant, whereas a high predicted amount can be used to substantiate an observed toxicity in the nursing infant. Maturational liability of the nursing infant modifies these assessments in some cases.

CONCLUSION

Our review of the literature pertaining to the general issue of infant effects from drugs or chemical toxins in milk have highlighted major deficiences in our knowledge. Simple criteria to validate a proposed effect on the nursing infant have not been fulfilled in the majority of reports. The relationship between infant effect and administered dose via milk is essential for documentation of the risk of breast feeding by medicated mothers. Estimates of this dose in milk using the traditional M/P ratio are flawed; they fail to control for variations associated with time after dosing and number of doses.

We recommend that future studies address both infant effect and drug dose in milk. Both objective and prospective observation of these infants is mandatory. Their exposure to drug or toxin from milk must be quantified by drug or toxin measurement in the infant or at least in the milk. The observed effect must be consistent with the known pharmacological effect of the drug or toxin in question. An exception is pharmacologic vulnerability exhibited by/in the developing organism. Prospective studies of drug excretion into breast milk must record sufficient concentration data points in maternal plasma and milk after both single and repeated dosing. These concentrations can be used by the predictive kinetic approach to generate reasonably accurate estimates of drug or toxin amounts in milk at any time that feeding occurs after a dose. Such predictions may be applied to other breast feeding women if sufficiently accurate population parameters are available. A combination of these approaches will substantiate observed or potential adverse effects in the nursing infant exposed to a drug or toxin in breast milk.

ACKNOWLEDGEMENT

We appreciate the high quality of manuscript preparation undertaken by Mrs. Carole Webb.

REFERENCES

1 J. T. Wilson, R. D. Brown, D. Cherek, J. W. Dailey, B. Hilman, P. C. Jobe, B. R. Manno, J. E. Manno, H. M. Redetzki, and J. J. Stewart, Drug Excretion in Breast Milk: Principles, pharmacokinetics and projected consequences, Clin. Pharmacokin. 5:1-66, (1980).

2 J. T. Wilson, R. D. Brown, D. Cherek, J. W. Dailey, B. Hilman, P. C. Jobe, B. R. Manno, J. E. Manno, H. M. Redetzki, and J. J. Stewart, Drugs in Breast Milk, ADIS Pres, Australia (1980).

3 J. T. Wilson. Determinants and consequences of drug excretion in breast milk, Drug Metab. Rev. 14:619-652 (1983).

4 J. T. Wilson. Contamination of human milk by drugs and chemicals, Nutrition and Health, 2:191-201 (1983).

5 J. T. Wilson. Chemical contaminants of cow's milk and human milk, in: "Health Hazards of Milk," D. L. J. Freed, ed., Bailliere Tindall, London, (1984).

6 J. T. Wilson, R. D. Brown, J.L. Hinson, and J. W. Dailey, Pharmacokinetic pitfalls in the estimation of the breast milk/plasma ratio for drugs, Ann. Rev. Pharmacol. 25:667-689 (1985).

7 J. T. Wilson, I. J. Smith, J. L. Hinson, T. W. Woods, V. A. Johnson, and R. D. Brown, Fundamental kinetics of drug excretion in breast milk, in: "Human Lactation 2," Plenum Publishing Corporation, New York, (1986).

8 J. T. Wilson, J. L. Hinson, R. D. Brown, and I. J. Smith, A comprehensive assessment of drugs and chemical toxins excreted in breast milk, in: "Human Lactation 2," Plenum Publishing Corporation, New York, (1986).

9 J. T. Wilson, J. L. Hinson, V. A. Johnson, Thomas W. Woods, I. J. Smith, and R. D. Brown, Pharmacologic factors contributing to variance in the milk to plasma (M/P) ratio for acetaminophen in the goat, Dev. Pharmacol. Ther. In Press (1986).

10 J. T. Wilson, R. D. Brown, and J. L. Hinson, A comprehensive assessment of drugs in breast milk, III World Conference on Clinical Pharmacology and Therapeutics, Stockholm, 1986.

11 I. Matheson, K. Kristensen, and P. K. M. Lunde, Breast feeding, maternal and infant drug use, III World Conference on Clinical Pharmacology and Therapeutics, Stockholm, 1986.

12 American Academy of Pediatrics (Committee on Drugs) The transfer of drugs and other chemicals into human breast milk, Pediatrics, 72:375-383 (1983).

13 R. D. Brown, and J. E. Manno, ESTRIP: A BASIC computer program for obtaining initial polyexponential parameter estimates, J. Pharm. Sci. 67:1687 (1978).

14 C. C. Peck, S. L. Beal, L. B. Sheiner, and A. I. Nichols, Extended least squares nonlinear regression: A possible solution to the choice of weights in analysis of individual pharmacokinetic data, J. Pharmacokin. Biopharm. 12:545-558 (1984).

15 C. C. Peck, L. B. Sheiner, and A. I. Nichols, The problem of choosing weights in nonlinear regression analysis of pharmacokinetic data, Drug Metab. Rev. 15:133-148, (1984).

16 K. Yamaoka, T. Nakagawa, and T. Uno, Statistical moments in pharmacokinetics, J. Pharmacokinet. Biopharm. 6:547-558 (1978).

VITAMIN K DEFICIENCY IN BREASTFED INFANTS

Rüdiger von Kries[1], Rudolf Tangermann[1], Martin Shearer[2], and Ulrich Göbel[3]

[1]University of Düsseldorf, Kinderklinik, Abt. für allgemeine Pädiatrie, Neonatologie und Gastroenterologie
[2]Department of Haematology, Guy's Hospital and United Medical and Dental Schools of Guy's and St. Thomas's Hospitals, London
[3]University of Düsseldorf, Kinderklinik, Abt. für pädiatrische Hämatologie und Onkologie

Vitamin K is required for gamma carboxylation of a number of proteins, most of which are involved in haemostasis. Bleeding is the main clinical problem observed in vitamin K deficiency.

Most cases of vitamin K deficiency bleeding occur in the newborn period and early infancy. In relation to the clinical picture and the onset of bleeding, three forms of vitamin K deficiency bleeding in infancy have been recognized[1]: early vitamin K deficiency bleeding, classical haemorrhagic disease of the newborn and late onset vitamin K deficiency haemorrhage.

Maternal drug ingestion is common in babies with early vitamin K deficiency bleeding. In such infants, bleeding usually occurs on the first day of life, and often presents as intracranial haemorrhage. Classical vitamin K deficiency haemorrhagic disease of the newborn is a haemorrhagic diathesis confined to the first week of life. Common bleeding sites for classical haemorrhagic disease are the gastrointestinal tract, skin, nose and bleeding following circumcision, whereas intracranial haemorrhage is rarely observed. Late onset vitamin K deficiency haemorrhage occurs in babies aged 3 - 26 weeks, and intracranial haemorrhage accounts for more than 50% of the bleeding episodes.

A pathogenetic role for both classical haemorrhagic disease of newborn and late onset vitamin K deficiency haemorrhage has been attributed to breastfeeding. In this paper, the role of breast feeding and other factors in the genesis of these disorders will be examined.

Classical Haemorrhagic Disease of Newborn

A 15 to 20 fold higher incidence of moderate and severe vitamin K deficiency bleeding in breastfed newborns as compared to those fed cow's milk based formula has been reported[2].

Subsequent coagulation studies on babies within an experimental setting revealed an influence of feeding on clotting factors[3]. A recent clinical study in a nursery setting[4] on healthy 5 and 6 day old newborns, not given vitamin K pro- phylaxis at birth showed significantly different distribution patterns of Factor II clotting times according to the feeding regimen. None of the babies on either supplementary or exclusive cow's milk formula (n = 36) had factor II clotting times longer than 34 seconds, detectable in 38 % of the exclusively breastfed babies (n = 168) . Factor II clotting times, however, do not solely represent the baby's vitamin K status, as they also reflect liver maturity and function.

Diagnosis of vitamin K deficiency with clotting assays requires demonstration of the corrective effect of vitamin K on vitamin K dependent coagulation factors. Immediate and more sensitive diagnosis of vitamin K deficiency may be attained with the demonstration of non-carboxylated precursors of vitamin K dependent proteins, 'Proteins Induced by Vitamin K Absence' (PIVKA), such as the prothrombin PIVKA: PIVKA II. The classical test for PIVKA II is crossed immunoelectrophoresis. We modified this test for capillary blood for use in studies of vitamin K deficiency in newborns[5].

The PIVKA II were detected in 95 out of 191 healthy babies aged 5 to 6 days. Most of the PIVKA II positive babies were exclusively breastfed. The PIVKA II detection rate in exclusively breastfed babies was significantly higher than in babies who received cow's milk formula feeding (Tab. 1).

Tab. 1: PIVKA II detectability in relation to the kind of feeding

FEEDING GROUPS	PIVKA II +	PIVKA II –	
EXCLUSIVELY BREASTFED	88	74	# (vs. breastfeeding + formula); * (vs. formula only)
BREASTFEEDING + FORMULA	6	9	+ (vs. formula only)
FORMULA ONLY	1	13	

(*: $p<0.0001$, +: $p>0.1$, #: $p<0.01$; chi^2-test)

Thus, exclusive breastfeeding appears to be a strong risk factor for vitamin K deficiency in the neonatal period. Lower vitamin K concentrations in human milk as compared to cows milk[6] have been postulated to explain this association[2,3].

Newborns in central Europe, however, will rarely be fed pure cow's milk, and little is known about the vitamin K concentrations of both human colostrum and cow's milk formula not supplemented with vitamin K. We therefore analysed the vitamin K concentrations in maternal milk in relation to the stage of lactation and in some infant formulas commonly used in nurseries in West Germany. For vitamin K determination, 2 ml of milk were extracted with hexane. The lipid extract was then purified on sep-pak cartridges and by semi-preparative HPLC. Reverse phase HPLC and dual electrode electrochemical detection of vitamin K were then used for subsequent chromatography. Using this method, vitamin K_1 was found to be the main vitamin K component of maternal milk[7].

Vitamin K_1 in milk samples from 9 mothers was quantified during the course of lactation. The vitamin K_1 concentrations in colostrum (day 1 to 5) were significantly higher as compared to milk samples collected at day 22 to 36 from the same mothers (Fig. 1). In one cow's milk formula the vitamin K1 concentration (3.8 ng/ml) was within the range of human colostrum, whereas the concentrations were much higher in two other preparations (7.8 and 10.8 ng/ml).

During the period, when the PIVKA II study was performed, babies were fed the formula with vitamin K1 concentrations within the range of human colostrum. Thus low concentrations of vitamin K in the maternal milk alone could not account for the high PIVKA II detection rate in breastfed babies.

The baby's vitamin K supply is determined by the milk intake and the vitamin K1 concentrations in the milk. Lactation often takes some days to establish, whereas formula is immediately available. To find out whether the milk intake was

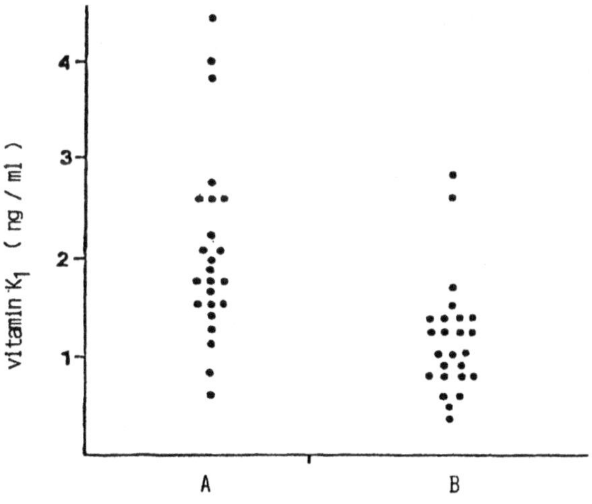

Fig. 1. Vitamin K_1 concentrations in human milk obtained from 9 mothers during the colostral phase (A): n=25, median 1.8 ng/ml and at day 22 to 36 (B): n=25, median 1.1 ng/ml. Differences in the concentrations of vitamin K_1 in the early and late samples of human milk were significant; $p < 0.01$ Wilcoxon U test.

related to the PIVKA II detection rates, we analysed the feeding records of 79 exclusively breastfed newborns, who had been weighed before and after each feed. The cumulative distributions revealed significantly lower mean daily milk intakes for PIVKA II positive babies than for those, in whom PIVKA II was not detected[8].

Feeding records of babies receiving either supplementary or exclusive cow's milk formula showed that PIVKA II was only detected, when formula feeding was initiated on day 3 or 4 of life as a consequence of insufficient maternal milk production. In these babies, the mean daily intake of human milk and cow's milk formula during the first 5 days of life was often even below that for PIVKA II positive breastfed babies.

It must be concluded that babies receiving little or no milk during their first days of life are at risk of vitamin K deficiency haemorrhage, unless prophylactic vitamin K is administered at birth.

Late Onset Vitamin K Deficiency Haemorrhage

Late onset vitamin K deficiency bleeding occurs at a period when lactation is fully established. Most of these babies are in their third to eighth week of life and appear healthy before the bleeding occurs. Still the disease is virtually confined to breastfed babies[1].

For German milk powder preparations, not supplemented with vitamin K_1, concentrations of the vitamin ranged between 5.7 and 10.8 ng/ml and were much higher than in mature human milk (range 0.4 - 2.8 ng/ml). Even higher vitamin K_1 concentrations have been detected in formula supplemented with vitamin K[9].

Consequently, breastfed infants receive less vitamin K_1 than those fed cow's milk formula. Due to low vitamin K_1 intake in exclusively breastfed infants, subclinical vitamin K deficiency might be more common than symptomatic vitamin K deficiency.

To assess the potential relevance of low vitamin K concentrations in maternal milk for hemostasis, we analysed factor II clotting times in 165 five and six week old healthy babies not given vitamin K prophylaxis at birth[4]. The distribution pattern of prothrombin clotting times in exclusively breastfed babies (n = 78) were very similar to those in babies receiving supplementary or exclusive formula feeding (n=87). The failure to detect subclinical vitamin K deficiency in exclusively breastfed babies might have been due to either absence of such deficiency or insufficient sensitivity of the method.

In a subsequent study on 202 healthy 5 and 6 week old infants, we used a more sensitive marker, crossed immuno-electrophoresis for PIVKA II, to detect vitamin K deficiency. One of 113 exclusively breastfed babies, but none of 89 babies on either exclusive or supplementary cow's milk formula showed evidence of subclinical vitamin K deficiency[10].

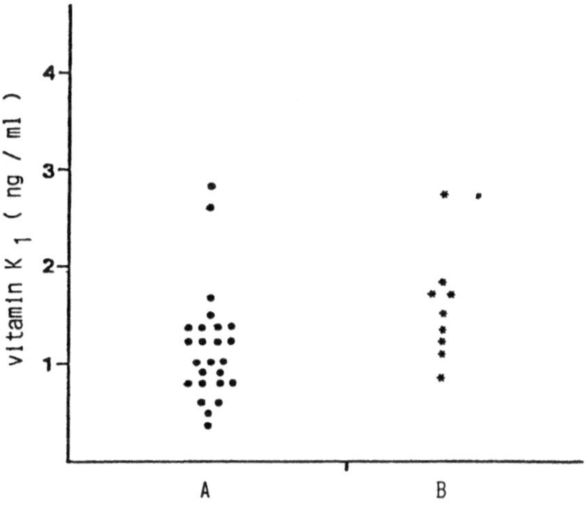

Fig. 2. Vitamin K_1 in mature human milk; controls, n=25 (A); milks fed to babies with late onset vitamin K deficiency haemorrhage, n=9 (B).

Even higher PIVKA II detection rates in exclusively breastfed infants were reported with a monoclonal PIVKA II antibody assay, which is more sensitive than crossed immunoelectrophoresis. With that assay in exclusively breastfed babies, PIVKA II detection rates were 5% on day 30, 2% on day 60 and 3% on day 90. PIVKA II was never detected in babies receiving cow's milk formula[14]. Although marginal vitamin K supply in exclusively breastfed infants appears to be more common, the incidence of vitamin K deficiency bleeding is low and the triggering factors accounting for haemorrhage in the individual baby are not well understood.

The vitamin K_1 concentrations in human milk were found to vary widely and some mothers might have consistently low vitamin K_1 concentrations in their milk. Analysis of maternal milk samples fed to 10 babies with late onset vitamin K deficiency haemorrhage in Japan[12] did not support this hypothesis. More than 20 cases of late onset vitamin K deficiency bleeding have been observed in Germany and all of the babies were exclusively breastfed. Milk of 9 of the mothers was available for vitamin K analysis (Fig. 2).

The vitamin K_1 concentrations in these milks, however, were within the range of controls, which consisted of human milk collected from women, whose infants did not have symptomatic vitamin K deficiency. Random milk samples had been used in Motohara's and our study. Investigations of the variability of the vitamin K_1 concentrations in human milk for 25 mature milk samples from one mother showed wide variations in the concentrations of vitamin K_1 (0.6 - 4.2 ng/ml). The vitamin K_1 concentrations in the random milk samples analyzed might not be representative for the baby's vitamin K_1 supply. Further studies to assess dietary deficiency in babies with late onset vitamin K deficiency bleeding should therefore be performed with milk collected over a longer period. Present data, however, suggest no evidence of an insufficient vitamin K supply due to low vitamin K concentrations in milks from mothers of babies with late onset vitamin K deficiency. Other pathogenetic factors should therefore be considered.

For some babies with late onset vitamin K deficiency bleeding underlying disease such as alpha-1-antitrypsin deficiency[13], bile duct atresia[14], hepatitis[15] and cystic fibrosis[16] has been demonstrated. Impaired bile acid circulation is a characteristic feature these diseases have in common, and bile acids are essential for vitamin K_1 absorption[17].

An insufficient vitamin K supply could also be a consequence of impaired absorption of the vitamin, and cholostasis, if present, could give a good explanation. Eleven out of 16 German babies with late onset vitamin K deficiency bleeding were found to have laboratory evidence of cholostasis[18]. Temporary malabsorption of vitamin K_1 associated with mild cholostasis has been described recently in a baby with late onset vitamin K deficiency bleeding[19].

Idiopathic late onset vitamin K deficiency bleeding has been differentiated from vitamin K deficiency bleeding secondary to underlying disease, and it has been suggested that idiopathic forms account for the majority of cases[1]. This certainly does not account for the German observations and several of the idiopathic cases may have been misclassified. Bhanchet et al.[20] reported a series of 96 "idiopathic" cases[1] from Thailand: 73 % had hepatomegaly and 27 % were jaundiced. Some kind of hepatopathy may well have been present in these babies. The clinical relevance of cholostasis and its relation to malabsorption of vitamin K_1 for late onset vitamin K deficiency bleeding should therefore be further evaluated.

In summary, low vitamin K_1 concentrations in mature human milk probably constitute a predisposing factor for late onset vitamin K deficiency haemorrhage. Subclinical vitamin K deficiency appears to be more common in exclusively breastfed infants. As yet there is no evidence, however, for symptomatic late onset vitamin K deficiency solely due to insufficient dietary intake of the vitamin. The potential role of vitamin K malabsorption due to liver disease deserves further studies.

References

1. P. A. Lane, W. E. Hathaway, Vitamin K in infancy, J. Pediatr. 106:351 (1985)

2. J. M. Sutherland, H. I. Glueck, G. Gleser, Hemorrhagic disease of the newborn: Breast feeding as a necessary factor in the pathogenesis, Am. J. Dis. Child 113:524 (1967)

3. W. J. Keenan, Th. Jewett, H. I. Glueck, Role of feeding and vitamin K in hypoprothrombinemia of the newborn, Am. J. Dis. Child 121:524 (1971)

4. U. Göbel, R. v. Kries, S. Bewersdorff, B. Henninghausen, E. Schmidt, Erniedrigte Prothrombin-Gerinnungsactivitäten bei gestillten Kindern? Klin. Pädiat. 198:13 (1986)

5. R. v. Kries, Ch. Zenses, U. Göbel, Immunoelectrophoretic determination of PIVKA II in capillary plasma, Haemostasis 15:42 (1985)

6. H. Dam, J. Glavind, H. E. Larsen, P. Plum, Investigations into the cause of physiological hypoprothrombinemia in newborn children. IV. The vitamin K content of woman's milk and cow's milk. Acta Med. Scand. 112:210 (1942)

7. R. v. Kries, M. J. Shearer, P. T. McCarthy, M. Haug, G. Harzer, U. Göbel, Vitamin K_1 content of maternal milk: Influence of the stage of lactation, lipid composition and vitamin K_1 supplements given to the mother, Pediatr. Res. (submitted for publication)

8. R. v. Kries, U. Göbel, B. Maase, Vitamin K deficiency in the newborn, Lancet II (letter):728 (1985)

9. Y. Haroon, M. J. Shearer, S. G. Gunn, G. McEnery, P. Barkhan, The content of phylloquinone (vitamin K_1) in human milk, cow's milk and infant fomula determined by high-performance liquid chromatography, J. Nutr. 112:1105 (1982)

10. R. v. Kries, B. Maase, A. Becker, U. Göbel, Latent vitamin K deficiency in healthy infants? Lancet II (letter):1421 (1985)

11. J. Widdershoven, K. Motohara, F. Endo, I. Matsuda, L. Monnens, Influence of the type of feeding on the presence of PIVKA II in infants, Helv. paediat. Acta, 41:25 (1986)

12. K. Motohara, M. Matskura, I. Matsuda, K. Iribe, T. Ikeda, Y. Kondo, A. Yonekubo, Y. Yamamoto, F. Tsuchiya, Severe vitamin K deficiency in breastfed infants, J. Pediatr. 105:943 (1984)

13. P. L. Hope, M. A. Hall, P. Millward-Sadler, Alpha$_1$-antitrypsin deficiency presenting as a bleeding diathesis in the newborn, Arch. Dis. Child 57:68 (1982)

14. Y. Fujimura, Y. Okubo, T. Sakai, M. Sugimoto, T. Takase, A. Yoshioka, H. Fukui, Precursor proteins PIVKA II, IX and X in the plasma of patients with 'hemorrhagic disease of the newborn', Haemostasis 14:211, (1984)

15. J. R. Poley, G. B. Humphrey, Bleeding disorder in an infant associated with anicteric hepatitis, Clin. Pediatr. 13:1045 (1974)

16. O. L. Torstensen, G. B. Humphrey, J. R. Edson, Cystic fibrosis presenting with severe hemorrhage due to vitamin K malabsorption: A report of 3 cases, Pediatrics 45:857 (1970)

17. M. J. Shearer, A. McBurney, P. Barkhan, Studies on the absorption and metabolism of phylloquinone (vitamin K_1), Vitam Horm. 32:523 (1974)

18. A. H. Sutor, H. Pollmann, R. v. Kries, H. Thais, F. Schindera, U. Göbel, W. Künzer, Vitamin K Mangelblutungen bei Säuglingen jenseits der Perinatalperiode, weitere Beobachtungen, Monatsschr. Kinderheilk. 132:608

19. R. v. Kries, A. Reifenhäuser, U. Göbel, P. McCarthy, M. Shearer, P. Barkhan, Late onset haemorrhagic disease of newborn with temporary malabsorption of vitamin K_1, Lancet I(letter): 1035, (1985)

20. P. Bhanchet, S. Tuchinda, P. Hathirat, P. Visudhiphan, N. Bhamarapharati, S. Bukkavesa, A bleeding syndrome in infants due to acquired prothrombin complex deficiency, Clin. Pediatr. 16:992 (1977)

TRANS-FATTY ACIDS IN HUMAN MILK AND INFANT PLASMA AND TISSUE

Berthold Koletzko*, Maria Mrotzek and Hans Joachim Bremer

Universitäts-Kinderklinik Düsseldorf, West Germany

INTRODUCTION

During the past decades, human dietary consumption of trans-isomers of unsaturated fatty acids (t-FAs) has increased considerably in industrialized countries. This increase is mainly due to the extensive use of the catalytic hydrogenation process of vegetable oils developed by Wilhelm Normann in 1902. Catalytic hydrogenation is performed in order to harden liquid oils and to improve their oxidative and flavour stability for their use in margarines, shortenings and a large variety of food products. Depending on the conditions of the process, usually positional and geometric isomerization of the naturally occurring cis-unsaturated FAs is favoured over saturation reactions; thus in margarines made with hydrogenated fats, 10-50% of all fatty acids can be t-FAs (1,2). The use of hydrogenated oils in visible fats and processed foods has become so widespread in North America and Western Europe, that it is almost impossible to avoid their consumption. They account for 44% of the average dietary fat intake in the USA (3). Dietary t-FAs are also found in body and milk fats of ruminants, as a result of microbial fat-hydrogenation in the animal's forestomach. However, these products appear to contribute only a small proportion (5% in the USA) of total t-FA-consumption (4).

In comparison to cis-FAs, such as the essential FAs and the non-essential FAs synthesized by man, t-FAs differ in tertiary structure (Fig. 1) and in physicochemical properties, and they lack any essential FA-activity. A number of untoward biological effects of t-FA-consumption have been proposed, although their significance is controversial. However, since the presence of t-FAs in human milk was first demonstrated (5) concern has risen as to their metabolic fate and side effects in the growing infant, which is especially sensitive to environmental influences.

* Present address: The Hospital for Sick Children, 555 University Ave., Toronto, Ontario, Canada, M5G 1X8

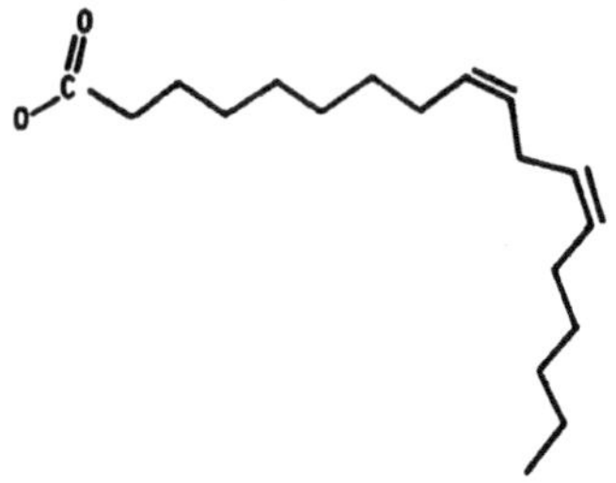

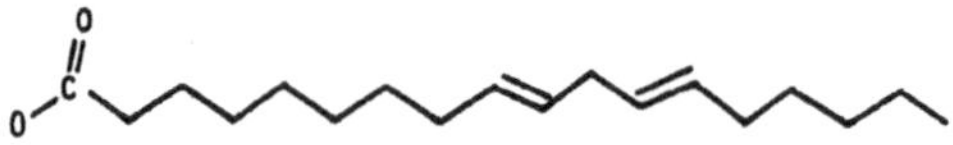

Linoleic acid
(18:2ω6, cis, cis)

Linolelaidic acid
(18:2ω6, trans, trans)

Figure 1
Tertiary structure of linoleic and linolelaidic acids.

BIOLOGICAL EFFECTS OF TRANS-FATTY ACIDS

Considerable differences in the metabolism and physiologic effects of cis- and t-FAs have been demonstrated in mammals (table 1). Not only is the rate of mitochondrial ß-oxidation of t-FAs only half that of their cis-isomers with a resulting lower mitochondrial ATP-production (6,7), t-FAs also impede the oxidation of the cis-FAs (8). The decreased mitochondrial oxidation of t-FAs appears to be partly compensated for by a higher proportion of peroxisomal chain-shortening (9), so that the overall oxidation in the rat is not decreased (7). T-FAs seem to increase the essential FA-requirements and interfere with the synthesis of desaturation-elongation products of essential FAs (10-13). These side effects could be especially serious in the growing infant, whose high requirements of essential FA for tissue synthesis contrast with marginal reserves and a low capacity for synthesis of the desaturation-elongation products (14). T-FAs also inhibit the synthesis of prostaglandins and thromboxanes through a reduction of cyclooxygenase activity (12) and resulting low serum and tissue levels of prostaglandins have been demonstrated (15, 16). The incorporation of t-FAs, which have strikingly high melting points, in cell membranes may decrease membrane fluidity and alter both permeability (17) and resistance to osmotic stress (18). Some studies suggest unfavourable effects of t-FAs on serum cholesterol and atherogenesis, but others do not confirm these effects (12). Based on epidemiological data, a correlation between t-FA consumption and carcinogenesis has been proposed (20).

Table 1
Biological effects of trans-fatty acids

↓	Mitochondrial ß-Oxidation and ATP-synthesis
↑	Essential fatty acid requirements
↓	Desaturation and elongation of dietary polyunsaturates
↓	Arachidonic acid content of body lipids
↓	Cyclooxygenase activity
↓	Serum and tissue prostaglandin levels
↓	Membrane fluidity, alteration of membrane functions
↑	Serum cholesterol levels?
↑	Atherogenesis?
↑	Carcinogenesis?

DETERMINATION OF TRANS-FATTY ACIDS

The analysis of trans fatty acids demands very careful preparation and handling of the samples. Spontaneous isomerization from trans- to cis-fatty acids and thus false low results may be induced by light, heat and long-term storage. Comparison of published analysis results on t-FAs from different laboratories is difficult, as the amounts detected vary considerably with the analytical technique used. Infrared spectrometry, formerly a widespread method for the determination of the degree of trans unsaturation in a sample, does not give any information about the position of the double bond and the configuration of the molecule. This method is particularly inaccurate at low t-FA-concentrations. Gaschromatographic analysis became the method of choice when the introduction of highly polar columns allowed isomer-separation. Packed columns, as we have emphasized earlier (21), tend to underestimate trans-fatty acids secondary both to detection of a smaller number of trans-isomers and to underevaluation of the individual trans-peak when the small and broad trans-peak is partly hidden under the tall cis-peak. More standardization of analytical procedures should be reached; it would be helpful if a generally acceptable quantitative reference standard was available.

For the results presented here sample preparation and capillary gaschromatography were performed as described earlier (22) yielding a very satisfying coefficient of variation of 3% for t-FA determination. The use of human milk and infant samples was approved by the Ethics Commission of the Medical Faculty of the University of Düsseldorf as part of different study protocols.

TRANS-FATTY ACIDS IN HUMAN MILK LIPIDS

We analyzed the t-FA content in total lipids of 24-hour collections of manually expressed hindmilk of 15 women from the area of Düsseldorf (23). All were on ad libitum diets typical for this country, as assessed by 24-hour dietary recall. Each women was in the third or fourth month of lactation and was fully breastfeeding one infant born at term. Hindmilk samples were chosen, because their collection were easier for the mothers. Moreover t-FA content of breastmilk does not change during feeding (22). The mean t-FA content of total lipids from the German milk samples was 4.14% of all FA detected (Table 2) and thereby is remarkably similar to the estimated content of approximately 4% t-FAs in the average German diet (24). The major proportion of all t-FAs was contributed by 18:1t, followed by 16:1t and 14:1t, again reflecting the composition of the diet. Only small amounts of the geometric isomers of linoleic acid (18:2) were found. The higher value for 18:2ct, as compared to 18:2tc, may result in part from consumption of ruminant fats; 18:2ct-production from linoleic acid is favoured by the rumen organism Butyrivibrio fibrosolvens (25). The presence of 20:1t in human milk was demonstrated for the first time.

In contrast to the samples collected in Germany, we found strikingly low values for both the sum of t-FAs as well as all isomers in breastmilk of 15 rural Sudanese women, each of whom manually expressed one late morning sample in the third and fourth month of lactation (26). The difference in t-FA content is presumably due to a very low consumption of technically hardened fats by the African women. It has been shown that the dietary t-FA content is the main

Table 2

Trans-fatty acid content (wt. %, mean ± SD) of total lipids in mature human milk of 15 West German and 15 Sudanese women.

Fatty acid	German women	Sudanese women
14:1t	0.21 ± 0.06	0.08 ± 0.04 *
16:1t	0.56 ± 0.23	0.10 ± 0.04 *
18:1t	2.94 ± 0.88	0.48 ± 0.26 *
18.2tt	0.15 ± 0.05	0.04 ± 0.02 *
18.2ct	0.14 ± 0.04	0.05 ± 0.01 *
18:2tc	0.07 ± 0.05	N.D.
20:1t	0.16 ± 0.10	N.D.
sum t-FAs	4.14 ± 1.17	0.75 ± 0.36 *

* significant at $p < 0.0005$

determinant of the trans-concentration in breastmilk. Significant changes of maternal t-FA intake were reflected in milk within 1-3 days (5,27,28). T-FA content of breastmilk doubled when nothing but the margarine used by the mother was changed from a low- to a high-trans product (29). A linear correlation between t-FA content of the mother's diet and the breastmilk has been suggested (28). However, maternal weight loss during lactation may also have an influence. Apparently significant amounts of t-FAs can be liberated from adipose tissue stores and used for mammary gland lipid synthesis (29).

When comparing our results with other published data on t-FA content of human milk (Table 3), differences in methodology must be taken into account (cf. above). Both in the USA and in the FRG, the 18:1t values obtained with packed gaschromatographic columns were lower than those obtained with capillary methods (23,28). The sums of t-FAs derived from between 2 and 6 isomers are also hardly comparable. The only study in the USA using capillary gaschromatography (28) revealed a mean 18:1t content 1.6-fold higher than in the German samples. Assuming a similar ratio of total t-FAs to 18:1t, as found

Table 3

Published trans-fatty acid content (%) in human milk lipids.

Country	Subjects	Method[1]	18:1t	Sum t-FAs	Reference
USA	3	p.GC	2.1-4.0	?	(30)
USA	11	p.GC	3.44 ± 0.67	3.88 ± 0.73	(27)
USA	11	p.GC	4.48 ± 1.33	4.87	(31)[2]
USA	10	p.GC	3.38 ± 1.78	?	(32)[2]
USA	8	c.GC	4.76 ± 2.06	?	(28)
USA	?	p.GC	3.67 ± 2.28	4.24	(33)
Canada	?	IR?	?	6-18	(34)
Canada	14	c.GC	2.1-4.1	2.3-4.7	(29)
FRG	5	IR	?	1-4	(5)
FRG/UK	14	p.GC	N.D.	0.59	(35)
FRG	15	c.GC	2.94 ± 0.88	4.14 ± 1.17	This study
Sudan	15	c.GC	0.48 ± 0.26	0.75 ± 0.36	This study

[1]p.GC and c.GC = Gaschromatography with packed and capillary columns respectively, IR = Infared spectrometry.
[2]Analysis of milk triglycerides.

in the German samples (1.4), the average total t-FA-content may be calculated from Craig-Schmidt's results as 6.7%. This value is similar to the estimated average dietary intake of t-FAs in the USA, i.e. approximately 6.5 - 8% (20,36,37). Extremely high t-FA values published for Canadian milk-samples in abstract form only (34) could not be confirmed in a later Canadian study (29).

After separation of the major lipid classes of human milk by thin-layer-chromatography, we found the highest content of t-FAs in triglycerides (Table 4). The value is almost equal to the total lipid content, because triglycerides contribute 98-99% to the milk lipids. Less than 40% of the t-FA content of triglycerides were detected in sterolesters and phospholipids, which are mainly found in the membrane of the milk fat globule. A similar trend with regard to the relative distribution of t-FAs between triglycerides and phospholipids was found by Hundrieser et al. (32), who also demonstrated differences in t-FA content between phosphatidylcholine and phosphoethanolamine. Thus, t-FAs are not randomly distributed within the various lipid classes in human milk.

Table 4
Sum of trans-fatty acids (wt. %, mean ± SD) in the main lipid fractions of mature human milk of 5 German women.

Total Lipid	Triglyceride	Sterolester	Phospholipid
3.80 ± 0.86	3.85 ± 0.96	1.35 ± 0.66*	1.45 ± 0.60*

*significant at $p < 0.005$ vs. total lipid and triglyceride

TRANS-FATTY ACIDS IN INFANTS' PLASMA LIPIDS

To our knowledge, no data have been published before on the absorption and metabolism of t-FAs by the human infant. We determined the t-FA content of plasma lipids of 3-week old, healthy premature infants, who had lipid analyses performed for other reasons. In 10 infants fed exclusively human milk, t-FAs were found in all major lipid fractions (Table 5) demonstrating that the human infant absorbs t-FAs and incorporates them into endogenous lipids. The highest proportion was found in triglycerides, where the relative distribution of the individual isomers was similar to breastmilk (cf. Table 2). Phospholipids showed a similar relative distribution of the individual t-isomers, but a somewhat lower content. In contrast, the percentage of most isomers in the cholesterol-esters was similar to triglycerides, but a far lower 18:1t value resulted in a lower sum of t-FAs. This finding suggests a negative selectivity for 18:1t-esterification with cholesterol.

In Table 5, the lipid composition of the infants fed human milk is compared with that of 7 infants raised with an adapted formula containing 1.94% t-FAs and thus consuming about half of the average content detected in human milk (cf. Table 2). The t-FA content of plasma triglycerides and sterolesters was significantly lower in the formula group. The relative isomer distribution was similar to the infants fed breastmilk, again showing a strikingly low 18:1t value in sterolesters. In contrast to triglycerides and sterolesters, no clear difference was found between the two groups in the trans-values of phospholipids. Obviously, the relative t-FA-content of plasma

Table 5
Trans-fatty acids (wt. %, mean ± SD) in plasma lipid fractions of 21-day old premature infants fed human milk (n =10) or formula (n=7).

	HUMAN MILK	FORMULA
Triglycerides		
14:1t	0.21 ± 0.04	0.17 ± 0.05
16:1t	0.33 ± 0.12	0.17 ± 0.03 *
18:1t	1.75 ± 0.41	0.94 ± 0.12 *
18:2tt	0.11 ± 0.06	0.06 ± 0.03
18:2ct	0.10 ± 0.04	0.14 ± 0.04 *
18:tc	0.05 ± 0.03	0.13 ± 0.05 *
sum t-FAs	2.55 ± 0.47	1.62 ± 0.11 *
Sterolesters		
14:1t	0.39 ± 0.15	0.26 ± 0.10 *
16:1t	0.27 ± 0.15	0.16 ± 0.12 *
18:1t	0.59 ± 0.23	0.30 ± 0.15
18:2tt	0.15 ± 0.11	0.19 ± 0.18
18:2ct	0.13 ± 0.06	0.18 ± 0.11 *
18:tc	0.03 ± 0.04	0.07 ± 0.12 *
sum t-FAs	1.55 ± 0.37	1.21 ± 0.57 *
Phospholipids		
14:1t	0.11 ± 0.07	0.15 ± 0.08
16:1t	0.17 ± 0.11	0.10 ± 0.07
18:1t	1.14 ± 0.31	1.06 ± 0.31
18:2tt	0.03 ± 0.02	0.03 ± 0.02
18:2ct	0.04 ± 0.03	0.04 ± 0.04
18:2tc	0.02 ± 0.02	0.09 ± 0.08
sum t-FAs	1.52 ± 0.30	1.46 ± 0.46

*significant at $p < 0.05$

sterolesters and phospholipids in the human infant is not a simple linear function of the dietary intake; rather it seems to be subject to a more complex metabolic regulation.

This assumption is further substantiated by the analysis of plasma lipid fractions from German and Nigerian infants and toddlers aged 5 to 24 months (38). The pattern of t-FA distribution among the main lipid classes in both groups (Table 6) is very similar to the one found in the premature infants. The significantly lower trans values in the Nigerian group presumably reflect a lower t-FA content in the breastmilk and weaning foods consumed by the African children.

Table 6
Sum of trans-fatty acids (wt. %, mean ± SD) in plasma lipids of 8 Nigerian and 17 West German infants and toddlers.

	Triglycerides	Sterolesters	Phospholipids
German	1.95 ± 0.69	1.37 ± 0.59	1.38 ± 0.60
Nigerian	1.20 ± 0.61*	0.76 ± 0.26*	0.83 ± 0.38*

* significant at $p < 0.05$ vs. German

TRANS-FATTY ACIDS IN ADIPOSE TISSUE

Adipose tissue contains the major portion of the body lipids. It has been suggested that the fatty acid composition of adipose tissue lipids correlates well with the average dietary intake of infants (39,40) but no data on its t-FA content in human infants has been published before. We analyzed the total lipid composition of specimens obtained from subcutaneous tissue of the abdominal wall during autopsy of 6 infants who had died at age 13.0 ± 1.6 weeks from Sudden Infant Death Syndrome (Table 7). Two of the children had been breastfed, the other four had been fed mixed diets. The t-FA content and the relative distribution of the individual isomers were similar to the values observed in the triglyceride fraction of the German infants and toddlers (Table 6), and lie between the triglyceride content of the two premature groups fed human milk and formula (Table 5). This finding suggests that t-FAs may be interchangeable between plasma triglycerides and adipose tissue without the presence of a selective process.

Table 7

Trans-fatty acids (wt. %, mean ± SD) in adipose tissue of 6 German infants having died from Sudden Infant Death Syndrome.

Fatty acid	
14:1t	0.12 ± 0.05
16:1t	0.26 ± 0.18
18:1t	1.46 ± 0.88
18:2tt	0.12 ± 0.07
18:2ct	0.13 ± 0.04
18:tc	0.06 ± 0.04
20:1t	0.02 ± 0.02
sum t-FAs	2.14 ± 1.17

The proportion of 18:1t in infant adipose tissue (1.46 ± 0.88 %) is just slightly lower than the 1.89 ± 0.75% detected in 16 German males aged 50 to 75 years (24). The similar degree of t-FA contamination in the two groups raises concern regarding the possible greater susceptibility of the growing infant to some untoward effects of t-FA (cf. above). On the other hand, we conclude from these data that t-FAs do not accumulate in adipose tissue over time, as might be expected from the cited data on lower mitochondrial t-FA oxidation. Rather, the t-FA content of adipose tissue seems to be correlated to the long term average dietary intake. Both the German values are clearly lower than the average 3.4 and 4.0% of 18:1t found in the subcutaneous tissue of American adults (41,42), again reflecting the higher consumption of t-FAs in the USA.

TRANS-FATTY ACIDS IN MEMBRANE LIPIDS

The infant synthesizes a considerable amount of membranes for its growing tissues and the FA composition of these membranes has been demonstrated to be of major importance for their fluidity, their permeability and the activity of membrane-bound enzymes. The incorporation of t-FAs into biological membranes has been demonstrated to alter membrane-functions (cf. above). The dietary intake of t-FA, therefore, may be of special importance during early life, when the introduction of t-FAs into newly synthesized structural lipids with a

Table 8
Trans-fatty acids (wt. %, mean ± SD) in erythrocyte total lipids of five 21-day old premature infants fed human milk.

Fatty acid	
14:1t	N.D.
16:1t	0.17 ± 0.02
18:1t	0.68 ± 0.10
18:2tt	0.09 ± 0.04
18:2ct	0.10 ± 0.02
18:tc	0.09 ± 0.02
sum t-FAs	1.13 ± 0.10

very low turnover rate (e.g. those of the central nervous system) might influence their composition for the following decades. As an accessible example of a cell membrane, we analyzed the total lipid composition of washed erythrocytes, in which practically all lipid is membrane bound, in five 3-week old premature infants fed human milk. As shown in Table 8, 1.13 % t-FAs in fact were incorporated into erythrocyte lipids. It can be assumed that membranes of other tissues also contain t-FAs. We do not know, whether a proportion of t-FAs as small as 1% in membrane lipids affects membrane function. However, it can not be excluded that some infants may reach concentrations in their structural lipids that alter membrane properties in animal experiments (18).

CONCLUSIONS

1. The content of trans-fatty acids in human milk is markedly higher in the USA than in W. Germany, and very low in Sudan.

2. The concentration of trans-fatty acids is higher in the core than in the membrane lipids of the milk fat globule.

3. The main determinant for the trans-fatty acid content in human milk is the mother's recent dietary intake.

4. Dietary trans-fatty acids are absorbed and metabolized by the human infant. In plasma their content in sterolesters and phospholipids is lower than in triglycerides and varies with dietary intake in a non-linear fashion.

5. A negative selectivity for 18:1t esterification with cholesterol is suggested.

6. Trans-fatty acids seem to be interchangeable between plasma triglycerides and adipose tissue lipids. There is no accumulation over time in adipose tissue; rather, concentration may be related to the long term average dietary intake.

7. Trans-fatty acids are incorporated into the infants' membrane lipids, as demonstrated by red cell analysis.

8. These data raise concern but do not permit definite conclusions concerning the nutritional safety of trans-fatty acid intake by the human infant.

ACKNOWLEDGMENT

Supported by the Ministry of Science and Research, Northrhine Westphalia, Düsseldorf, FRG (V B5-FA 6).

REFERENCES

1. H. Heckers, F.W. Melcher, Trans-isomeric fatty acids present in West German margarines, shortenings, frying and cooking fats. Am J Clin Nutr 31: 1041 (1978)

2. J.L. Beare-Rogers, L.M. Gray, R. Hollywood, The linoleic and trans-fatty acids of margarines. Am J Clin Nutr 32: 1805 (1979)

3. Agricultural Statistics. US Govt Print Off Cat No AL 47, 977: 146 (1982)

4. J.B. Ohlrogge, Distribution in human tissues of fatty acid isomers from hydrogenated oils. In: Dietary fats and health, E.G. Perkins, W.J. Visek, eds., American Oil Chemists Society, Champaign, 359 (1983)

5. H.P. Kaufmann, F. Volbert, G. Mankel, Anwendung der IR-Spektrographie auf dem Fettgebiet V: Untersuchung von Milchfetten auf trans-ungesättigte Fettsäuren. Fette, Seifen, Anstrichmittel 63: 261 (1961)

6. R. de Schrijver, O. S. Privett, Energetic efficiency and mitochondrial function function in rats fed trans fatty acids. J Nutr 144: 1183 (1984)

7. T. Ide, M. Sugano, Strain dependence of the metabolism of cis- and trans-isomers of 9-octadecenoic acid in perfused liver and cell-free preparation in rats. Biochim Biophys Acta 877: 104 (1986)

8. F.A. Kummerow, Effects of isomeric fatty acids on animal tissue, lipid classes and atherosclerosis. In: Geometrical and positional fatty acid isomers, E.A. Emken, H.J. Dutton, eds., The American Oil Chemists Society, Champaign, 151 (1979)

9. H. Osmundsen, Peroxisomal ß-oxidation of long fatty acids: effects of high fat diets. Ann N Y Acad Sci 386: 13 (1982)

10. R.R. Brenner, R.O. Peluffo, Regulation of unsaturated fatty acids biosynthesis. Biochim Biophys Acta 176: 471 (1969)

11. J. Chern, J.E. Kinsella, The effects of unsaturated fatty acids on the synthesis of arachidonic acid in rat kidney cells. Biochim Biophys Acta 750: 465 (1983)

12. E.A. Emken, Nutrition and biochemistry of trans and positional fatty acid isomers in hydrogenated oils. Ann Rev Nutr 4: 339 (1984)

13. M.M. Mahfouz, T.L. Smith, F.A. Kummerow, Effect of dietary fats on desaturase activities and the biosynthesis of fatty acids in rat-liver microsomes. Lipids 19: 214 (1984)

14. B. Koletzko: Essentielle Fettsäuren: Bedeutung für Medizin und Ernährung. Akt Endokr Stoffw 7: 18 (1986)

15. J.E. Kinsella, D.H. Hwang, D.H. Yu et al., Prostaglandins and their precursors in tissues from rats fed on trans, trans-linoleate. Biochem J 184: 701 (1979)

16. D.H. Hwang, P. Chanmugam, R. Anding, Effects of dietary 9-trans, 12-trans linoleate on arachidonic acid metabolism in rat platelets. Lipids 17: 307 (1982)

17. F.A. Kummerow, Nutritional effects of isomeric fats: Their possible influence on cell metabolism or cell structure. In: Dietary fats and health, E.G. Perkins, W.J. Visek, eds., American Oil Chemists Society, Champaign, 391 (1983)

18. W.J. Decker, W. Mertz, Effects of dietary elaidic acid on membrane function in rat mitochondria and erythrocytes. J. Nutr 91: 324 (1967)

19. A. Csordas, K. Schauenstein, Temperature-dependant specificity of cis-trans isomeric fatty acid interaction with the erythrocyte membrane. Biochim Biophys Acta 856: 212 (1986)

20. M.G. Enig, R. J. Munn, M. Keeney, Dietary fat and cancer trends - a critique. Fed Proc 37: 2215 (1978)

21. B. Koletzko, M. Gwosdz, H.J. Bremer, Trans isomeric fatty acids in human milk in West Germany. In: Composition and physiological properties of human milk, J. Schaub, ed., Elsevier Science Publishers, Amsterdam, 225 (1985)

22. B. Koletzko, M. Mrotzek, H.J. Bremer, Fat content and cis- and trans-isomeric fatty acids in human fore- and hindmilk, In: Human lactation 2, maternal and environmental factors, M. Hamosh, A.S. Goldman, eds., Plenum Press, New York, 589 (1986)

23. B. Koletzko, M. Mrotzek, H.J. Bremer, Fatty acid composition of mature human milk in Germany, submitted

24. H. Heckers, F.W. Melcher, K. Dittmar, Zum täglichen Verzehr trans-isomerer Fettsäuren. Eine Kalkulation unter Zugrundelegung der Zusammensetzung handelsüblicher Fette and verschiedener menschlicher Depotfette. Fette, Seifen, Anstrichmittel 81: 217 (1979)

25. C.R. Kepler, W.P. Tucker, S.B. Tove, Biohydrogenation of unsaturated fatty acids. J Biol Chem 245: 3612 (1970)

26. Abbas Omer El Karib, Eisa Osman El Amin, H.J. Bremer et al., Fatty acid composition of human milk in healthy Sudanese women, in preparation

27. J.M. Aitchison, W.L. Dunkley, N.L. Canolty et al., Influence of diet on trans fatty acids in human milk. Am J Clin Nutr 30: 2006 (1977)

28. M.C. Craig-Schmidt, J.D. Weete, S.A. Faircloth et al., The effect of hydrogenated fat in the diet of nursing mothers on lipid composition and prostaglandin content of human milk. Am J Clin Nutr 39: 778 (1984)

29. J.E. Chappell, M.T. Clandinin, C. Kearney-Volpe, Trans fatty acids in human milk lipids: influence of maternal diet and weight loss. Am J Clin Nutr 42: 49 (1985)

30. M.F. Picciano, E.G. Perkins, Identification of trans-isomers of octadecenoic acid in human milk. Lipids 12: 407 (1977)

31. R.M. Clark, A.M. Ferris, N. Key et al., The identity of the cholesteryl esters in human milk. Lipids 15: 972 (1980)

32. K.E. Hundrieser, R.M. Clark, P.B. Brown, Distribution of trans-octadecenoic acid in the major glycerolipids of human milk. J Pediatr Gastroenterol Nutr 2: 635 (1983)

33. D.A. Finley, B, Lönnerdal, K.G. Dewey et al., Breast milk composition: fat content and fatty acid composition in vegetarians and non-vegetarians. Am J Clin Nutr 41: 787 (1985)

34. J.L. Beare-Rogers, E.A. Nera, Some nutritional aspects of partially hydrogenated oils. J Am Oil Chem Soc 53: 467A (1976)

35. G. Harzer, M. Haug, I. Dieterich et al., Changing patterns of human milk lipids in the course of lactation and during the day. Am J Clin Nutr 37: 612 (1983)

36. J.E. Kinsella, G. Bruckner, J. Mai et al., Metabolism of trans-fatty acids with emphasis on the effects of trans, trans octadecadienoate on lipid composition, essential fatty acid, and prostaglandins: an overview. Am J Clin Nutr 34: 2307 (1981)

37. M.M. van den Reek, M.C. Craig-Schmidt, J.D. Weete et al., Fat in the diets of adolescent girls with emphasis on isomeric fatty acids. Am J Clin Nutr 43: 530 (1986)

38. B. Koletzko, P.O. Abiodun, M.D. Laryea et al., Comparison of fatty acid composition of plasma lipid fractions in well-nourished Nigerian and German infants and toddlers. J Pediatr Gastroenterol Nutr 5:581 (1986)

39. M.J. Sweeney, J.N. Etteldorf, L.J. Throop et al., Diet and fatty acid distribution in subcutaneous fat and in the cholesterol-triglyceride fraction of serum of young infants. J Clin Invest 42: 1 (1963)

40. D. Francis, N. Koster, F. Quaade et al., Fatty acid composition of brown and white fat in children and adults. Dan Med Bull 18: 143 (1971)

41. J.B. Ohlrogge, E.A. Emken, R.M. Gulley, Human tissue lipids: occurrence of fatty acid isomers from dietary hydrogenated oils. J Lipid Res 22: 955 (1981)

42. R.O. Adlof, E.M. Emken, Distribution of hexadecenoic, octadecenoic and octadecadienoic acid isomers in human tissue lipids. Lipids 21: 543 (1986)

REPORT OF EPODEMIOLOGY WORKSHOP: INTRODUCTION

J-P. Habicht, K.M. Rasmussen, and A.S. Goldman*

Division of Nutritional Sciences, Cornell University
Ithaca, New York
*Department of Pediatrics and Human Biological Chemistry
and Genetics, The University of Texas Texas Medical Branch
Galveston, Texas

As a result of a growing body of knowledge about variations in the volume and composition of human milk, many questions have arisen concerning the determinants of lactational performance, the effects of human milk upon the recipient infant, and the effects of breast feeding on both infant and mother. Inasmuch as information from the basic and clinical sciences does not suffice to answer these questions, epidemiologic studies in populations living in different environments are needed both for testing hypotheses evolving from basic and clinical studies and for developing public health methods to improve the performance, health, and survival of infants and their mothers.

A necessary first step is the description of breast-feeding patterns in terms that have biological and behavioral meaning, because these descriptions may identify salient differences within and among populations that can serve as natural experiments. They also may reveal puzzling traditions that challenge knowledge from basic and clinical sciences. For instance, colostrum, which basic and clinical science judges to be good for the infant, is often replaced by fluids that are apparently less appropriate. The basis of such practices is unknown, but such traditions raise questions about whether they result in positive effects that are as yet not understood or, conversely, whether they have a negative effect on the health of the recipient infant.

Descriptive epidemiology also may identify the patterns and relative importance of the incidence, severity, and duration of disease and behavioral outcomes that are affected by breast feeding. So far, such descriptions reveal as groundless many claims of potential danger from breast feeding in that the expected deficiencies and toxicities never or rarely occur even in high-risk populations. Such unlikely claims are sometimes made to justify research because of the potential public health importance of the outcomes, whereas in reality the studies are motivated and better justified by scientific reasons for which lactation and breast feeding provide useful models.

As basic and clinical scientists and epidemiologists move to relate breast feeding and the components of human milk to health, behavioral, and social outcomes in infants and mothers, special problems arise because breast feeding is only one of a myriad of factors that can affect outcomes. Epidemiologic approaches and methods to infer causality in such cases are not conceptually difficult, but do require precision of expression. Kramer describes in the following paper how final proof of causality is obtained by intervention studies for which epidemiologists have devised strict conditions. These conditions sometimes are impossible to achieve, but their careful consideration often results in new ways of considering the problem. For example, one condition, which is not feasible with breast feeding, is randomization. However, when breast feeding is not common, randomization of support for breast feeding in otherwise comparable groups who wish to breast feed can be accomplished, and may result in very different breast feeding patterns and potentially different outcomes in the infants.

Intervention trials are essential, but they are difficult and costly. Therefore, they should only be undertaken within an orderly process of scientific investigation. For example, it has been proposed that isolated components of human milk, such as lactoferrin or secretory IgA, be added to infant formula as a means of preventing diarrhea. Their use for this purpose should be tested sequentially in appropriate animal models, in selected groups of infants, and only then in larger populations to establish their public health effects. In such studies, it will be essential to establish control groups to minimize confounding variables and to utilize the most appropriate statistical methods to determine sample size and the validity of outcome analysis. This process not only prevents costly program failures but also assures that major side effects, if any, are identified as quickly as possible. This is important because banked milk or isolated components may not behave in the same fashion as whole, fresh human milk. Furthermore, the benefits sought may be lost when breast milk is fed other than through breast feeding, because breast feeding entails more than the transfer of milk from mother to child.

Finally, a different kind of intervention study is required to establish the feasibility of public health interventions. It is important to note that intervention trials to establish causality of effect and those to determine feasibility of intervention usually need to be done separately because their different objectives almost always require different designs.

The above discussion shows how the epidemiology of breast feeding and of its determinants and consequences may give insights to basic scientists and clinicians. Of equal importance, basic and clinical research provide the basis for the formulation of epidemiological hypotheses and provide methods necessary for their investigation. Examples, many of which would not have been recognized without this conference, are found in the following paper and the notes from this Workshop.

BREAST FEEDING AND CHILD HEALTH: METHODOLOGIC ISSUES IN EPIDEMIOLGIC RESEARCH

Michael S. Kramer

Department of Pediatrics and of Epidemiology and Biostatistics
McGill University Faculty of Medicine
Montréal, Québec

INTRODUCTION

A remarkable renaissance of interest in breast feeding has occurred over the last 10-15 years. As if rediscovering the wheel, scientists are only beginning to understand the extent to which millions of years of evolution have succeeded in ways that modern technology cannot yet approach. The child health effects of breast feeding has become a fertile field for research by both basic laboratory investigators and epidemiologists.

Ideally, these two major branches of clinical research should complement one another. Population-based epidemiologic studies may provide the original clue to a health effect whose physiologic mechanism will be subsequently revealed by laboratory studies. Conversely, demonstration by basic scientists of a biochemical or immunologic substance in breast milk, or even demonstration that such a substance confers health advantages in laboratory animals, requires epidemiologic study of free-living populations to confirm corresponding benefits in human infants and children. Dosage (concentration) differences, inter-species variation, and effect modification by other factors operating in the intact human organism prevent the automatic inference of human health effects based on laboratory studies.

Four of the potential health benefits that have received the widest attention are reduction in overall mortality, protection against infection, prevention of allergic disease, and improvement in growth. In this paper, I shall review epidemiologic studies published since 1970 bearing on these alleged health benefits. For each, I shall briefly summarize previous research findings, highlight the major methodologic pitfalls in these studies, and suggest priorities for future research. But before beginning this review, I shall first examine the important methodologic issues relevant for all epidemiologic research in this domain.

METHODOLOGIC ISSUES

In investigating the relationship between infant feeding and some alleged health effect, we are particularly interested in the extent to which the feeding practice causes the health outcome. In epidemiologic parlance, the question is: Does exposure cause the outcome? Mere association without causation is less useful. The distinction is important, because changes in infant feeding practices will be ineffective in changing outcomes whose association with feeding is not a causal one.

Adequate demonstration of causality requires that a study be internally valid. Internal validity, in turn, is predicated on the absence of analytic bias and proper statistical inference. Furthermore, since other investigators and clinicians may be interested in the extent to which a study's result can be generalized to other populations, external validity is also an issue. Each of these major categories of methodologic concern will be discussed in turn, with specific indications of how they pertain to studies of infant feeding practices and their alleged health effects.

Internal Validity

A. Analytic Inference. A research study will be internally valid to the extent that it is able to eliminate, or at least substantially reduce, analytic bias. Analytic bias can occur through any of four mechanisms: (1) information bias, (2) selection bias, (3) confounding bias, and (4) reverse causality ("cart-vs-horse") bias.

In information bias, misclassification (i.e., erroneous measurement) of either exposure (here, type of infant feeding) or health outcome leads to a biased estimate of the association between the two. If measurement errors are random, as would be the case with merely "sloppy" measurements, the bias would be toward finding no association between infant feeding and the health outcome. Such a situation might arise from inadequate questionnaires, poor history taking, or reliance on prolonged maternal recall.

Of potentially greater import is nonrandom (differential) misclassification, in which outcome measurements are systematically different in those infants exposed to different feeding regimens or, conversely, when exposure histories differ according to outcome status. This is particularly likely to happen when study subjects or observers are not "blind" to the research hypothesis. In forward-directional (cohort) studies, observers who are measuring the health outcome, or mothers who are reporting it, might be influenced (even if unconsciously) by knowledge of a child's prior feeding history. In backward-directed (case-control) studies, the feeding history reported may be altered by knowledge of the child's health status. It is important to emphasize that the blinding discussed here does not entail the ethical difficulties often encountered with "double-blinding" in experimental (randomized) trials of treatment. Obviously, mothers choose how to feed their infants, and the infants' physicians need to monitor both feeding and health status to provide optimal care. But research studies should make every attempt to blind study observers to infant feeding history (cohort studies) or health status (case-control studies). Ideally, therefore, the observer for any study child should not also be that child's physician.

Another major source of systematic information bias in infant feeding studies stems from how breast feeding is defined, e.g., exclusive vs predominant vs any breast feeding. Obviously, the inclusion of partially breast-fed infants with either the exclusively bottle- or breast-fed groups will reduce the magnitude of any differences attributable to the two modes of feeding.

In selection bias, the exposure-outcome association is biased by the way study subjects are selected (or subsequently lost to follow-up) from the target population of which they are presumably representative. If the relationship between exposure and outcome in those selected for study is different from those not selected, bias will obviously occur. Selection bias is particularly common in hospital-based studies, since physicians may be reluctant to hospitalize breast-feeding infants for fear of separating the nursing mother and infant.[1] It is also likely when feeding status is defined as feeding at the time the outcome is determined or based on a minimum duration of breast feeding. Adverse health outcomes in infancy often lead to changes in feeding practices. A child who is being breast-fed may be switched to bottle feeding, but a bottle-fed infant who develops a health problem cannot usually be switched to breast feeding. Thus any group of infants selected by their current feeding status, or based on a minimum threshold of duration of breast feeding, will automatically lead to a selection bias whereby breast-fed infants are healthier.[2]

The third type of analytic bias is confounding bias, in which the relationship between exposure and outcome is altered by the existence of a third factor that is associated with both exposure and, independently of exposure, with the study outcome. The potential for confounding is the main rationale for the randomized clinical trial, the epidemiologic study design that comes closest to the laboratory experiment. By randomizing study subjects to the study exposure, any third factor that affects outcome should no longer be associated with exposure. Unfortunately, infants cannot be randomized to different feeding regimens. A mother's decision or a baby's ability to initiate or continue breast feeding depends on a variety of biological, attitudinal, behavioral, and economic considerations that are often difficult to conceptualize and even more difficult to measure.

Mothers (or communities or regions) could, however, be randomized to receive or not receive some breast feeding promotion intervention. The two experimental groups could then be compared for one or more child health outcomes. The success of such trials, of course, depends on the efficacy of the intervention. If only small differences are achieved in breast feeding rates, the health effects may not be detectable. Because of this and other practical (e.g., cost) considerations, most epidemiologic studies in this domain have used an observational (nonexperimental) design, and confounding bias is therefore a major methodologic concern.

One of the main ways confounding can occur in infant feeding studies is through the use of historical controls. Here the confounding factor is calendar time; if mortality, morbidity, or growth are improving in a population for reasons other than feeding, a before-and-after study of the effect of any change in infant feeding practices will show those practices to be beneficial.

Low-birth-weight infants are at increased risk for both mortality and morbidity during infancy. They are also more likely to be unable to breast feed in the immediate neonatal period. Consequently, in settings (especially developing countries) where low birth weight prevalence is high, investigators should control for this factor to avoid biasing the results toward showing a beneficial effect of breast feeding.

Another source of confounding is that of sociodemographic differences between mothers who breast feed and those who bottle feed, or among those who breast feed for different durations. Variables such as age, parity, racial/ethnic origin, and socioeconomic status are potential confounders, because these factors are associated with both infant feeding practices and infant and child health.

The direction of this confounding bias is likely to be opposite in developed and developing countries. In developing countries, mothers who engage in prolonged, exclusive breast feeding are more likely to be poor and multiparous, with short intervals between successive births. They are more likely to live in crowded dwellings with less access to clean water and sanitation. All these factors may lead to adverse health outcomes, and a simple comparison between breast- and bottle-fed infants in such a setting will lead to an underestimate of the health benefits of breast feeding. In developed countries, prolonged breast feeding is more common among the privileged, and failure to control for this difference may lead to an overestimate of the beneficial effects of breast feeding. In either setting, however, sociodemographic variables are fairly easy to measure and control for through matching, stratification, or multivariate statistical techniques. Consequently, the investigator who is aware of the potential for confounding will be able to eliminate or reduce confounding bias through adequate design or statistical analysis.

But other confounding variables present more difficult analytic problems. Particularly prominent among these are motivational differences that often exist between mothers engaging in different feeding practices. Since these motivational differences may have an independent effect on many infant and child health outcomes, they are likely to be potent confounders. And since adequate methods for measuring motivation to breast feed have not been developed, control for this source of confounding may be extremely difficult to achieve.

Finally, another source of confounding concerns treatment "contamination" by care givers. In the context of a research study, physicians, nurses, and other care givers may provide more support or encouragement to women who engage in the "favored" feeding practice (usually breast feeding), thus leading to additional treatments or exposures other than infant feeding and thereby confounding the feeding effect. Although it is often impossible to blind care givers to the feeding status of the subjects, they can often at least be blinded to the research hypothesis.

The fourth type of analytic bias is reverse causality bias, in which the study exposure and outcome are causally related, but the temporal sequence is the reverse of that indicated by the research hypothesis. In other words, the cart and the horse are reversed; the hypothesized outcome actually preceded and caused the exposure. Rather than being caused by bottle feeding, illness or poor growth may lead to cessation of breast feeding. Reverse causality bias is a major concern in cross-sectional studies, since it is often difficult to know the true temporal sequence between two variables when the two are determined at a single point in time. When feeding status is defined at the same point in time as the health outcome is determined, it is impossible to know whether the outcome actually led to the current feeding practice, or vice-versa. Infants who develop gastroenteritis or fail to thrive, for example, may be switched from breast feeding to formula or some nonmilk-containing liquid. If the problem persists, the investigator will find that fewer affected infants are being breast-fed than are healthy control infants and may erroneously infer that breast feeding confers protection against gastroenteritis.

B. Statistical Inference. Statistical inference involves an examination of the role of chance (random variation) in explaining a research study's findings. The actual study sample is presumed to be representative (even if not randomly selected) of some larger target population. The observed association between exposure and outcome is compared to the null hypothesis of no such association in the target population. (Note that the null hypothesis of no association is usually quite different from the investigator's research hypothesis, which often posits a beneficial effect of breast feeding.) The statistical inference may be erroneous for one of two reasons: (1) the null hypothesis is rejected when in fact it is true (Type I error); or (2) the null hypothesis is not rejected when it is actually false (Type II error).

The P value that results from most statistical tests is an expression of the probability of obtaining the sample result if the null hypothesis is in fact true for the target population. The P value thus reflects the probability of a Type I error. In other words, if the P value associated with a given association is .05, such a result would be obtained 1 out of 20 times in samples of the same size randomly selected from a target population in which the null hypothesis is true.

Type I errors are not generally a major problem for either breast feeding studies or other research studies when a single _a priori_ hypothesized association is being tested. Occasionally, however, multiple health outcomes are investigated simultaneously, and many of the associations analyzed arise _post hoc_, i.e., after the data are examined. In this setting, the measured P values underestimate the true probability of a Type I error (the null hypothesis is more likely to be falsely rejected). This is a common statistical pitfall in so-called "fishing expeditions", in which infant feeding practices are compared for numerous health outcomes. Of course, when several studies net the same "fish" (find the same health effects to be "significant"), the probability of error is reduced.

Another source of Type I error relates to the unit of statistical analysis, i.e., the group vs the individual. For reasons of feasibility or ethics, many health promotion interventions are carried out _en bloc_ on entire communities or regions. Most investigators then analyze their results based on the individuals within these communities as the units of statistical analysis. But this is often inappropriate, since individuals within communities tend to be more similar to each other than they are to individuals in other communities, and the extent of within-group dependence will reduce the effective sample size.[3-5] If the sample size is taken to be the number of individuals, the resulting P value will then be too low (too "significant"), and the null hypothesis may be falsely rejected. If an investigator wishes to study the effect of a breast feeding promotion intervention in one community and compares health outcomes in the "experimental" community to those in a "control" community without such intervention, the true sample size is not the total number of individuals living in the two communities. In fact, if the communities were perfectly homogeneous, it could be as a low as 2! Investigators need either to demonstrate substantial independence of individuals within groups or to include a sufficient number of groups to ensure adequate statistical power.[5]

The second type of erroneous statistical inference, Type II error, can arise from inadequate statistical power whenever small samples are used.[6] In this case, the null hypothesis may be false but is accepted simply because the sample size is inadequate to reject it. The absence of an association in small studies makes it hazardous to infer that such an association does not exist in the target population. This is particularly important in infant feeding studies whenever the health outcome under investigation is rare. Small studies are simply incapable of adequately excluding a true effect on these outcomes.

External Validity

Many clinicians and investigators will be interested in the extent to which the results of a given epidemiologic study, even if internally valid, are generalizable (externally valid) to other populations. External validity may be limited by factors related to the study subjects or study procedures.

Differences in study subjects can arise either because of selective participation in a study, the study of special subgroups, or regional differences between one population and another. Selective participation means subjects who agree to participate in a study may not be representative even of their own community. If study mothers are highly motivated, for example, their infants and children may experience better health outcomes and thus may show less benefit from breast feeding than would the general community. In this kind of scenario, a study could produce a false negative result; breast feeding would appear ineffective, whereas its effectiveness would be much higher in the general population. Other kinds of special subgroups, defined by a variety of clinical or sociodemographic criteria, might be either particularly susceptible or particularly resistant to the health benefits of breast feeding, and generalizability beyond the subgroup studied would be hazardous. Finally, regional differences may exist so that study results in one setting cannot be applied to those in another. This is a crucial limiting factor in attempting to extrapolate results from developed to developing countries or vice versa.

External validity can also be compromised by a study's procedures. One potential threat is the Hawthorne effect: taking part in a research project may itself affect outcome. If, for example, infant or child health improves as a result of participation in an infant feeding study, such improvement may make it difficult to demonstrate any additional benefit of breast feeding (the so-called "ceiling" effect).[5] Once again, the result will be a false negative one, i.e., breast feeding will appear to be ineffective, whereas it may be quite effective in a real-world, nonstudy situation.

MORTALITY

In developing countries, most (but not all) epidemiologic studies indicate that breast feeding leads to lower infant and early childhood mortality than bottle feeding.[7-14] Several of these studies have adequately controlled for potential confounding by socioeconomic status and age and have still found the protective effect of breast feeding to be highly significant.[8,10,12-14] The best of these have controlled for possible selection bias and reverse causality bias either by defining feeding status as ever vs never breast-fed or by using feeding status at a time sufficiently prior to the child's death to ensure that an illness leading to death was not the actual cause of breast feeding discontinuation.[12-14]

In developed countries, no recent epidemiologic studies have appeared to investigate the relationship between infant feeding and overall infant or early childhood mortality. Most of the better epidemiologic studies indicate no association between infant feeding and the sudden infant death syndrome (SIDS).[15-18] In industrialized societies, the vast majority of infant mortality occurs in the perinatal period, when infant feeding might be expected to have relatively little impact. Gastrointestinal and respiratory infections are not frequent causes of death in children from developed countries, and a large case-control study would probably be necessary to demonstrate a statistically significant decreased risk of mortality from these causes attributable to breast feeding.

INFECTION

A number of epidemiologic studies from developing countries indicate that breast feeding strongly protects against the development and severity of diarrhea, in general,[19-27] and a variety of pathogen-specific gastrointestinal infections.[28-31] Several of these have adequately controlled for confounding factors,[22,25,31] although reasons for initial choice of bottle feeding, or from switching from breast to bottle feeding, have rarely been considered. Since infants of low birth weight or those with other neonatal problems may be less able to breast feed, and since such infants may be at greater risk for subsequent gastrointestinal infection, part of the protective effect of breast feeding could be artificial. On the other hand, many studies (including one with negative results[32]) have included partially breast-fed infants in either the breast- or the bottle-feeding group, leading to misclassification (information) bias that would tend to reduce the protective effect of breast feeding. Furthermore, the magnitude of the protective effect is so high in several studies that it seems unlikely that any form of analytic bias could adequately explain it.

Even in developed countries, most epidemiologic studies,[33-47] including those controlling for the higher socioeconomic status of breast-feeding mothers in such settings,[34,36,38,41,43,45-47] have concluded that breast feeding does indeed have a protective effect against gastroenteritis. This has been shown in both advantaged subgroups[34,38] and disadvantaged native North American subgroups.[39,44,45] Negative findings in five other studies[48-52] are probably explained by their large potential for Type II error (i.e., inadequate statistical power) or information bias caused by inclusion of partially breast-fed infants in the breast-fed group.

As to respiratory infections and otitis media, the evidence that breast feeding is protective is somewhat weaker than for gastroenteritis. Although many studies from both developing[22,26,53] and developed[22,38,39,46,54-61] countries suggest that breast feeding may be protective, several of those from developed countries do not adequately control for confounding socioeconomic differences between breast-feeding and formula-feeding mothers.[22,54,55,59] On the other hand, some of the negative findings may be explained by information bias stemming from classification of partially breast-fed infants as either breast- or bottle-fed.[23,32,36,41,42,48,62] As to specific pathogens, the one most frequently investigated has been respiratory syncytial virus (RSV).[51,56,58,62] Here, too, the findings have been mixed, and some of the same methodologic problems may well explain the discrepancies.

Isolated reports also indicate that breast feeding may protect against invasive *Hemophilus influenzae* type B infection in Alaskan Inuit

(Eskimo) children[63] and against urinary tract infection,[64] although inadequate control for confounding factors may account for at least part of the reported effects. Finally, breast feeding also appears to decrease the risk of neonatal sepsis,[65] especially among low-birth-weight infants.[66-69] Although reverse causality bias (sicker or more immature newborns may be less able to breast feed) could explain the results in those studies using an observational design,[65,66] randomized trials from India comparing human milk vs formula feedings have confirmed the protective effect of breast milk.[67-69]

ALLERGIC DISEASE

A number of epidemiologic studies from developed countries have investigated the potential protective role of breast feeding in preventing atopic eczema,[70-84] asthma,[73,75-78,82,83,85,86] hay fever,[73,77] or food allergy.[72,87,88] Of the studies bearing on atopic eczema, only four[71,78,82,84] found a significant protective effect, and their findings are likely to be explained by information bias due to nonblinding of the physicians diagnosing eczema among the study infants. Most of the large-scale epidemiologic studies with adequate blinding to reduce information bias have failed to demonstrate such a protective effect, despite their larger sample sizes and enchanced statistical power. Methodologically superior studies have also indicated no protective effect against either hay fever or food allergy (with the possible exception of cow milk allergy[88]).

The best evidence (with adequate blinding) concerning asthma is also generally negative.[73,75-78,82,83] A prophylactic effect of breast feeding against subsequent asthma could be explained by protection against early RSV or other respiratory infections. In support of this possibility, one population-based study reported that initially breast-fed infants were less likely to develop ≥1 episode of wheezing in the first year,[73] but no less likely to develop ≥2 episodes by age 4.[83] Unfortunately, most epidemiologic studies in this domain have not controlled for parental smoking, which current evidence indicates to be a risk factor for bronchospasm and wheezing in childhood.

It should be pointed out that cow milk and other foreign protein antigens ingested by the mother are secreted in breast milk.[89] None of the large-scale epidemiologic studies have controlled for differences in maternal diet. Maternal dietary restriction and exclusive breast feeding might conceivably provide protection against allergic disease. The difficulty in achieving such a strict regimen, however, renders this theoretical benefit of limited practical public health importance.

GROWTH

The main methodologic difficulty in population-based studies relating infant feeding and growth is the lack of an appropriate standard. Virtually all cohort studies from both developing[90-97] and developed[98-101] countries have shown that exclusively breast-fed infants begin to "fall off" their weight-for-age growth charts after three or four months of age. But these percentile charts are almost always based on a predominantly bottle-fed population from a developed country [the U.S. National Center for Health Statistics (NCHS) standards[102] have recently been adopted by the World Health Organization[103]]. The "fall off" might just as easily indicate overnutrition by bottle-fed babies as undernutrition of breast-fed babies. Even if the latter is true, subsequent "catch-up" may occur, and the long-term consequences for growth and development are unknown. Furthermore, the potential for ultimate growth is at least in part genetically determined, and part of the "fall off" in some developing countries may represent the normal postnatal attainment of that genetic potential.

Although there is some disagreement,[32,104-116] several epidemiologic studies from developed countries indicate that bottle-fed infants are at greater risk for infant and childhood obesity.[117-123] Many of the studies in this area, however, have not controlled for the confounding differences between breast- and bottle-feeding mothers nor for age at introduction of solid foods. Furthermore, the long-term outlook for adult obesity and consequent morbidity remains unclear.

By contrast, epidemiologic studies from developing countries have found that breast-fed infants grow better than bottle-fed infants, at least during the first six months of life.[19,20,27,32,92,124-126] Although many of the latter studies did not consider potentially confounding factors, most of the confounding influences in such settings would probably bias the results against breast feeding. Selection and reverse causality biases probably represent more serious methodologic drawbacks. The difference in findings between developing and developed countries may reflect either a protective effect of breast feeding against diarrheal disease, with its consequent malnutrition, or a tendency to use nutritionally inadequate milk (due to dilution) or nonmilk liquids in the artificially-fed infants.

After six months of age, the data are conflicting, but most of the evidence from developing countries suggests that breast-fed infants continue to grow better than artificially-fed infants in such settings,[27,32,92,125,127-131] especially in poor socioeconomic groups.[131] Differences in results probably reflect, at least to some degree, differences in the nutritional adequacy or microbial contamination of supplemental foods, as well as problems of confounding bias (particularly by gastroenteritis or other infection that can lead to breast feeding cessation and malnutrition) and reverse causality bias (poorly growing infants are more likely to receive supplements).

CONCLUSION

Based on the best epidemiologic data available, breast feeding appears to confer several important benefits for infant and child health. In developing countries, it leads to lower mortality and lower infectious disease (especially gastroenteritis) morbidity. Breast-fed babies in these settings appear to grow better than their artificially-fed counterparts, despite a progressive "fall-off" from Western standards after 3-6 months. In developed countries, the healths benefits are less marked but still demonstrable for gastroenteritis and possibly for respiratory and other infections. Breast feeding may also offer some protection against childhood obesity. Prevention of allergic disease, however, seems more dubious based on the best and largest studies. Further research is required in all of these areas to confirm the existence of some benefits and better define the magnitude of others.

Future epidemiologic studies in this domain should give greater attention to the internal validity of their design and analysis. Better measurement methods need to be devised both for infant feeding practices and health outcomes. Blinding of subjects, care givers, and observers needs to be routinely instituted. Historical controls should be avoided wherever possible. Breast feeding promotion interventions in defined communities should be compared using the community as the unit of statistical analysis, unless outcomes in individuals are shown to be independent of community membership. Potentially confounding variables should be measured and controlled for appropriately. Although scientific perfection may be unattainable, gross violation of the design and statistical guidelines outlined here is likely to yield misleading research findings. The consequence of such misleading findings may be the establishment of premature or otherwise inappropriate public health programs that result in unfulfilled promises of health benefits and considerable waste of human and financial resources.

REFERENCES

1. Sauls HS: Potential effect of demographic and other variables in studies comparing morbidity of breast-fed and bottle-fed infants. Pediatrics 1979;64:523-527.
2. Hill AB: A Short Textbook of Medical Statistics. London: Hodder and Stoughton, 1977, p.27.
3. Cornfield J, Mitchell S: Selected risk factors in coronary disease: possible intervention effects. Arch Environ Health 1969;19:382-394.
4. Buck C, Donner A: The design of controlled experiments in the evaluation of non-therapeutic interventions. J Chronic Dis 1982;35:531-538.
5. Kramer M, Shapiro S: Scientific challenges in the application of randomized trials. JAMA 1984;252:2739-2745.
6. Feinstein AR: Clinical biostatistics XXXIV: The other side of statistical significance: Alpha, beta, delta, and the calculation of sample size. Clin Pharmacol Ther 1975;18:491-505.
7. Cantrelle P, Leridon H: Breastfeeding: Mortality in childhood and fertility in a rural zone of Senegal. Popul Stud 1971;25:505-533.
8. Plank SJ, Milanesi ML: Infant feeding and infant mortality in rural Chile. Bull WHO 1973;48;203-210.
9. Dugdale AE: Infant feeding, growth and mortality: A twenty year study of an Australian aboriginal community. Med J Aust 1980;2:380-385.
10. Janowitz B, Lewis JH, Parnell A, et al: Breast-feeding and child survival in Egypt. J Biosoc Sci 1981;13:287-297.
11. Schmidt BJ: Breast-feeding and infant morbidity and mortality in developing countries. J Pediatr Gastroenterol Nutr 1983;2(suppl 1): S127-S130.

12. Goldberg HI, Rodrigues W, Thome M, et al: Infant mortality and breastfeeding in Northeastern Brazil. Popul Stud 1984;38:105-115.

13. Butz WP, Habicht J-P, DaVanzo J: Environmental factors in the relationship between breastfeeding and infant mortality: The role of sanitation and water in Malaysia. Am J Epidemiol 1984;119:516-525.

14. Habicht J-P, DaVanzo J, Butz WP: Does breastfeeding really save lives, or are apparent benefits due to biases? Am J Epidemiol 1986;123:279-290.

15. Froggatt P, Lynas MA, Mackenzie G: Epidemiology of sudden unexpected death in infants ('cot death') in Northern Ireland. Br J Prev Soc Med 1971;25:119-134.

16. Naeye RL, Ladis B, Drage JS: Sudden infant death syndrome: A prospective study. Am J Dis Child 1976;130:1207-1210.

17. Biering-Sorensen F, Jorgensen T, Hilden J: Sudden infant death in Copenhagen 1956-1971: I. Infant feeding. Acta Paediatr Scand 1978;67:129-137.

18. Valdes-Dapena MA: Sudden infant death syndrome: A review of the medical literature 1974-1979. Pediatrics 1980;66:597-614.

19. Grantham-McGregor SM, Back EH: Breast feeding in Kingston, Jamaica. Arch Dis Child 1970;45:404-409.

20. Kanaaneh H: The relationship of bottle feeding to malnutrition and gastroenteritis in a pre-industrial setting. J Trop Pediatr Environ Child Health 1972;18:302-306.

21. Schoub BD, Greef AS, Lecatsas G, et al: A microbiological investigation of acute summer gastroenteritis in Black South African infants. J Hyg 1977:78:377.

22. Chandra RK: Prospective studies of the effect of breast feeding on the incidence of infection and allergy. Acta Paediatr Scand 1979;68:691-694.

23. Urrutia JJ, Sosa R, Kennell JH, et al: Prevalence of maternal and neonatal infections in a developing country: Possible low-cost preventive measures. Ciba Found Symp 1980;77:171-186.

24. Clavano NR: Mode of feeding and its effect on infant mortality and morbidity. J Trop Pediatr 1982;28:287-293.

25. Mittal Sk, Kanwar A, Varghese A, et al: Gut flora in breast and bottlefed infants with and without diarrhea. Indian Pediatr 1983;20:21-26.

26. Lepage P, Munyakazi C, Hennart P: Breastfeeding and hospital mortality in children in Rwanda. Lancet 1981;2:409-411.

27. Bravo IL, Cabiol L, Arcuch S, Rivera E, Vargas S: Breast-feeding, weight gains, diarrhea, and malnutrition in the first year of life. Bull Pan Am Health Organ 1984;18:151-163.
28. Mata LJ, Urrutia JJ: Intestinal colonization of breast-fed children in a rural area of low socioeconomical level. Ann NY Acad Sci 1971;176:93-109.
29. Gunn RA, Kimball AM, Pollard RA, et al: Bottlefeeding is a risk factor for cholera in infants. Lancet 1979;2:730-732.
30. Stoll BJ, Glass RI, Huq MI, et al: Epidemiologic and clinical features of patients infected with Shigella who attended a diarrheal disease hospital in Bangladesh. J Infect Dis 1982;146:177-183.
31. Clemens JD, Stanton B, Stoll B, Shahid NS, Bana H, Chowdbury AKMA: Breast feeding as a determinant of severity in shigellosis: Evidence of protection throughout the first three years of life in Bangladeshi children. Am J Epidemiol 1986;123:710-720.
32. Dugdale AE: The effect of the type of feeding on weight gain and illness in infants. Br J Nutr 1971;26:423-432.
33. Ironside AG, Tuxford AF, Heyworth B: A survey of infantile gastroenteritis. Br Med J 1970;3:20-24.
34. Cunningham AS: Morbidity in breast-fed and artificially fed infants. J Pediatr 1977;90:726-729.
35. Tripp, JH, Wilmers MJ, Wharton BA: Gastroenteritis: A continuing problem of child health in Britain. Lancet 1977;2:233-236.
36. Fergusson DM, Horweed LJ, Shannon FT, et al: Infant health and breast-feeding during the first 16 weeks of life. Aust Pediatr J 1978;14:254-258.
37. Larsen SA, Homer DR: Relation of breast versus bottle feeding to hopsitalization for gastroenteritis in a middleclass US population. J Pediatr 1978;992-417-418.
38. Cunningham AS: Morbidity in breast-fed and artificially fed infants: II. J Pediatr 1979;95:685-689.
39. Ellestad-Sayed J, Coodin FJ, Dilling LA, et al: Breast-feeding protects against infection in Indian infants. Can Med Assoc J 1979;120:295-298.
40. France GL, Marmer DJ, Steele RW: Breast-feeding and Salmonella infection. Am J Dis Child 1980;134:147-152.
41. Fergusson DM, Horweed LJ, Shannon FT, et al: Breast-feeding, gastrointestinal and lower respiratory illness in the first two years. Aust Paediatr J 1981;17:191-195.

42. Paine R, Coble RJ: Breast-feeding and infant health in a rural US community. Am J Dis Child 1982;136:36-38.
43. Taylor B, Wadsworth J, Golding J, et al: Breast-feeding, bronchitis, and admissions for lower-respiratory illness and gastroenteritis during the first five years. Lancet 1982;1:1227-1229.
44. Schaefer O, Spady DW: Changing trends in infant feeding patterns in the Northeast Territories 1973-1979. Can J Public Health 1982;73:304-309.
45. Forman MR, Graubard B, Hoffman HJ, et al: The Pima infant feeding study: Breastfeeding and gastroenteritis in the first year of life. Am J Epidemiol 1984;119:335-349.
46. Palti H, Mansbach I, Pridan H, Adler B, Palti Z: Episodes of illness in breast-fed and bottle-fed infants in Jerusalem. Isr J Med Sci 1984;20:395-399.
47. Duffy LC, Byers TE, Riepenhoff-Talty M, La Scolea LJ, Zielezney M, Ogra PL: The effects of infant feeding on rotavirus-induced gastroenteritis: A prospective study. Am J Pub Health 1986;76:259-263.
48. Adebonojo, FO: Artificial vs breast-feeding: Relation to infant health in a middle class American community. Clin Pediatr 1972;11:25-29.
49. Research Sub-committee of the South-East England Faculty of the Royal College of General Practitioners: The influence of breast-feeding on the incidence of infectious illness during the first year of life. Gen Practitioner Forum 1972;209:356-363.
50. Cushing AH, Anderson L: Diarrhea in breast-fed and non-breast-fed infants. Pediatrics 1982;70:921-925.
51. Weinberg RJ, Tipton G, Klish WJ, Brown MR: Effect of breast-feeding on morbidity in rotavirus gastroenteritis. Pediatrics 1984;74:250-253.
52. Eiger MS, Rausen AR, Silverio J: Breast- vs. bottle-feeding: A study of morbidity in upper middle class infants. Clin Pediatr 1984;23:492-495.
53. Elliott RB: An epidemiologic study of pulmonary disease in preschool Rarotongan children. NZ Med J 1975;81:54-57.
54. Schaefer O: Otitis media and bottle-feeding: An epidemiological study of infant feeding habits and incidence of recurrent and chronic middle ear disease in Canadian Eskimos. Can J Public Health 1971;62:478-489.

55. Hildes JA, Schaefer O: Health of Igloolik Eskimos and changes with urbanization. J Hum Evolution 1973;2:241-246.

56. Downham MAPS, Scott R, Sims DG, et al: Breast-feeding protects against respiratory syncytial virus infections. Br Med J 1976;2:274-276.

57. Watkins CJ, Leeder SR, Corkhill RJ: The relationship between breast and bottle feeding and respiratory illness in the first year of life. Epidemiol Community Health 1979;33:180-182.

58. Pullan CR, Toms GL, Martin AJ, et al: Breast-feeding and respiratory syncytial virus infection. Br Med J 1980;281:1034-1036.

59. Timmermans FJW, Gerson S: Chronic granulomatous otitis media in bottle-fed Inuit children. Can Med Assoc J 1980;122:545-547.

60. Cunningham AS: Otitis and breast-feeding. J Pediatr 1984;105:854-855.

61. Forman MR, Graubard BI, Hoffman HJ, Beren R, Harley EE, Bennett P: The Pima infant feeding study: Breastfeeding and respiratory infections during the first year of life. Int J Epidemiol 1984;13:447-453.

62. Frank AL, Taber LH, Glezen WP, et al: Breast-feeding and respiratory virus infection. Pediatrics 1982;70:239-245.

63. Lum MK, Ward JI, Bender TR: Protective influence of breast feeding on the risk of developing invasive H influenza type B (HIB) disease. Pediatr Res 1982;16:151A.

64. Cheong CIT: Human breast milk in Escherichia coli urinary tract infection. Milit Med 1982;147:202-204.

65. Winberg J, Wessner G: Does breast milk protect against septicemia in the newborn? Lancet 1971;1:1091-1094.

66. Patel RB, Khanna SA, Lahiri K, et al: Breast milk in low birth weight babies. Indian Pediatr 1981;48:195-196.

67. Narayanan I, Bala S, Prakash K, et al: Partial supplementation with expressed breast-milk for prevention of infection in low-birth-weight infants. Lancet 1980;2:561-563.

68. Narayanan I, Prakash K, Gujral VV: The value of human milk in the prevention of infection in the high-risk low-birth-weight infant. J Pediatr 1981;99:496-498.

69. Narayanan I, Prakash K, Prabhakar AK, et al: A planned prospective evaluation of the anti-infective property of varying quantities of expressed human milk. Acta Paediatr Scand 1982;71:441-445.

70. Halpern SR, Sellars WA, Johnson RB, Anderson DW, Saperstein S, Reisch JS: Development of childhood allergy in infants fed breast, soy, or cow milk. J Allergy Clin Immunol 1973;51:139-151.
71. Matthew DJ, Taylor B, Norman AP, et al: Prevention of eczema. Lancet 1977;42:613-622.
72. Saarinen UM, Kajosaari M, Backman A, et al: Prolonged breast-feeding as prophylaxis for atopic disease. Lancet 1979;2:163-166.
73. Hide DW, Guyer BM: Clinical manifestations of allergy related to breast and cows' milk feeding. Arch Dis Child 1981;56:172-175.
74. Kramer MS, Moroz B: Do breast-feeding and delayed introduction of solid foods protect against subsequent atopic eczema? J Pediatr 1981;98:546-550.
75. Gordon RR, Noble DA, Ward AA, Allen R: Immunoglobin E and the eczema-asthma syndrome in early childhood. Lancet 1982;1:72-74.
76. Golding J, Butler NR, Taylor B. Breastfeeding and eczema/asthma. Lancet 1982;1:623.
77. Taylor B, Wadsworth J, Golding J, et al: Breastfeeding, eczema, asthma, and hayfever. J Epidemiol Community Health 1983;37:95-99.
78. Businco L, Marchetti F, Pellegrini G, Cantani A, Perlini R: Prevention of atopic disease in "at-risk newborns" by prolonged breast-feeding. Ann Allergy 1983;51:296-299.
79. Pratt HF: Breastfeeding and eczema. Early Hum Dev 1984;9:283-290.
80. Taylor B, Wadsworth J, Wadsworth M, Peckham C: Changes in the reported prevalence of childhood eczema since the 1939-45 war. Lancet 1984;2:1255-1257.
81. Peters T, Golding J. Butler NR: Breast-feeding and childhood eczema. Lancet 1985;1:49-50.
82. Chandra RK, Puri S, Cheema PS: Predictive value of cord blood IgE in the development of atopic eczema and role of breast-feeding in its prevention. Clin Allergy 1985;15:517-522.
83. Hide DW, Guyer BM: Clinical manifestations of allery related to breast- and cow's milk-feeding. Pediatrics 1985;76:973-975.
84. Moore WJ, Midwinter RE, Morris AF, Colley JRT, Soothill JF. Infant feeding and subsequent risk of atopic eczema. Arch Dis Child 1985;60:722-726.
85. Blair H: Natural history of childhood asthma: A twenty-year follow-up. Arch Dis Child 1977;52:613-619.
86. Fergusson DM, Horwood LJ, Shannon FR: Asthma and the infant diet. Arch Dis Child 1983;58:48-51.

87. Gerrard JW, Mackenzie JWA, Goluboff N, et al: Cow's milk allergy: Prevalence and manifestations in an unselected series of newborns. Acta Paediatr Scand 1973;234(suppl):1-21.

88. Stintzing G, Zetterstrom R: Cow's milk allergy, incidence and pathogenetic role of early exposure to cow's milk formula. Acta Paediatr Scand 1979;68:383-387.

89. Kilshaw PJ, Cant AJ: The passage of maternal dietary protein into human breast milk. Int Arch Allergy Appl Immunol 1984;75:8-15.

90. Mata LJ: The Children of Santa María Cauqué: A Prospective Field Study of Health and Growth. Cambridge: Massachusetts Institute of Technology Press, 1978.

91. Lauber E, Reinhardt M: Studies on the quality of breast milk during 23 months of lactation in a rural community of the Ivory Coast. Am J Clin Nutr 1979;32:1159-1173.

92. Asha Bai PV, Leela M, Subramaniam VR: Adequacy of breast milk for optimal growth of infants. Trop Geogr Med 1980;32:158-161.

93. Khan M: Infant feeding practices in rural Meheran, Comilla, Bangladesh. Am J Clin Nutr 1980;33:2356-2364.

94. Waterlow JC, Ashworth A, Griffiths M: Faltering in infant growth in less developed countries. Lancet 1980;2:1176-1177.

95. Rowland MGM, Paul AA, Whitehead RG: Lactation and infant nutrition. Br Med Bull 1981;37:77-82.

96. Whitehead RG, Paul AA: Growth standards for early infancy. Lancet 1981;2:419-420.

97. Zhi-chien H: Breast-feeding in Xinhui district in South China. Food Nutr Bull 1981;3:42-48.

98. Chandra RK: Breast feeding, growth and morbidity. Nutr Res 1981;1:25-31.

99. Hitchcock NE, Gracey M, Owles EN: Growth of healthy breast-fed infants in the first six months. Lancet 1981;2:64-65.

100. Chandra RK: Physical growth of exclusively breastfed infants. Nutr Res 1982;2:275-276.

101. Hitchcock NE, Gracey M, Gilmour AI: The growth of breast fed and artificially fed infants from birth to twelve months. Acta Paediatr Scand 1985;74:240-245.

102. National Center for Health Statistics: NCHS Growth Curves for Children: Birth to 8 years. US Dept of Health, Education and Welfare, Publication No. (PHS) 78-1650, 1977.

103. World Health Organization: A Growth Chart for International Use in Maternal and Child Health Care: Guidelines for Primary Health Care Personnel. Geneva: World Health Organization, 1978.

104. Eid EE: Follow-up study of physical growth of children who had excessive weight gain in first six months of life. Br Med J 1970;2:74-76.

105. Holley D, Cullen D: A comparison of weight gain in breast fed and bottle fed babies. Public Health 1977;91:113-116.

106. de Swiet M, Fayers P, Cooper L: Effect of feeding habit on weigh in infancy. Lancet 1977;1:892-894.

107. Oakley JR: Differences in subcutaneous fat in breast- and formula-fed infants. Arch Dis Child 1977;52:79-80.

108. Saarinen UM, Siimes MA: Role of prolonged breast feeding in infant growth. Acta Paediatr Scand 1979;68:245-250.

109. Dubois S, Hill DE, Beaton GH: An examination of factors believed to be associated with infantile obesity. Am J Clin Nutr 1979;32:1997-2004.

110. Dine MS, Gartside PS, Glueck CJ, et al: Where do the heaviest children come from? A prospective study of white children from birth to 5 years of age. Pediatrics 1979;63:1-7.

111. Ferris AG, Beal VA, Laus MJ, e al: The effect of feeding on fat deposition on early infancy. Pediatrics 1979;64:397-401.

112. Ferris A, Laus MJ, Hosmer DW, et al: The effect of diet on weight gain in infancy. Am J Clin Nutr 1980;33:2635-2642.

113. Yeung DL, Pennel MD, Leung M, et al: Infant fatness and feeding practices: A longitudinal assessment. J Am Diet Assoc 1971;79:531-535.

114. Köhler L, Meeuwisse G, Mortensson W: Food intake and growth of infants between six and twenty-six weeks of age on breast milk, cow's milk formula, or soy formula. Acta Paediatr Scand 1984;73:40-48.

115. Salmenperä L, Perheentupa J, Siimes MA: Exclusively breast-fed healthy infants grow slower than reference infants. Pediatr Res 1985;19:307-312.

116. Birbeck JA, Buckfield PM, Silva PA: Lack of long-term effect of the method of infant feeding on growth. Hum Nutr Clin Nutr 1985;39C:39-44.

117. Taitz LS: Infantile overnutrition among artificially fed infants in the Sheffield region. Br Med J 1971;1:315-316.

118. Shukla A, Forsyth HA, Anderson CM, et al: Infantile overnutrition in the first year of life: A field study in Dudley, Worcestershire. Br Med J 1972;4:507-515.

119. Sveger T, Lindberg T, Weibull B, Olson UL: Nutrition, overnutrition, and obesity in the first year of life in Malmo, Sweden. Acta Paediatr Scand 1975;64:635-640.

120. Neumann CG, Alpaugh M: Birthweight doubling time: A fresh look. Pediatrics 1976;57:469-473.

121. Kramer MS: Do breast-feeding and delayed introduction of solid foods protect against subsequent obesity? J Pediatr 1981;98:883-887.

122. Kramer MS, Barr RG, Leduc DG, Boisjoly C, McVey-White L, Pless IB: Determinants of weight and adiposity in the first year of life. J Pediatr 1985;106:10-14.

123. Kramer MS, Barr RG, Leduc DG, Boisjoly C, Pless IB: Infant determinants of childhood weight and odiposity . J Pediatr 1985;107:104-107.

124. Kanawati AA, McLaren DS: The epidemiology of protein-calorie malnutrition in Jordan: II. Results of a questionnaire. Trans R Soc Trop Med Hyg 1970;64:761-768.

125. Lambert J, Basford J: Port Moresby infant feeding survey. Papua New Guinea Med J 1977;20:175-179.

126. Victora CG, Vaughan JP, Martines JC, Barcelos LB: Is prolonged breast-feeding associated with malnutrition? Am J Clin Nutr 1984;39:307-314.

127. Antrobus ACK: Child growth and related factors in a rural community in St. Vincent. J Trop Pediatr Environ Child Health 1971;17:188-210.

128. Kanawati AA, McLaren DS: Failure to thrive in Lebanon: II. An investigation of the causes. Acta Paediatr Scand 1973;62:571-576.

129. Zeitlin M, Masangkay Z, Consolacion M, et al: Breast feeding and nutritional status in depressed urban areas of Greater Manila, Philippines. Ecol Food Nutr 1978;7:103-113.

130. Greiner T, Latham MC: Factors associated with nutritional status among young children in St. Vincent. Ecol Food Nutr 1981;10:135-141.

131. Young HB, Buckley AE, Hamza B, et al: Milk and lactation: Some social and developmental correlates among 1,000 infants. Pediatrics 1982;69:169-175.

REPORT OF EPIDEMIOLOGY WORKSHOP: WORKSHOP NOTES

K. Dewey[1], C. Garza[2], R. Martorell[3], M.S. Kramer[4], L.A. Hanson[5], and R.K. Chandra[6]

[1]College of Agricultural, Univ. of Calif. Davis, Davis CA
[2]Dept. of Pediatrics, Baylor College of Med., Houston, Texas
[3]Food Research Institute, Stanford University, Stanford, CA
[4]Dept. of Pediatrics McGill Univ., Faculty of Med., Montreal, Quebec
[5]Dept. of Clinical Immunology, Univ. of Göteborg, Göteborg, Sewden
[6]Dept. of Pediatrics, Memorial Univ. of Newfoundland, St. John's, Newfoundland, Canada

An epidemiologic workshop was convened at this conference and consisted of epidemiologists, basic scientists, and clinical investigators who had previously studied certain facets of human lactation and the composition and effects of human milk. It was evident that the participants from these several disciplines learned a great deal from each other, and at the end there was a consensus that additional meetings of this type would be highly desirable to plan for future investigations in the field. We hope that these future meetings will result not only in further formulations of epidemiologic research concerning these matters but also in the presentation of the fruits of such collaboration.

Four major research areas in which epidemiologic approaches would be helpful to help answer questions about the effects of breast feeding emerged from the Workshop. These four areas were: (1) growth and nutrient utilization, (2) potentially harmful effects of breast milk, (3) host resistance and morbidity, and (4) breast milk and atopic disease. They are directly related to concerns that were raised during the planning and execution of the plenary sessions and other workshops at this meeting. They also covered some of the issues dealt with at the NICHD-supported conference on maternal and environmental influences upon breast feeding that was held in January, 1986 in Oaxaca, Mexico (HUMAN LACTATION 2, Maternal and Environmental Factors, Margit Hamosh and Armond S. Goldman, eds., Plenum Press, New York, 1986).

A. Growth and nutrient utilization (Rapporteurs: K. Dewey and C. Garza)

1. Growth of breast-fed infants

Recent studies in developed countries document that the growth pattern of breast-fed infants differs from international standards of growth. Exclusively breast-fed infants one to three months of age grow as well as the children of the World Health Organization's reference (predominantly bottle-fed) population, but they grow more slowly after that age. These different patterns of growth raise several issues: are they a reflection of demographic characteristics of populations with dissimilar infant feeding practices or do they reflect fundamental differences in absolute nutrient intake, the nutrient balance of the diet, and/or behaviors associated with breast and bottle feeding? The hypothesis that different growth patterns of breast- and bottle-fed infants are not associated with consequences for other functional competencies--such as neurodevelopmental indices, immunological function, and physical activity--merits careful investigation from both biological and epidemiological perspectives.

These issues also extend to the development and use of criteria for assessing the adequacy of infant feeding. Decisions to initiate the supplementation or complementation of diets based exclusively on human milk most often are based on growth performance. The significance of this decision is clear when sanitation is poor and contaminated food or fluids pose a serious health risk to young infants. This decision also is important even in less hostile environments. Health risks are expected to fall in progressively less contaminated environments, but a continuum of risk is likely to exist. The consequences of promoting potentially excessive intakes also are of concern.

2. Nutrient utilization by breast-fed infants

The effects of iron fortification of formulas on morbidity, the short- and long-term consequences of not meeting intrauterine rates of bone mineralization in infants born prematurely, and vitamin K deficiency are issues that merit close scrutiny.

Iron deficiency remains a problem in the very young. The high bioavailability of iron from human milk protects the breast-fed infant against this deficiency, but the mechanism permitting the efficient utilization of iron is poorly understood. Formula-fed infants therefore often are fed products high in iron to compensate for the low bioavailability of iron in commercial formulas. Iron fortification of formula results in significant amounts of free iron in the gastrointestinal tract that may promote the growth of pathogens.

Present schemes for feeding human milk to very low birthweight infants do not supply sufficient calcium or phosphorus to achieve fetal rates of mineral accretion. The functional significance of this condition may be assessed by such indices as bone density in infants at risk or recumbent length in older infants who received varying amounts of calcium and phosphorus from diverse sources that differ in net bioavailability.

Vitamin K deficiency and hemorrhagic disease are more frequent in breast-fed infants. Although hemorrhagic disease of the newborn is quite rare, it is common practice in most developed countries to supplement breast-fed infants with vitamin K soon after birth. Attempts to relate vitamin K levels in milk to maternal diet, to the stage of lactation, and to the infant's utilization of the vitamin have been hampered by inadequate techniques to measure compounds with vitamin K activity in milk. Current advances in analytical technologies suggest that the capability may be available for doing the requisite population studies in less developed countries where vitamin K supplementation at birth is difficult. In these countries, maternal consumption of vegetable sources of the vitamin may be great; however, the content of the vitamin in breast milk may be limited by poor absorption arising from a low consumption of fat. Thus, descriptive investigation of these aspects in developing countries is desirable.

B. Potentially harmful effects of breast milk (Rapporteurs: R. Martorell, M.S. Kramer, and L.A. Hanson)

The benefits of breast feeding for infant health, growth, and development as well as for child spacing are well documented. Breast milk remains the food of choice for infants, especially in developing countries but also in developed countries. In addition to considering these benefits, conference participants reviewed harmful effects that might be associated with human milk. Information on the extent and nature of these "problems" was judged to be generally inadequate. It was felt that descriptive epidemiological studies are required to assess properly any dangers that might be associated with breast feeding.

1. Viruses in human milk

It is known that human milk may contain viruses, including both cytomegalovirus (CMV) and human immunodeficiency virus (HIV), the cause of AIDS. More is known about the passage of CMV into human milk, and the hypothesis that breast feeding plays a major role in the transmission of infection should be investigated further. Infection of the breast-fed infant with CMV may not be undesirable because these infants may be less likely to express symptoms than bottle-fed infants. It may even be that the breast-fed infant of a seropositive mother is efficiently immunized via the milk. However, the issue requires research inasmuch as data relating mode of feeding to clinical symptoms and complications of CMV infection are not available.

Concern for AIDS, although it is not known whether or not it is transmitted by breast feeding or breast milk, will probably result in the routine screening of donors to milk banks not only for HIV but also for other microbiological organisms. CMV might also be one of the chosen organisms because milk from milk banks is used for premature infants who are at high risk of developing complications from CMV infection.

2. Drugs and toxins

A variety of pesticide residues, drugs, and toxins may be present in human milk, and the analysis of milk samples may be desirable as part of surveillance systems for environmental contaminants. These data would enable researchers to identify potentially dangerous situations. Of great interest to participants was the basic question of whether breast-fed infants ingest more or less drugs and toxins than bottle-fed infants and, more importantly, whether this exposure has any implications for performance, health or survival of these infants.

Contamination of human milk with DDT is well documented, and in areas such as Guatemala the levels have been reported to be very high. No studies have apparently been carried out to assess the consequences. Inasmuch as DDT has been in wide use in developing countries since World War II, there is the possibility of carrying out studies of its long-term effects on both infant and child health and maternal lactational performance.

Establishing the possible harmful effects of maternal drugs on the breast-fed infant involves three types of assessments of the causal link between a given drug and some observed adverse event: Can It?, Will It?, and Did It? "Can It?" addresses the question of whether the suspected drug leads to any increased risk of the adverse event. It is best answered from large-scale epidemiologic studies using conventional clinical trial, observational cohort or case-control designs. "Will It?" is the probability that the next exposed infant will develop a drug-caused adverse event. Description of the conditions under which effects occurred in the "Can It?" studies are essential for answering the "Will It?" question, which is the most important causality question for assessing overall future public health impact and, therefore, risk-benefit ratios. "Did It?" refers to a specific infant in which the adverse event has been observed and represents the probability of drug causation in that case. The "Did It?" causality assessment problem is the chief problem facing the pediatrician who observes an adverse event in a specific patient. It is also of great importance to drug regulatory agencies in their interpretation of reports on cases of adverse reactions received through their post-marketing surveillance activities, although it is the answer to "Will It?" that determines their regulatory actions.

3. Other issues

A number of other contaminants of breast milk were discussed, some of which may require epidemiological studies to establish their prevalence and their effect on the infant. An example is *trans* fatty acids in human milk. Western diets contain high amounts of *trans* fatty acids resulting from the process of hydrogenation of fats (e.g. margarine). Breast milk in countries such as the United States is known to contain high amounts of these substances. Formulas, on the other hand, do not often contain *trans* fatty acids. Some investigations have shown that breast-fed infants accumulate *trans* fatty acids in their tissues to a greater extent than bottle-fed infants. However, the significance of these observations for long-term health are unknown. Also, it is not known whether exposure to *trans* fatty acids in the diet once weaning has occurred overwhelms initial differences in *trans* fatty acid tissue composition. If this occurs, the possibility of health consequences arising from breast feeding becomes even more remote.

C. Host resistance and morbidity (Rapporteurs: R. Martorell)

Considerable progress has been made in recent years in the analysis of breast milk composition and in the isolation of those properties likely to be associated with protection against infection. Much could be learned from the systematic application of epidemiologic methods and population-based studies with respect to the role played by breast milk and breast feeding in non-laboratory settings. A number of salient issues can be proposed for investigation.

1. Diarrheal disease

In the past most studies of the relationship between infant feeding and infection have examined incidence patterns. These studies have generally concluded that breast feeding is associated with reduced risk of illness. To understand the role played by breast feeding better in protecting the infant against diarrheal diseases, more attention now needs to be given to a careful analysis of the impact of feeding mode on both the severity and duration of such episodes. More attention also needs to be given to dose-response effects (how much breast feeding and for how long) and to extended protection beyond the age of weaning. Finally, studies should consider whether the protective effects of human milk differ by causal agent (e.g. rotavirus, *Shigella*, etc.).

It is not clear from existing data whether the protective role played by breast feeding in enteric infections results from its intrinsic qualities or to the fact that the breast-fed infant, especially when exclusively breast-fed, has a reduced exposure to pathogenic organisms. Although the answer is likely to be that both characteristics are at play, it is important from the point of view of research that these be isolated and their relative impact be described. From the point of view of public health planning, it is only important if the two causes have a differential impact in varying environmental circumstances.

The evidence that breast milk contains certain glycolipids and glycoproteins that may inhibit binding of specific bacteria and bacterial toxins is intriguing and deserves further study before presently envisaged public health trials are undertaken. Firstly, these glycolipids and glycoproteins are present in much lower concentrations than are IgA and other antibodies in breast milk, and their relative contributions *in vivo* to the protective effects of breast milk are unknown because the necessary epidemiologic studies have not yet been done. Adding these glycoproteins and glycolipids in large quantities to infant formula may have unexpected effects on the infant. Therefore, public health supplementation trials of these compounds in infants should await clinical trials under controlled conditions to test for safety and possibly for efficacy, which in turn should await experimental studies in animals.

2. Other specific infections

The role of breast feeding in protecting against respiratory infection has not been as clearly defined as in the case of diarrheal diseases. Nevertheless, the literature suggests that the incidence of respiratory infection is lower among breast-fed than among artifically fed infants. Similar studies to those in the area of diarrheal diseases--in which etiology, incidence, severity (including complications), duration, and specific pathogenic agent are looked at separately--are called for.

The protective effect of the intrinsic qualities of breast milk appears largely due to humoral factors. Less is known about the degree to which cell-mediated immunity is transmitted by breast milk. It is recommended that research be focused on tuberculosis and that the following aspects be considered: a) protection against the occurrence or severity of tuberculosis, b) success of BCG vaccination, and c) interpretability of TB skin testing in areas endemic for tuberculosis.

Evidence was presented indicating that breast milk may stimulate local immune factors (IgA, lactoferrin) in the gut and urinary tract. This suggests a possible mechanism for a protective effect of breast feeding against urinary tract infection (UTI). Inasmuch as very few epidemiologic studies have focused on the possible protective effect of breast feeding on UTI, such studies should be undertaken. Emphasis should be placed on UTI occurring in the first 12-24 months of life in otherwise healthy infants.

3. Use of colostrum

Although colostrum is a potentially important immunologic resource, women in many societies do not use it in the feeding of infants and they delay breast feeding until more mature milk becomes available. The consequences of this widespread practice remain largely unstudied. In view of the potential value of colostrum and the fact that little work has been done, it is recommended that studies be carried out to examine the role of colostrum in helping protect against infection and its complications, and especially in looking for any disbenefits to mother or child of breast feeding colostrum.

4. Use of expressed breast milk

Little attention has been given in previous research to the possibly differential impact played by expressed breast milk as opposed to breast feeding. Possible changes in the composition of breast milk once it has been expressed, and the implications of this for nutritional and immunologic function, thus have tended to be overlooked. Similarly, the psychobiological dimensions of breast feeding and the role played by the interaction between mother and infant during the act of suckling have not been considered with respect to the protective effects of breast feeding against morbidity. In view of the advice increasingly given to working mothers that they should express and store their milk for later feeds, these issues merit further study.

D. Breast milk and atopic disease (Rapporteurs: M.S. Kramer and R.K. Chandra)

Further studies are required on the effect of infant feeding on the incidence, severity, and age at onset of atopic disease in infants and children. Standardized criteria for the diagnosis of atopic eczema, asthma, allergic rhinitis, and food allergy should be used. The major concern is the early introduction of allergens to the infant's diet. It also has been suggested that infants may be exposed to allergens through breast milk. In studying this issue one must differentiate between *in utero* sensitization to proteins in the maternal diet and postnatal sensitization to substances in breast milk. Epidemiologic studies of the above issues should focus on children at risk (those with first-degree relatives with atopic disease) and experimental trials should attempt to influence and measure maternal diet as well as receipt of foods other than breast milk by the breast-fed infant. In cohort studies, observers examining children for atopic disease should be blinded to the feeding history; in case-control studies, interviewers obtaining the feeding history should be blinded to the case vs. control status of the children whose mothers are interviewed. Smoking should be considered an important potential confounding variable in studies of respiratory allergy.

REPORT OF EPIDEMIOLOGY WORKSHOP: RECOMMENDATIONS REGARDING FUTURE RESEARCH CONCERNING THE EFFECTS OF HUMAN MILK UPON INFANT RECIPIENTS

J-P. Habicht, M.W. Woolridge, K.M. Rasmussen, R. Martorell, M.S. Kramer, F. Jalil, L.A. Hanson, A.S. Goldman, C. Garza, K. Dewey, R.K. Chandra and M. Carballo

A. Breast milk and resistance to infection

Rationale

Investigations on the potential mechanisms of protection offered by breast feeding against infections and other diseases are sufficiently advanced to raise many questions concerning the *in vivo* advantages of banked and fresh human milk, components of human milk, and other factors related to breast feeding in populations living in different environments. Inasmuch as *in vitro* studies and studies in clinical settings cannot provide definitive answers to the following questions, epidemiological studies could contribute both scientific and public health insights to research.

1. What are the mechanisms by which breast feeding protects against diarrheal pathogens? Knowledge is now sufficient so that intervention trials are possible to test the efficacy of some of the protective factors. Questions of interest include:

   a. Are all pathogens equally susceptible?

   b. For similar pathogens, is there a different effect on acute and chronic diarrhea?

   c. In cases where breast feeding does not prevent infection, does it influence severity and duration of infection?

   d. What are the maternal health and behavioral factors that influence the putative protective effects of breast feeding? In contrast to bottle feeding, is breast feeding essential for conveying some of this protection, either because of feedback mechanisms or other factors?

   e. Do isolated components of human milk have the same effects as those in fresh milk?

   f. What are the relationships between pathogen loads and the amounts of protective factors in human milk? Very few infants are fed breast milk exclusively, even in traditional societies; other contaminated fluids often also are given.

g. Do the responses to the protective factors also depend upon the age (maturation) of the recipients? The increasing use of human milk in the feeding of very low birthweight infants who cannot be breast fed underscores the need to consider these questions.

h. To what extent are there differences in these protective factors in the breast milk of mothers with premature babies as compared to those with term infants? Are these differences beneficial to the premature infant?

2. Does breast feeding protect against diseases other than enteric infections? The most relevant illness that requires study is infectious respiratory disease.

3. What is the role of colostrum in disease prevention? Does colostrum have harmful effects?

Recommendation

A workshop should be convened with epidemiologists and with basic and clinical scientists, all of whom are experts in this area, to specify in detail epidemiologic research approaches to the above questions.

B. Breast feeding and growth

Rationale

Differences in growth between breast-fed and bottle-fed infants in developed countries present a dilemma in interpreting the meaning of the slower growth rate of many breast-fed infants after a few months. Historically, higher growth velocity has been assumed to be good because less-than-optimal growth is caused by infections and malnutrition. It is unclear, however, that the observed slower growth rate in breast-fed, term infants is the result of any pathologic process or has bad effects. In fact, it may be an indicator of a more desirable state of health.

Further studies are needed regarding the basic issue of the warranty implied in the development of growth criteria. At present little can be said with confidence about any relationships between small differences in growth and performance, health or survival of the infant. Understanding this issue more thoroughly requires examining basic metabolic and behavioral data.

The following questions need to be addressed specifically in relationship to breast feeding, although the questions also have broader implications for infant feeding. These questions should be asked in the context of newly available methods for measuring body composition, energy expenditure, and activity patterns as well as motor and cognitive development. For example, advances in stable isotope methodologies and other novel approaches for assessing body composition and energy expenditure may permit testing of hypotheses not previously possible because of technological limitations.

1. What practical approaches can be recommended to determine the timing for the introduction of supplementary foods to the diets of individual infants?

2. What indices can be used to relate specific functional competencies (e.g. infant mental and motor development, immunological responses,

and physical activity) to different growth patterns or to levels of nutrient intake that influence growth?

3. In view of the impossibility of randomly assigning breast or bottle feeding, what strategies can be developed to examine functional outcomes of differing growth patterns of breast- and bottle-fed infants?

4. Should environmental and socioeconomic characteristics (e.g. sanitation, availability of safe weaning foods, and education) influence our interpretation of acceptable growth patterns?

5. What behavioral and metabolic facets of infant feeding are most promising for investigating the causes of apparent differences in intake and growth patterns of breast- and bottle-fed infants?

6. Are distinct patterns of growth due to differences in infant feeding associated with differences in partitioning of energy expenditure and consequent differences in body composition, performance, health and survival in infancy and childhood?

7. Do the stresses of different environments moderate the effects of differences in infant feeding on growth and on associated differences in performance, health and survival?

Recommendation

A workshop should be convened of experts in growth, nutrient requirements, measurements of performance, health and survival of infants and children as well as epidemiology to examine the potential of recent technical advances and to design epidemiologic research to address the questions listed above.

C. Classification of breast feeding experience

Rationale

There is some doubt as to whether present methods of measuring and classifying breast feeding patterns are adequate to address the questions listed above as well as other epidemiologic questions related to the determinants of breast feeding and lactation. It is unlikely that a single group of measurements and classifications will prove useful for all purposes. For instance, strong effects of breast feeding--as in the prevention of death in areas with poor environmental sanitation--can be identified with much cruder measurements and classifications of infant feeding patterns than weaker effects. As another example, the characteristics of breast feeding that are important for protecting against diarrheal disease are different from those necessary to identify the ingestion of foreign allergens in studies of atopic disease.

On the other hand, it may be possible to develop objective classifications for categorizing the "intensity" of breast feeding. Discussion with epidemiologists would help to determine whether a simple scale would suffice for many studies. This would foster the collection of useful information about breast feeding in situations when it would otherwise not be collected.

Recommendations

1. In each of the above workshops appropriate and feasible schemes for measuring and classifying breast feeding and infant feeding practices must be reviewed specifically for each epidemiologic study proposed.

2. The lessons learned in implementing the above recommendation should be reviewed to decide whether to recommend to some normative body, such as the World Health Organization, the convening of an expert group to develop measuring and classification procedures for breast and infant feeding patterns that can be recommended for cohort studies and for studies that depend upon recall.

D. Advice to practitioners on healthy breast feeding practices

Rationale

There is an urgent need for practical advice to give to clinical and public health practitioners on healthy breast feeding practices. This may require the development of criteria of normal growth for breast-fed infants. Comparison of international standards with available growth data obtained in studies of exclusively breast-fed infants raises the concern that primary health care professionals may be counseling women to introduce solid foods or formula prematurely. This is because longitudinal assessments of the growth of individual infants are likely to indicate that the child is failing to maintain an "expected growth trajectory". This raises issues of defining standards and their application. These concerns are focused at the core of the conceptualization and application of growth monitoring.

Recommendation

Convene as soon as possible a group of experts under the aegis of a normative body, such as the World Health Organization or the U.S. National Academy of Sciences, to establish recommendations for the growth monitoring of breast-fed infants and associated recommendations for the introduction of complementary and weaning foods in developed and developing countries.

E. Criteria for atopic diseases

Rationale

Much of the ambiguity in interpreting studies on the effect of breast feeding in preventing allergic diatheses results from a lack of clarity in defining allergic syndromes and their diagnoses. This is a prerequisite for both basic laboratory and epidemiologic research; the alternative is continued chaos and controversy.

Recommendation

Clinical immunologists/allergists, dermatologists, and pneumonologists should develop objective clinical criteria for diagnosis of atopic eczema, asthma, allergic rhinitis, and food allergy. Once such criteria have been accepted, further epidemiologic studies should be undertaken, with incorporation of the specific research design features mentioned above in the Workshop Notes.

PREGASTRIC LIPASE TRIGGERS FAT DIGESTION

Stefan Bernbäck, Olle Hernell and Lars Bläckberg

Department of Physiological Chemistry
University of Umeå
S-901 87 Umeå, Sweden

Adults have a high capacity for digestion and absorption of dietary fat. For this sufficient concentration of colipase-dependent pancreatic lipase for digestion, and micellar concentrations of bile salts for product absorption, are required. Relative deficiency of both are therefor considered the major cause of fat malabsorption seen in some newborn, especially preterm, infants. Therefor the observation that fat digestion is initiated in the stomach by pregastric/gastric lipase has gained interest mainly in the neonatal period. To study the properties and physiological function of a pregastric lipase we purified the enzyme from calf pharyngeal tissue. The electrophoretically pure enzyme is a single-chain glycoprotein with Mr ∿50,000 D. It crossreacts immunochemically with a lipase of similar size in human gastric juice and, since the two have similar properties also in other respects, we consider the bovine enzyme as useful model for the human counterpart. Since pregastric lipase has an acid pH-optimum and is extensively resistant to proteolysis by pepsin it has properties required for a function in stomach contents. In contrast, it is rapidly inactivated by physiological concentrations of pancreatic proteases and bile salts. Hence, it is not likely that it contributes to fat digestion in duodenal contents. Pregastric lipase has no known cofactor and in contrast to other lipases involved in fat digestion it can by itself hydrolyze human milk triglycerides. This property is thus unique and probably the most essential function for pregastric lipase. On the other hand its activity is susceptible to product inhibition. Thus, *in vitro* hydrolysis of 2-5 % of the substrate (milk) triglyceride is enough to cause complete inhibition. This observation suggests that gastric lipolysis is not mainly of quantitative importance. Interestingly however the low concentration of free fatty acid released by pregastric lipase has an opposite effect on milk triglyceride digestion by colipase-dependent lipase. The latter lipase can by itself not hydrolyze human milk fat globule triglyceride *in vitro* not even in presence of its cofactor colipase and bile salt. After hydrolysis of a few per cent of milk triglyceride by pregastric lipase the remaining triglyceride is rapidly hydrolyzed by colipase-dependent lipase. This triggering effect of pregastric lipase can be replaced by direct addition of a low concentration of free fatty acid.

THE SOURCE OF "LOST CALORIES" FROM FORTIFIED BREAST MILK

Nitin R. Mehta, Margit Hamosh, Joel Bitman, and D. Larry Wood

Georgetown University Medical Center, Washington, DC

USDA Agricultural Research Service, Beltsville

MCT[1] oil is frequently chosen as fortifier for human milk (HM) feeds for the very premature infant because it improves caloric density without increasing the osmotic load of the feed. In practice, however, these infants who apparently are receiving adequate calories do not demonstrate the expected growth. We have therefore investigated whether fat might be lost because of adherence to feeding sets in the prevailing methods of MCT supplementation. Forty feeding sets, collected from infants on feeds of 3 different compositions, were analyzed for amount of residual fat (measured as % of total fat in feed by gravimetry). The fatty acid composition (FA) of the fat residue was analyzed by gas chromatography. The results mean $\pm$ SEM show:

Feed Composition	MCT Supplementation Method	N	RESIDUAL FAT Amount % Total Fat	FA (% of total) C8:0	C10:0	>C12:0
HM	--	5	1.90$\pm$0.69	0.0	0.4	99.6
HM & MCT	MCT followed by HM	12	2.63$\pm$1.59	31.6	16.0	52.4
HM & MCT	MCT & HM mixed & fed	23	15.77$\pm$10.8*	45.5	30.5	24.0

MCT[1]: medium chain triglyceride oil. Mead-Johnson (C8:0=64%,C10:0=34%).
*t-test p 0.001

Conclusion: 1) When MCT is delivered prior to milk, the lipid loss in the feeding set is not different from that occurring with unfortified milk feeding; 2) however, when MCT is mixed with milk prior to feeding, as much as 20% of lipid calories may be lost in the feeding set; 3) the analysis of the fat residue indicates that it is composed mainly of the added MCT; 4) up to 90% of MCT added is lost as residue when MCT and breast milk are premixed and fed.

We therefore question the rationale for adding MCT to human milk. Should we instead consider human milk fat, or a fat blend similar to human milk fat, when there is a need for higher caloric density?

GLYCOPROTEINS OF THE HUMAN MILK FAT GLOBULE MEMBRANE: ULTRASTRUCTURE AND RELATION TO FAT ABSORPTION

Stuart Patton, Wolfgang Buchheim*, and Ulrich Welsch**

School of Medicine, University of California San Diego, La Jolla, CA; *Bundesanstalt für Milchforschung, Kiel, FRG; **Anatomische Anstalt, Universität München, München, FRG

Evidence is presented that unique high molecular weight glycoproteins of human milk fat globules may be involved in milk fat absorption by the infant. Milk fat globules from a number of species were evaluated for surface structure by freeze-etch and thin-section electron microscopy, and for glycoprotein composition by SDS-gel electrophoresis. Globules of the human, but not of cow, goat, sheep and rat, showed a filamentous surface coat. These filaments could be removed from the globule by heating (80°C-10 min), and were identified by gel electrophoresis as 3 high molecular weight glycoproteins of the globule membrane. This group of proteins was missing from globules of the other species. Several studies (eg. Atkinson et al. J. Pediatr. 99:617, 1981) show that heating milk lowers absorption of its fat in the preterm infant. We suggest that heat destruction of bile-salt-stimulated lipase may not be the explanation of this phenomenon since fat in an infant formula containing no lipase was well absorbed (Shenai et al. Pediatrics 66:233, 1980). Heat-induced changes in the globule surface, such as loss of high molecular weight glycoproteins, affords an alternative or additional explanation.

COMPARISON OF THE DEUTERIUM DILUTION AND TEST-WEIGHING TECHNIQUES FOR THE DETERMINATION OF HUMAN MILK INTAKE

N.F. Butte, W.W. Wong, B.W. Patterson, C. Garza, and P.D. Klein

USDA Children's Nutrition Research Center, Baylor College of Medicine, Department of Pediatrics
Houston, TX

The deuterium dilution and test-weighing techniques to determine human milk intake were compared in 5 exclusively breast-fed infants, mean age 93 days, and in 4 supplemented infants, mean age 112 days. Deuterium oxide was administered orally to the mothers at 100 mg/kg body weight. The monoexponential decay curve of ^{2}H from the mother's milk was defined from milk sampled on days 1, 2, 6, 10, 13, 14 of the experiment. Milk samples were defatted by centrifugation and the milk water was reduced to hydrogen gas for hydrogen isotope ratio measurements by gas-isotope ratio mass spectrometry (GIRMS). Total body water (TBW) of the infants was determined by dilution after the administration of 60 mg ^{18}O/kg body weight. Infant urine was sampled on days 1, 2, 4, 6, 8, 10, 12, 14 of the experiment and analyzed for ^{2}H and ^{18}O enrichment by GIRMS. The SAAM27 computer program for compartmental analysis was used to estimate fractional rate constants and predict the average human milk intake over the 14 days of the experiment. The test-weighing procedure was conducted for 5 consecutive days. All supplemental formula and solid foods consumed during the 14 days were measured.

The mean ± SD intake of human milk was 648 ± 63 g/d estimated by deuterium dilution and 636 ± 84 g/d estimated by test-weighing. The mean difference between the two methods of 12 ± 32 g/d was not statistically different from zero (paired-t-test). The compartmental model also estimates water intake from sources other than breast milk. For the supplemented infants the predicted water intake was 159 g/d compared with the measured amount of 130 g/d. For the exclusively breast-fed infants (presumably receiving no supplemental water), the predicted water intake was 24 g/d.

In conclusion, the deuterium dilution and test-weighing techniques provide similar estimates of daily human milk intake.

INFANT SELF-REGULATION OF BREAST MILK INTAKE

Kathryn G. Dewey and Bo Lönnerdal

Department of Nutrition
University of California
Davis, CA 95616

There is a very wide range in milk intake among normal, exclusively breast-fed infants. It is unknown to what extent this range in intake is attributable to maternal "supply" vs. infant "demand." Do intakes at the low end of the range (500-650 g/day) represent cases of "insufficient milk," or are such infants able to meet their needs on relatively low energy intakes? The present study was developed to determine 1) whether breast milk production can be increased through regular expression of extra milk, and 2) whether infants will increase breast milk intake if maternal milk supply is augmented.

Eighteen mothers of exclusively breast-fed infants 6 to 21 weeks of age participated in the experiment, in which milk supply was stimulated by daily expression of extra milk (at least 100 g/day) for 2 weeks. The study was divided into four phases: baseline, expression phase, 48-hr post-expression, and follow-up (5-16 days after the expression phase). During each phase, infant milk intake was recorded by test-weighing for 2-4 days, and 1-2 milk samples were collected.

All but 4 mothers increased milk production by >73 g/day over baseline, with an average increase of 124 g/day. On the average, the 14 infants of mothers who increased milk production took in significantly more milk immediately post-expression (849 vs. 732 g/day), but about half of them returned to near baseline levels of milk intake after 1-2 weeks. There were no significant changes in mean milk concentrations of protein, fat or lactose during the study. Net change in infant intake at the end of the study (a measure of infant "response") was positively correlated with infant weight-for-length (r=0.59) and age (r=0.58), and was unrelated to baseline milk intake (r=-0.06). Therefore, the wide range in breast milk volume in well-nourished populations is due more to variation in infant "demand" than to inadequacy of milk production.

(in press, Acta Paediatrica Scandinavica, 1986)

X-RAY STRUCTURAL STUDIES OF LACTOFERRIN

Sylvia V. Rumball, Bryan F. Anderson, Heather M. Baker,
Gillian E. Norris, Joyce M. Waters and Edward N. Baker

Department of Chemistry and Biochemistry
Massey University
Palmerston North
New Zealand

Knowledge of protein tertiary structure is fundamental to a proper understanding of function. Currently there is little detailed structural information available on human milk proteins. The structure of bovine milk β-lactoglobulin at 2.8 A has recently been determined and this has led to speculation about its function, but this protein is apparently not found in human milk. Preliminary crystallographic data on α-lactalbumin has been published but no high resolution structure has yet been forthcoming.

Lactoferrin is a member of the transferrin family of proteins and a prominent protein in human milk. There has been much speculation about its function. The three-dimensional structure will allow a description of iron sites in lactoferrin (and by analogy in other transferrins) and the nature and location of other metal-binding sites, thus leading to a better understanding of the function of lactoferrin in human milk, leucocytes and various bodily secretions.

The tertiary structure of human lactoferrin is presently being studied by X-ray diffraction. An electron-density map at 3.2 A has been calculated and the path of the polypeptide chain traced. The molecule is folded into two globular lobes with each lobe further sub-divided into two equal-sized domains. Each of the two iron atoms is located at the interface between the domains of each lobe.

Crystals of Cu_2-Lf, apo-Lf and bovine Fe_2-Lf have also been prepared. Future studies should allow both a between species comparison (important considering the proposal to add bovine lactoferrin to infant formulae) and a comparison of the structure with and without iron and copper.

WHEY PROTEINS IN FECES OF PRETERM INFANTS RECEIVING PRETERM MILK AND INFANT FORMULA

Sharon M. Donovan*, Stephanie A. Atkinson** and Bo Lönnerdal

University of California, Davis* and McMaster University
Hamilton, Ontario, Canada**

Human milk whey contains proteins thought to have functional and nutritional roles for the infant. These proteins include lysozyme(Lys), lactoferrin(Lf), secretory IgA(sIgA), serum albumin(SA) and α-lactalbumin(α-LA). SA and α-LA primarily have nutritional importance as sources of amino acids. Lys,Lf and sIgA may have both nutritional and physiological functions. Lys and Lf have been suggested to have bacteriostatic effects within the infant's gastrointestinal(GI) tract. SIgA acts immunologically within the gut of the infant. For these proteins to exert their functions, they must remain largely intact during passage through the upper GI tract. We have demonstrated (AJCN 41: 852(1985)) that LF and sIgA are detected in feces of exclusively breast-fed term infants. We were interest in quantitating these proteins in feces of preterm infants for two reasons. First, preterm milk contains higher quantities of these whey proteins. Secondly, the proteins may serve a greater protective role for preterm infants, due to immature immune function in such infants.

Three-day balance studies were carried out in 14 infants at the neonatal unit of McMaster University Medical Centre. One group received only preterm milk)PTM); the other group received 50% PTM and 50% infant formula (PTM+f). PTM samples (24 h) were collected from each mother. Fecal samples were collected from the infants during the 3-day period. Milk and fecal samples were analyzed for total N by Kjeldahl analysis, and soluble protein by the Lowry assay. Whey proteins were quantitated immunological. Statistical analysis was performed using a 2-tailed Student's t-test.

The average intakes and fecal losses of whey proteins are in Table I. PTM infants received less total N but more soluble whey proteins than PTM+F infants. Whey proteins were observed in the following percentages of fecal samples analyzed: Lf and sIgA, 100%; Lys, 86%; SA, 57% and α-LA, 0%. Fecal losses of Lf and Sa were significantly greater in PTM-fed infants, but when fecal losses were expressed as % of intake, the relative amounts of Lf and SA excreted in both groups were similar. PTM+F-fed infants excreted a greater % of their intake of sIgA and Lys.

In summary, PTM-fed infants received and metabolized more whey proteins than infants fed PTM+F. We observed lower absolute quantities of sIgA in feces of preterm infants than we have observed in term infants, despite a greater incidence a fecal SA. The presence of SA may indicate impaired digestion or mucosal breakdown. If the digestion capacity of preterm infants is less than that of term infants, we would expect higher levels of whey proteins in feces. The present data suggest that macromolecular absorption may be occurring in the preterm infant.

Table I: Average Intakes and Fecal Losses of Whey Proteins (mg/kg/day)

Whey Protein	PTM-fed Infants		PTM+F-fed infants	
	Intake	Fecal Loss	Intake	Fecal Loss
Total N	373 ± 81[a]	128 ± 32	467 ± 79 *	128 ± 12
Lactoferrin	554 ± 190	16 ± 9	282 ± 110*	11 ± 7 *
α-Lactalbumin[b]	454 ± 104	not detected	171 ± 32 *	not detected
SIgA	242 ± 244	23 ± 40	78 ± 11 **	13 ± 12 ***
Serum Albumin[b]	83 ± 66	0.6 ± 0.3	45 ± 18 **	0.1 ± 0.2*
Lysozyme	31 ± 23	1.1 ± 1.2	11 ± 8 *	2.5 ± 1.6***

a. Mean ± SD. b. Human protein only; bovine proteins not accounted for.
* (p=.001) ** (p>.05) *** (p<.05)

FORTIFIED HUMAN MILK FOR VERY LOW BIRTH WEIGHT INFANTS: CORRECTION OF MINERAL INADEQUACIES

Richard J. Schanler and Cutberto Garza

USDA/ARS Children's Nutrition Research Center, Dept. of Pediatrics, Baylor College of Medicine, Houston, TX

Very low birthweight (VLBW) infants fed human milk fortified with pasteurized, lyophilized skim and cream fractions of mature human milk developed signs of calcium and phosphorus inadequacy despite retentions of nitrogen and energy that were similar to intrauterine references. Sixteen VLBW infants were fed a revised preparation of fortified human milk (Group FM2, fresh mother's milk plus human skim milk, calcium lactate, and mono- and dibasic phosphates) and compared to 10 similar infants fed cow milk-based formula (Group CM). Birthweight and gestation were similar (<1.35 kg and 30 weeks, respectively). Both groups received similar intakes (per kg/d) of N (479 mg), energy (129 kcal), and P (69 mg); Ca intake was greater in group CM (144 vs 128 mg). Ninety-six hour balance studies were performed at 25 (B1) and 38 (B2) days of age. N retention, 330 vs 317 mg/kg/d, and metabolizable E, 106 vs 108 kcal/kg/d were similar between groups, FM and CM, respectively. Despite greater intakes of calcium in CM and similar intakes of P in both groups, absorption and retention of Ca and P were greater in FM2.

	FM2		FORMULA (CM)	
(mg/kg/d)	B1	B2	B1	B2
Ca absorption	70±10*	93±23	52±13	82±19
Ca retention	66±20	86±21	49±13	78±18
P absorption	66±11	69±10	55± 4	60± 3
P retention	58± 8	56±14	37± 3	49± 4

* M±SD.

No differences were observed between groups for gain in weight, length, head circumference, or serum Ca, P, Mg, PTH, CT, and D metabolites. Differences in nutrient utilization in groups fed human milk or cow milk preparations persist despite efforts to humanize commercial products. The net mineral retention in group FM was below estimates for fetal Ca and P accretion. This may lead to differences in mineralization especially as a consequence of the transition from feeding fortified to unfortified human milk. Follow-up evaluations are necessary.

RELATIONSHIPS AMONG MATERNAL SIZE AND CARCASS COMPOSITION, LACTATIONAL PERFORMANCE, AND GROWTH AND COMPOSITION OF THE YOUNG: COMPARISONS ACROSS VARYING DEGREES OF CHRONIC MALNUTRITION

Kathleen M. Rasmussen

Division of Nutritional Sciences
Cornell University
Ithaca, N.Y. 14853

To describe the relationships among dietary intake, lactational performance, and litter growth in chronically underfed Sprague-Dawley rats, data from four studies were combined. In all experiments, animals were assigned at random to their dietary treatment group at 42 d of age and were bred over a 3- or 4-wk period beginning at 70 d of age. Throughout the experiments, rats were fed diet AIN-76A ad libitum (AL) or were fed 75%, 60%, 50% or 40% of AL intake. Litter size was adjusted to 8 pups at day 1 of lactation (L). At day 14 of L, milk yield was determined among half the dams by the tritiated water method. Milk and carcass composition were determined among the remaining dams and litters. At day 0 of L, maternal weight was a positive, linear function of food intake. At day 14 of L, milk yield and energy content were positively related to maternal body weight and carcass fat values; both increased at decreasing rates as body weight and carcass fat increased. Only within the AL range were decreases in maternal food intake not associated with compromised lactational performance. Pup weight gain also was compromised by restriction of maternal food intake but not by changes in maternal food intake within the AL range. At any given milk energy intake, pups of the AL-fed dams grew better than those of the restricted dams. Weight gain of the pups increased at a decreasing rate as a function of milk energy intake. The pups of restricted dams grew better as milk energy increased, whereas the pups of AL-fed dams grew at similar rates across a wide range of milk energy intake. A potential explanation for this is that as maternal food intake increased pups of the AL-fed dams deposited more of their new tissue as fat; however, the limited data available here on litter carcass composition are insufficient to support this hypothesis. This is the first time that these relationships have been explored across varying degrees of chronic underfeeding. These data support the existence of a non-linear relationship between chronic dietary intake and lactational performance. They also provide clear evidence of adaptation in milk yield and pup growth over the range of intake characteristic of AL-fed rats. However, this adaptation was overwhelmed by maternal dietary restriction with a resultant compromise in lactational performance and pup weight gain. (Supported by NIH Grant HD-14953.)

HUMAN ß-CASOMORPHIN-8 IMMUNOREACTIVE MATERIAL IN THE PLASMA OF WOMEN DURING PREGNANCY AND AFTER DELIVERY

G. Koch, E. Drebes, K. Wiedemann, W. Zimmermann*, G. Link* and H. Teschemacher

Rudolf Buchheim-Institut für Pharmakologie, Frankfurter Straße 107, D-6300 Gießen and *Zentrum für Frauenheilkunde und Geburtshilfe, Klinikstraße 32, D-6300 Gießen

Human ß-casein contains fragments with opioid activity, which have been termed "human ß-casomorphins"[1,2].

The present study was undertaken to investigate whether ß-casomorphins or fragments thereof may be cleaved from ß-casein in the organism of pregnant or nursing women and may eventually appear in the blood. Thus, for the determination of $ß_H$-casomorphin-8 immunoreactive material in human plasma an extraction method and a radioimmunoassay were developed and blood samples were collected from 35 pregnant women and from 138 women after delivery; in addition samples were taken from four men and sixteen nonpregnant women at different stages of the ovarian cycle as controls.

No $ß_H$-casomorphin-8 immunoreactive material was detectable in the plasma of the four men and the sixteen nonpregnant women. However, $ß_H$-casomorphin-8 immunoreactive material was present in the plasma of 26 out of 35 pregnant women. After delivery, in 100 out of 138 women $ß_H$-casomorphin-8 immunoreactive material was found. Plasma levels of immunoreactive $ß_H$-casomorphin-8 seem to rise during pregnancy and obviously continue to rise during the first week after delivery.

It is suggested that the presence of $ß_H$-casomorphin-8 immunoreactive material in the plasma of women during pregnancy and after delivery reflects certain secretory changes which the mammary gland undergoes during that time.

REFERENCES

1. V. Brantl, Novel opioid peptides derived from human ß-casein: human ß-casomorphins, Eur. J. Pharmacol. 106:213 (1984).
2. G. Koch, K. Wiedemann and H. Teschemacher, Opioid activities of human ß-casomorphins, Naunyn-Schmiedeberg's Arch. Pharmacol. 331:351 (1985).

EFFECT OF BREAST MILK INGESTION UPON THE THYROXINEMIA OF SUCKLING RAT PUPS

Linda V. Oberkotter

Department of Biological Sciences
Florida Institute of Technology
Melbourne, Florida 32901

The purpose of our studies was to develop a method for separation of milk iodothyronines in order to define the thyroid hormone metabolites present both in expressed milk from lactating dams and the stomach contents of their sucklings. Breast milk from *in vivo* radioiodine-labelled and -unlabelled lactating dams was subjected to chloroform:methanol extraction followed by $CaCl_2$ precipitation; the lyophilized extract was then further purified using a Bio-Rad ion-exchange resin column sequentially eluted with increasing concentrations of acetic acid. The iodothyronine eluate was then evaporated under nitrogen and a reconstituted aliquot injected onto a DuPont C-8 column and absorbance measured at 300 nm. Extracts of dams' milk contained only 8-10% of a $Na^{125}I$ cpm initially present in the whole milk. Recovery was dramatically increased by enzymatic digestion using either pepsin, trypsin, or pancreatin. Of these enzymes, we chose to investigate further the activity of pepsin, which most effectively liberated the iodothyronines of interest. Five litters of 12- and 14-day old suckling rats were separated from their dams for 6 hrs, after which time one animal per litter was sacrificed ("pre-suckling"animals). The remaining pups were permitted to suckle for 90 min; at 0, 1, 2, 3, and 4 hrs post-suckling, one animal per litter was sacrificed, trunk blood collected, and the stomachs removed. The latter were homogenized and then extracted as described above. The most marked changes in thyroxine (T_4) and triiodothyronine (T_3) levels occurred between Hr 0 and Hr 2 post - suckling: T_3 levels in Hr 0 stomach contents (180 ng) fell to 80 ng by Hr 2; T_4 levels decreased from 120 ng at Hr 0 to less than 60 ng at 2 hrs. Evidence that these decreases reflect rapid absorption rather than metabolic breakdown was found by examining serum T_4 levels from these same animals. Pre-suckling (2.71 ± 0.16 µg/dl, n=4) vs Hr 0 (2.47 ± 0.20, n=9) values did not differ significantly; at Hr 1, however, T_4 levels displayed a significant increase to 3.10 ± 0.37, n=9, $p < 0.001$ (all values expressed as the mean $\pm$ 1 SD). T_4 levels plateaued after Hr 2 post-suckling. Increases in serum T_3 concentrations were also observed between Hrs 0 and 2 post-suckling, but only in the older (day 14) group of animals. Our data indicate that (a) rat milk contains levels of T_3 and T_4 measurable by HPLC and spectrophotometric detection; (b) the bioavailability of these substances appears to be enhanced within the stomach of the suckling neonate; and (c) within 1 - 2 hrs of milk ingestion, a significant increase in neonatal circulating T_4 is observable, suggesting the possibility of rapid direct absorption of biologically-effective doses of intact hormone.

MOTILITY OF HUMAN MILK LEUKOCYTES IN COLLAGEN GELS

Fatih Ozkaragoz, Helen E. Rudloff, Frank C. Schmalstieg, and Armond S. Goldman

The Departments of Pediatrics and Human Biological Chemistry and Genetics, University of Texas Medical Branch, Galveston Texas

We previously reported that adherence, orientation and directed movement of human milk leukocytes (HMLs) were less than peripheral blood leukocytes (PBLs). Since the diminished motility may have been due to a decrease in adherence, a type 1 collagen gel system where leukocyte movement is less dependent on adherence was used to explore these questions.

Unfractionated HMLs or PBLs fractionated by Ficoll-hypaque density gradients were placed on collagen gels in microwells and the leading edge of migration was determined by inverted phase microscopy. The mean rates of invasion of HMLs, blood neutrophils and mononuclear blood leukocytes were 20, 200 and <1 μ/H, respectively ($p<0.01$). We then examined the characteristics of the motile HMLs by immunoperoxidase techniques using antibodies to selected cell markers. The HMLs in the matrix were positive for a neutrophil and monocyte marker Mac-1 and a specific macrophage marker cathepsin B, but were negative for markers for lymphocytes (Leu1), NK cells (Leu7) and neutrophils (cathepsin G). Furthermore, the motility of these cells was inhibited (<1 μ/H; $p<0.01$) when unfractionated HMLs were stimulated with a T cell lectin, phytohemagglutinin, to produce lymphokines, eg. migration inhibitory factor.

Thus, the diminished motility of human milk neutrophils does not appear to be due to decreased adherence per se, but may be due to their inability to change shape in the matrix. Furthermore, those HMLs which are motile are macrophages. The findings suggest a dichotomy for the function of HMLs. Neutrophils may be relegated to the lumen of the alimentary tract, whereas macrophages may penetrate into mucosal sites for the defense of the recipient.

ISOLATION OF LYMPHOCYTE

ACTIVATING FACTORS FROM HUMAN MILK

Olof Söder

Dept of Histology and Dept. of Pediatrics, St Göran's Children's Hospital, Karolinska Institute
S-104 01 Stockholm, Sweden

The protective effect of breast-feeding against infections has mainly been attributed to the presence in human milk of factors such as secretory immunoglobulins, lactoferrin and lysozyme. However, milk is also a rich source of regulatory molecules with putative immunomodulating functions in the mammary gland, and in the recipient infant. Such factors include many hormones, peptide growth factors and interferon, but others still remain to be characterized. The present study explores human milk in regard to the presence of lymphocyte activating factors. Ammonium sulphate precipitated protein from cell-free and defatted mature human milk was tested for interleukin-1 (IL-1), interleukin-2 (IL-2), thymocyte growth peptide (TGP) as well as T and B lymphocyte stimulating activity in various lymphocyte culture assays. The following resul were obtained. Employing the murine thymocyte proliferation assay, human milk protein displayed an IL-1-like activity, which eluted in three distinct peaks after Sephadex G 150 gel chromatography, with apparent M_r's of 14K, 30K and 60K, respectively. Chromatofocusing revealed charge heterogeneity, with an isoelectric point (pI) of 5 of the major peak. No pI 7 activity was detected, indicating that milk-derived IL-1 is similar to macrophage-derived IL-1α, and that no IL-1β-like material is present. Studies in progress on the cellular origin of milk IL-1 indicate that it is produced in high amounts by milk macrophages. Also T and B lymphocyte stimulating factors were detected. Thus, milk protein was comitogenic in combination with dextran sulphate for rat splenic B lymphocytes and in combination with phytohemagglutinin for guinea pig lymph node T cells. Gel filtration showed four peaks of B cell stimulating activity with apparent M_r's of less than 10K, 14K, 30K, and 70K, whereas the T cell stimulating activity was even more heterogeneous. IL-1 could be responsible fo some, but not all, of this activity. Human milk protein displayed no IL-2 or TGP activity. Pasteurization at 72°C for 15 sec. abolished all activity, except for a slight B cell stimulating effect.

The results demonstrate the presence in intact human milk of lymphocyte activating factors, one of which has been identified as IL-1. These factors might influence immune reactions in the mammary gland, and contribute to the anti-infectious effect of breast-feeding by stimulating immune functions in the recepient infant. Milk-derived IL-1 can be one of the factors behind the well-known pyrogenic activity of milk.

(Supported by grants from First of May Flowers Annual Campaign for Childrens Health, Swedish Society of Medicine, and Karolinska Institute)

PARTICIPANTS

Jean-Pierre Habicht, Chairman; Division of Nutritional Sciences, Cornell University, Ithaca, NY

Manuel Carballo, Maternal and Child Health Unit, World Health Organization, Geneva, Switzerland

Ranjit K. Chandra, Janeway Child Health Center, Memorial University of Newfoundland, St. John's, Newfoundland, Canada

Kathryn Dewey, Department of Nutrition, University of California, Davis, CA

Cutberto Garza, Department of Pediatrics, Baylor College of Medicine, Houston, TX

Armond S. Goldman, Department of Pediatrics, University of Texas Medical Branch, Galveston, TX

Lars A. Hanson, Department of Immunology, University of Goteborg, Goteborg, Sweden

Fehmida Jalil, Department of Preventive and Social Paediatrics, King Edward Medical College, Lahore, Pakistan

Michael S. Kramer, Departments of Pediatrics and Epidemiology, McGill University, Montreal, PQ, Canada

Reynaldo Martorell, Food Research Institute, Stanford University, Stanford CA

Kathleen M. Rasmussen, Division of Nutritional Sciences, Cornell University, Ithaca, NY

Michael W. Woolridge, Department of Child Health, University of Bristol, Bristol, UK

CONTRIBUTORS

Steven A. Abrams, Department of Pediatrics, Baylor College of Medicine, Houston, Texas 77030 USA

Sarah Alvarez, Department of Pediatrics, UCLA School of Medicine, Harbor-UCLA Medical Center, Torrance, California 90509 USA

Bryan F. Anderson, Department of Chemistry and Biochemistry, Massey University, Palmerston North, New Zealand

Stephanie Atkinson, Department of Pediatrics, McMaster University Medical Centre, Faculty of Health Sciences, Hamilton, Ontario, Canada L8N 3Z5

Edward N. Baker, Department of Chemistry and Biochemistry, Massey University, Palmerston North, New Zealand

Heather M. Baker, Department of Chemistry and Biochemistry, Massey University, Palmerston North, New Zealand

Janet K. Baltzell, Department of Food Science and Human Nutrition, University of Florida, Gainesville, Florida 32611 USA

Ch. Barth, Institut für Physiologie und Biochemie der Ernahrung, Bundesanstalt für Milchforschun, Kiel, Federal Republic of Germany

Henry S. Bayley, Department of Nutrition, University of Guelph, Guelph, Ontario, Canada

Alan Bedrick, Department of Pediatrics, University of Arizona College of Medicine, Tucson, Arizona 85724 USA

Graeme I. Bell, Chiron Labs, Emeryville, California 94608 USA

John Benson, Pediatric Nutritional Research, Ross Laboratories, Columbus, Ohio 43216 USA

Stefan Bernbäck, Department of Physiological Chemistry, University of Umeå, S-901 87 Umeå, Sweden

Joel Bitman, Milk Secretion and Mastitis Laboratory, United States Department of Agriculture, Beltsville, Maryland 20705 USA

Lars Bläckberg, Department of Physiological Chemistry, University of Umeå, S-901 87 Umeå, Sweden

Kurt J. Block, Departments of Pediatrics and Medicine, Massachusetts General Hospital, Boston, Massachusetts 02114 USA

Peggy R. Borum, Department of Food Science and Human Nutrition, University of Florida, Gainesville, Florida 32611 USA

Hans Joachim Bremer, Universitäts-Kinderklinik, Düsseldorf, Federal Republic of Germany

Marc Brown, Departments of Pediatrics and Medicine, Massachusetts General Hospital, Boston, Massachusetts 02114 USA

R. Don Brown, Departments of Pharmacology and Pediatrics, Clinical Pharmacology, School of Medicine in Shreveport, Louisiana State University Medical Center, Shreveport, Louisiana 71130-3932 USA

Wolfgang Buchheim, Institute for Chemistry and Physics, Federal Dairy Research Center, Kiel, Federal Republic of Germany

Robert J. Buczek, Department of Pediatrics, Georgetown University Medical Center, Washington, D.C. 20007 USA

Nancy Butte, Department of Pediatrics, Baylor College of Medicine, Houston, Texas 77030 USA

Manuel Carballo, Maternal and Child Health, Division of Family Health, World Health Organization, 1211 Geneva 27, Switzerland

Ranjit K. Chandra, Department of Pediatrics, Memorial University of Newfoundland, St. John's, Newfoundland, Canada A1B 3VD

Soter Dai, Department of Pharmacology, Faculty of Medicine, University of Hong Kong, Hong Kong

Ralph B. Dell, Department of Pediatrics, College of Physicians and Surgeons of Columbia University, New York, New York 10032 USA

Kathryn G. Dewey, College of Agricultural and Environmental Sciences, Agricultural Experiment Station, University of California, Davis, Davis, California 95616 USA

Sharon M. Donovan, Department of Nutrition, University of California, Davis, Davis, California 95616 USA

Klaus Dörner, Department of Pediatrics, University of Kiel, D-2300 Kiel, Federal Republic of Germany

Erika Drebes, Rudlof Buchheim-Institut für Pharmakologie der Justus Liebig-Universität Gießen D-6300, Gießeb, Federal Republic of Germany

Stefan Dziadzka, Department of Pediatrics, University of Kiel, D-2300 Kiel, Federal Republic of Germany

Johan Ek, National Hospital of Norway, Pediatric Research Institute, Oslo, Norway

Thorsten A. Fjellstedt, Endocrinology, Nutrition, and Growth Branch, Center for Research for Mothers and Children, National Institutes of Child Health and Human Development, Bethesda, Maryland 20014 USA

Gunn-Britt Fransson, Nutritional Research Department, Kabi Vitrim Nutr., AB, Stockholm, Sweden

Susan M. Gale, Department of Biochemistry, University of Adelaide, Adelaide, South Australia, Australia 5000

Cutberto Garza, Department of Pediatrics, Baylor College of Medicine, Houston, Texas 77030 USA

Carlos George-Nascimento, Chiron Corporation, Emeryville, California 94608 USA

George Giacoia, Department of Pediatrics, University of Oklahoma at Tulsa, Tulsa, Oklahoma 74136 USA

Ulrich Göbel, Zentrum für Kinderheilkunde, Abt. für pädiatrische, Hämatologie und Onkologie, University of Düsseldorf, Düsseldorf, Federal Republic of Germany

Randall M. Goldblum, Departments of Pediatrics and Human Biological Chemistry and Genetics, The University of Texas Medical Branch, Galveston, Texas 77550 USA

Armond S. Goldman, Departments of Pediatrics and Human Biological Chemistry and Genetics, The University of Texas Medical Branch, Galveston, Texas 77550 USA

Lawrence J. Grylack, Division of Neonatology, Columbia Hospital for Women, Washington, D.C. 20037 USA

Jean-Pierre Habicht, Department of Nutritional Sciences, Cornell University, Ithaca, New York 14853 USA

Leif Hambraeus, Institute of Nutrition, University of Upsulla, Upsulla, Sweden

Margit Hamosh, Department of Pediatrics, Georgetown University Medical Center, Washington, D.C. 20007 USA

Paul Hamosh, Departments of Physiology and Biophysics and Pediatrics, Georgetown University School of Medicine and Dentistry, Washington, D.C. 20007 USA

James W. Hansen, Mead Johnson Nutritional Group, Evansville, Indiana 47721-0001 USA

Donald G. Hanson, Departments of Pediatrics and Medicine, Massachusetts General Hospital, Boston, Massachusetts 02114 USA

Lars Å. Hanson, Department of Clinical Immunology, University of Göteborg, S-413 46 Göteborg, Sweden

Paul R. Harmatz, Departments of Pediatrics and Medicine, Massachusetts General Hospital, Boston, Massachusetts 02114 USA

Gerd Harzer, MILUPA Aktiengesellschaft, Friedrichsdorf/Taunus, Federal Republic of Germany

William C. Heird, Department of Pediatrics, College of Physicians and Surgeons of Columbia University, New York, New York 10032 USA

Olle Hernell, Department of Pediatrics, University of Umeå, 901 85 Umeå, Sweden

James L. Hinson, Departments of Pharmacology and Pediatrics, Clinical Pharmacology, School of Medicine in Shreveport, Louisiana State University Medical Center, Shreveport, Louisiana 71130-3932 USA

Jan Holmgren, Department of Medical Microbiology, University of Göteborg, S-413 46 Göteborg, Sweden

Fehmida Jalil, Department of Preventive and Social Paediatrics, King Edward Medical College and Mayo Hospital, Lahore, Pakistan

Robert G. Jensen, Department of Nutritional Sciences, University of Connecticut, Storrs, Connecticut 06268 USA

Sudha Kashyap, Department of Pediatrics, College of Physicians and Surgeons of Columbia University, New York, New York 10032 USA

Margaret Keller, UCLA School of Medicine, Harbor-UCLA Medical Center, Torrance, California 90509 USA

William R. Kidwell, Laboratory of Tumor Immunology and Biology, National Cancer Institute, Bethesda, Maryland 20892 USA

Michael Klagsbrun, Departments of Biological Chemistry and Surgery, Harvard Medical School and The Children's Hospital, Boston, Massachusetts 02115 USA

Peter D. Klein, Department of Pediatrics, Baylor College of Medicine, Houston, Texas 77030 USA

Ronald E. Kleinman, Departments of Pediatrics and Medicine, Massachusetts General Hospital, Boston, Massachusetts 02114 USA

Gertrude Koch, Rudlof Buchheim-Institut für Pharmakologie der Justus Liebig-Universität Gießen, D-6300 Gießen, Federal Republic of Germany

Otakar Koldovský, Department of Pediatrics, University of Arizona, College of Medicine, Tucson, Arizona 85724 USA

Berthold Koletzko, Universitäts Kinderklinik, Düsseldorf, Federal Republic of Germany

Michael S. Kramer, Departments of Pediatrics and of Epidemiology and Biostatistics, McGill University Faculty of Medicine, Montréal, Québec, Canada H3A 1B4

Idamarie Laquatra, Heinz Corporation, Pittsburgh, Pennsylvania 15230 USA

Ruth A. Lawrence, Department of Pediatrics, Obstetrics and Gynecology, The University of Rochester, Medical Center, Rochester, New York 14642 USA

Teresa H. Liao, Department of Pediatrics, Georgetown University Medical Center, Washington, D.C. 20007 USA

Marianne Lindblad, Department of Medical Microbiology, University of Göteborg, S-413 46 Göteborg, Sweden

Gerold Link, Rudlof Buchheim-Institut für Pharmakologie der Justus Liebig-Universität, D-6300 Gießen, Gießen, Federal Republic of Germany

Bo Lönnerdal, Departments of Nutrition and Internal Medicine, University of California, Davis, Davis, California 95616 USA

Reynaldo Martorell, Food Research Institute, Stanford University, Stanford, California 94305 USA

Lukas Matter, Department of Clinical Immunology, Kinderspital, St. Galen, Switzerland

Nitin R. Mehta, Department of Pediatrics, Georgetown University Medical Center, Washington, DC 20007 USA

Stanley G. Miguel, Medical Affairs, Mead Johnson Nutritional Group, Evansville, Indiana 47721-0001 USA

Iolanda Minoli, Department of Perinatal Pathology, Provincial Maternity Hospital, Milano, Italy

S. Mohanam, Laboratory of Tumor Immunology and Biology, National Cancer Institute, Bethesda, Maryland 20892 USA

Guido Moro, Department of Perinatal Pathology, Provincial Maternity Hospital, Milano, Italy

Maria Mrotzek, Universitäts-Kinderklinik, Düsseldorf, Federal Republic of Germany

Audrey Naylor, Wellstart - San Diego Lactation Program, San Diego, California 92138 USA

Margaret C. Neville, Department of Physiology, University of Colorado School of Medicine, Denver, Colorado 80262 USA

Gillian E. Norris, Department of Chemistry and Biochemistry, Massey University, Palmerston North, New Zealand

Linda V. Oberkotter, Department of Biological Sciences, Florida Institute of Technology, Melbourne, Florida 32901 USA

Clive W. Ogle, Department of Pharmacology, Faculty of Medicine, University of Hong Kong, Hong Kong

Jean Oliva-Rasbach, Departments of Physiology and Pediatrics, University of Colorado School of Medicine, Denver, Colorado 80262 USA

Anne-Brit Otnaess, Division of Molecular Cell Biology, Department of Biology, University of Oslo, Oslo, Norway

George M. Owen, Bristol-Myers International Group, New York, New York 10154 USA

Fatih Ozkaragoz, Department of Pediatrics, The University of Texas Medical Branch, Galveston, Texas 77550 USA

Robert F. Pass, Department of Pediatrics, University of Alabama at Birmingham, Birmingham, Alabama 35294 USA

Alesia Patera, Department of Food Science and Human Nutrition, University of Florida, Gainesville, Florida 32611 USA

Bruce W. Patterson, Department of Pediatrics, USDA Children's Nutrition Research Center, Baylor College of Medicine, Houston, Texas 77030 USA

Stuart Patton, Department of Neurosciences, University of California, San Diego, La Jolla, California 92093 USA

Mary Frances Picciano, Division of Nutritional Sciences, School of Human Resources and Family Studies, University of Illinois, Urbana, Illinois 61801 USA

Tom Picone, Medical Department, Ross Laboratories, Columbus, Ohio 43216 USA

Frank Pohlandt, Sektion Neonatologie, Universitäts-Kinderklink, Ulm, Federal Republic of Germany

Paul Pollack, Department of Pediatrics, University of Arizona College of Medicine, Tucson, Arizona 85724 USA

Guy Putet, Department of Neonatology, Hopital Edouard Herriot, 69003 Lyon, France

Lori K. Racaniello, Department of Biological Sciences, Florida Institute of Technology, Melbourne, Florida 32901 USA

Niels C.R. Raiha, University of Lund, Department of Pediatrics, Malmo General Hospital, Malmo, Sweden

Rajasekhar Ramakrishnan, Department of Pediatrics, College of Physicians and Surgeons of Columbia University, New York, New York 10032 USA

R.K. Rao, Department of Pediatrics, University of Arizona College of Medicine, Tucson, Arizona 85724 USA

Kathleen M. Rasmussen, Division of Nutritional Sciences, Cornell University, Ithaca, New York 14853 USA

Leanna C. Read, Department of Animal Sciences, University of Adelaide, Waite Agricultural Research Institute, Glen Osmond, South Australia, Australia 5064

Diane Reisinger, Diagnostic Systems Group, Beckman Instruments, Inc., Brea, California 92621

Christian Rieger, Zentrum für Kinderheilhunde der Philipps-, Universität Marburg, Marburg, Federal Republic of Germany

Jacques Rigo, Department of Pediatrics, Universite de Liège, Liège, Belgique

Annette L. Rodriguez, Department of Pediatrics, UCLA School of Medicine, Harbor-UCLA Medical Center, Torrance, California 90509 USA

Helen E. Rudloff, Department of Pediatrics, The University of Texas Medical Branch, Galveston, Texas 77550 USA

Sylvia Rumble, Department of Chemistry, Biochemistry, and Physics, Massey University, Palmerston North, New Zealand

Bernard Salle, Department of Neonatology, Hopital Edouard Herriot, 69003 Lyon, France

David S. Salomon, Laboratory of Tumor Immunology and Biology, National Cancer Institute, Bethesda, Maryland 20892 USA

Pieter Sauer, Sofia Children's Hospital, Rotterdam, The Netherlands

Richard J. Schanler, Department of Pediatrics, Baylor College of Medicine, Houston, Texas 77030 USA

Frank C. Schmalstieg, Jr., Departments of Pediatrics and Human Biological Chemistry and Genetics, The University of Texas Medical Branch, Galveston, Texas 77550 USA

Karl F. Schulze, Department of Pediatrics, College of Physicians and Surgeons of Columbia University, New York, New York 10032 USA

Jacques Senterre, Department of Pediatrics, Universite de Liège, Hôpital de la Citadelle, Liège, Belgique

Martin J. Shearer, Department of Haematology, Guy's Hospital and United Medical and Dental Schools of Guy's and St. Thomas's Hospitals, London, United Kingdom

Yuen W. Shing, Departments of Biological Chemistry and Surgery, Children's Hospital, Boston, Massachusetts 02115 USA

Erika Sievers, University Children's Hospital, D-2300 Kiel, Federal Republic of Germany

Jack C. Sinclair, Department of Nutrition, University of Guelph, Guelph, Ontario, Canada

Anne M. Smith, Division of Food and Nutrition, University of Utah, Salt Lake City, Utah 84112 USA

Olof Söder, Departments of Histology and Pediatrics, St. Göran's Children's Hospital, Karolinska Institute, S-10401 Stockholm, Sweden

Gerald Strecker, Laboratory of Biological Chemistry, University of Lille, F-59655 Villeneuve-Cedex, France

Janice Stuff, Department of Pediatrics, Baylor College of Medicine, Houston, Texas 77030 USA

Ann-Mari Svennerholm, Department of Medical Microbiology, University of Göteborg, S-413 46 Göteborg, Sweden

Rudlolf Tangermann, Zentrum für Kinderheilkunde, Abt. für allgemeine Pädiatrie, Neonatologie and Gastroenterologie, University of Düsseldorf, Düsseldorf, Federal Republic of Germany

Hansjörg Teschemacher, Rudlof Buchheim-Institut für Pharmakologie, Justus Liebig-Universität, D-6300 Gießen, Federal Republic of Germany

M. Rita Thomas, Grants Administration, Mead Johnson Nutritional Group, Evansville, Indiana 47721 USA

William Thornburg, Department of Pediatrics, University of Arizona College of Medicine, Tucson, Arizona 85724 USA

Rudiger von Kries, Kinderklinik für allgemeine, pädiatrie, Neonatologie, und Gastroenterologie, University of Düsseldorf, Düsseldorf, Federal Republic of Germany

W. Allan Walker, Departments of Pediatrics and Medicine, Massachusetts General Hospital, Boston, Massachusetts 02113 USA

Joyce M. Waters, Department of Chemistry and Biochemistry, Massey University, Palmerston North, New Zealand

Ulrich Welsch, Anatomy Institute, University of Munich, Munich, Federal Republic of Germany

Noel C. Wheeler, Department of Biomathematics, UCLA School of Medicine, Los Angeles, California 90024 USA

Robin K. Whyte, Department of Pediatrics, McMaster University Medical Centre, Hamilton, Ontario, Canada 48N 3Z5

Klaus Wiedemann, Rudlof Buchheim-Institut für Pharmakologie der Justus Liebig-Universität Gießen, D-6300 Gießen, Federal Republic of Germany

John T. Wilson, Department of Pharmacology and Pediatrics, Clinical Pharmacology, School of Medicine in Shreveport, Louisiana State University Medical Center, Shreveport, Louisiana 71130-3932 USA

William W. Wong, Department of Pediatrics, Baylor College of Medicine, Houston, Texas 77030 USA

D. Larry Wood, Milk Secretion and Mastitis Laboratory, United States Department of Agriculture, Beltsville, Maryland 20705 USA

Michael W. Woolridge, Department of Child Health, University of Bristol, Bristol, United Kingdom

Wilfried Zimmermann, Zentrum für Frauenheilkunde und Geburtshilfe, Justus-Liebig-Universität, D-6300 Gießen, Federal Republic of Germany

Christine L. Zucker, Department of Pediatrics, College of Physicians and Surgeons of Columbia University, New York, New York 10032 USA

INDEX

GPSR Compliance

The European Union's (EU) General Product Safety Regulation (GPSR) is a set of rules that requires consumer products to be safe and our obligations to ensure this.

If you have any concerns about our products, you can contact us on ProductSafety@springernature.com

In case Publisher is established outside the EU, the EU authorized representative is:

Springer Nature Customer Service Center GmbH
Europaplatz 3
69115 Heidelberg, Germany

Batch number: 10761599

Printed by Printforce, the Netherlands